WORKING IN WOOD

The Illustrated Manual of Tools, Methods, Materials and Classic Constructions

WORKING IN WOOD

The Illustrated Manual of Tools, Methods, Materials and Classic Constructions

Ernest Scott

G. P. Putnam's Sons   New York

Copyright © 1980 by Mitchell Beazley Publishers Limited
All rights reserved.
Published in the United States of America by
G. P. Putnam's Sons
200 Madison Avenue
New York, NY 10016
Reprinted 1984, 1988
Library of Congress Catalog Card Number 80-81148
ISBN 0-399-12550-7

Working in Wood was edited and designed by
Mitchell Beazley Publishers Limited,
Artists House, 14-15 Manette Street, London W1V 5LB
Phototypeset by Filmtype Services Limited, Scarborough
Origination by Gilchrist Bros, Leeds
Printed in Italy by New Inter Litho Limited, Milan

Art editor
John Ridgeway

Designers
Ayala Kingsley
Niki Overy
Nick Paul
Val Hill

Production controller
Barry Baker

Executive editor
Lawrence Clarke

Editor
Joanna Chisholm

Consultant editor
Pamela Tubby

Researchers
Louise Egerton
Pam Taaffe

Editorial assistant
Charlotte Kennedy

Contents

The author and publishers wish to acknowledge the contributions made by these specialist mastercraftsmen in the following subject areas:

Carcasses and frames
Mike Farrow
Lecturer in furniture studies at the London College of Furniture and Member of the Society of Industrial Artists and Designers. Served his apprenticeship as a cabinetmaker in his grandfather's workshop.

Classic constructions
John Coles
Lecturer in restoration and conservation studies at the London College of Furniture. Served his apprenticeship to an antiques restorer and cabinetmaker. Is a practicing restorer as well as a consultant to an international furniture group.

Design
Alan Tilbury
Design consultant and tutor in the School of Furniture and Design at the Royal College of Art.

John Barden
Furniture specialist at the Design Council. Is a practicing design consultant.

Materials
Bill Hallworth
Senior lecturer in wood science, technology and timber economics at the London College of Furniture and Associate of the Institute of Wood Science. Practiced as an importer in the timber trade and surveyor in timber preservation and is now a consultant to the furniture trade.

Parquetry and marquetry
Ernie Ives
Fellow of the Marquetry Society, and Editor of the *Marquetarian*. Practiced as a boatbuilder for 12 years and then spent 20 years teaching craft and marquetry.

Restoration
Roger Woods
Fellow of the College of Craft Education and lecturer in furniture subjects at Rycotewood College.

Alan Dyer
Professional restorer; family craftsmen since 1824.

Surface finishes
Bert Burrell
Lecturer in wood finishing and French polishing at the London College of Furniture. Served his apprenticeship as a polisher-improver and has practiced as a polisher for 35 years.

Veneering
Ernest Jones
Lecturer in fine craftsmanship and cabinetmaking at Rycotewood College. Served his apprenticeship to a handcraft furnituremaker; has practiced as a cabinetmaker and taught cabinetmaking for 37 years.

Wood carving
Paul Ferguson
Part-time lecturer in wood carving and gilding at the London College of Furniture. Served his apprenticeship to a Master Carver and now runs his own wood carving and gilding business.

Wood turning
William Wells
Member of the College of Craft Education. Has practiced as a woodworker and furniture designer for 10 years and has spent 20 years teaching both adults and children.

Foreword

Handling wood and sensing its textures and fragrances brings to mind the pleasures of a bygone age, when the woodworking skills of master craftsmen were commonplace. In our technological age, as these traditional skills are required less and less in the pursuit of a living, they are coveted more and more for the sheer pleasure inherent in their mastery.

The object of this book is to make the secrets of these skills available to all who seek them. It provides the essential sound basis upon which successful woodworking is founded: how to select the most suitable material and best method of working, how to maintain and use the correct tool for the job, and how to apply the principles of design and construction.

Working in Wood is divided into six sections: Classic constructions; Design; Tools; Methods; Materials; and Fixtures and fittings. Each section is treated separately and linked by an extensive index. All technical terms are fully explained in the comprehensive glossary.

The **Classic constructions section** introduces the subject of woodworking through some of the finest furniture ever made. Each piece has been selected for its universal appeal and as an example of quality craftsmanship. Even with the development of sophisticated power tools the principles underlying the best traditional techniques are wholly applicable to fine woodworking today. All the key methods and materials relating to each piece are contained in individual cross-reference panels.

The **Design section** is an analysis of all the problems that need to be resolved before starting any project. The appropriate design tools are illustrated and their uses demonstrated in the context of planning a project.

The **Tools section** is an analysis of all the hand and machine tools required in producing quality work, from the basic kit to specialized fixed power tools. The tools are divided according to their uses, such as sawing, planing, smoothing and chiselling. A complete description is supplied with each tool, and on the same or accompanying pages there are precise instructions on how to work with each.

The **Methods section** is a comprehensive demonstration of all the essential techniques in woodworking: joining; working manufactured boards; frame- and carcass-making; bending; laminating; wood turning; wood carving; veneering; finishing surfaces and restoring. Each technique is described through illustrations of the key actions by the specialist craftsman and precise step-by-step text.

The **Materials section** is an extensive study of the most sympathetic of all materials. It provides a complete understanding of wood, from how the tree grows, to how wood is converted and seasoned as well as the considerations to make when selecting and buying timber. More than 50 woods are illustrated in full color, and there is a detailed table defining the strengths, screwing, nailing and gluing properties and uses of the world's major timbers. Manufactured boards and veneers are also described.

Finally, the **Fixtures and fittings section** is a detailed compendium of all the universal woodworking accessories from intricate cranked and cylinder hinges to space-saving wardrobe rails and cupboard corner fittings.

As the scope of this book is so vast, Ernest Scott has worked in association with a team of specialist mastercraftsmen, each an expert in his chosen branch of woodworking. As a result, *Working in Wood* contains the accumulated knowledge of, literally, hundreds of years of practical experience. That experience is presented in a way that makes *Working in Wood* a unique book. The craftsman best communicates his expertise by demonstrating his skills and describing his actions as he works. This is exactly how *Working in Wood* has been created. Photographers and researchers have been specially commissioned to witness and record the author and the team of specialist mastercraftsmen, each demonstrating their particular craft. In this way their unique knowledge has been captured and passed on.

Classic Constructions

An introduction to the subject of woodworking through some of
the finest and most enduring furniture ever made

The Tudor chest

The chest may be described as the predecessor of all basic furniture forms and for many centuries it fulfilled a unique multipurpose role within the medieval and Tudor home. In this environment it served as table, chair, storage unit and often, owing to the highly mobile way of life, it was a packing case as well.

The earliest chests were crudely constructed from sections of solid timber that were hollowed out and bound with iron straps to minimize the effects of shrinkage. These dug-out chests were made from the trunks of fallen trees — the contemporary usage of the word "trunk" being a reminder of their origin.

It was not until technology was developed to convert timber into board form in the early fourteenth century that the now familiar chest structures appeared. These early planked chests consisted of six or more boards nailed and pegged together to form a lidded box onto which elaborately scrolled iron strapwork was applied to provide both strength and decoration.

During the latter half of the sixteenth century the chest underwent a fundamental change: the boards used to construct the front and back sections were joined, not edge to edge as before but at right angles to each other, forming a simple H frame. This represented an important technological step as, instead of relying on restraint to control shrinkage, efforts were now being made to construct a form that would be tolerant of timber movement.

As time progressed the two uprights became narrower and longer until they extended well below the central board to form feet. The central board itself gave way to the use of top and bottom rails connected by muntins, within which panels were contained loosely in grooves to enable them to shrink and expand without detriment to the structure as a whole. This form of construction is well demonstrated in the main illustration, which clearly shows the skeletal nature of the framework left by the evolution of the earlier H frames, in which thinner solid panels are contained. This form of construction proved immensely practical — so much so that it formed the basis of all furniture production until the end of the seventeenth century.

The chest was subject to many decorative treatments, of which the most international was the linenfold panel, which may be found on many items of furniture of the late sixteenth and early seventeenth centuries. This style of decoration, attributed to fifteenth-century Flemish architectural carvers, is often said to have been used on chests to denote their contents. However, this presupposes a selective use of the chest in medieval homes, which is contrary to their utilitarian evolution.

Lid in oak made from several narrow boards glued together — a technique that demonstrates the maker's awareness that several narrow boards were more stable than one or two wide boards

Top and bottom front rails in oak, decorated with guilloche carving. This device has been extensively used by furniture makers in numerous countries over many periods; it is, however, relatively uncommon on furniture of this period

Iron strap hinge — later known as a garnet hinge. This type of hinge would commonly be fixed with iron nails, often driven right through the back and then clinched

Panel in oak with scroll and roundel carving. This panel would be loose fitting within a groove to allow free movement, and have its grain direction running vertically

Bearer supporting the lid and providing protection against cupping. Bearers, however, tend to be self-defeating as, unless an allowance for shrinkage is made, they can cause severe splitting

Front rail tenoned into leg stile and secured with drawbore pegs. This technique was widely used to compensate for the inadequacies of available adhesives

Dug-out chests were hewn with axe and adze from hardwood trees that were roughly cut to shape. Commonly, these chests were bound with wrought iron straps to protect them from excessive deformation when shrinkage occurred.

The boarded chests of the thirteenth and fourteenth centuries were made from sawed hardwood boards held together by wooden pegs and iron straps. In many cases these straps consisted of elaborate scrolls that incorporated the nail positions as part of the overall design.

This joined chest of the late fifteenth century is an excellent example of an early piece of furniture that had incorporated the then newly-discovered jointing principles. Notice the wide uprights forming stiles that extend well below the center board.

This arcaded chest of the sixteenth century is evidence of the success of jointed structures. The upright stiles have become narrower, making them less vulnerable to shrinkage and giving them a more leg-like appearance. The arcading is typical.

The early-seventeenth-century mule chest is considered to be the first piece of furniture to contain drawers, or tills as they were first known. This example highlights the refined use of frame-and-panel construction to form a light portable storage unit.

Stile extending below the lower rail line to form the foot. This is a common technique in frame-and-panel construction

Bottom rail tenoned between the stile, and grooved to take both the panel and bottom boards

Muntin dividing two panels and providing resistance to racking. The prominence of the muntins in this example has been greatly reduced by addition of roundel carving

The split-carcassed chest of drawers of the seventeenth century represents the final development of the chest. The elaborate use of applied molding and l split turnings almost completely obliterated the frame-and-panel construction.

The sideboard

The sideboard, or dresser as it was then called, originated in the medieval period, when a trestle table or board was set aside for the purpose of dressing meat. The tables became more utilitarian when open shelves were added for the storage and display of utensils. However, during the Renaissance the communal halls disappeared as a more refined way of life evolved, and the dressing board became transformed into the more respectable and decorative court cupboard. This was simply an arrangement of tiered shelves and closed cupboards. It was not until the late seventeenth century that the dresser, or sideboard, as it is now recognized, made an appearance.

Because of their functional background sideboards were well suited to the rural way of life, and the basic form of graduated shelves over potboard or cupboard seems to have been simultaneously adopted over most of northern Europe. Many different styles existed but specific claims by some individual areas to a particular type or form were largely unfounded. Certainly, the term Welsh dresser, as some sideboards are called, holds no significance as to its origin.

As with most country-made articles the sideboard is found in a wide range of woods, because the rural craftsman would have used only those that were indigenous — oak, elm and some of the harder fruit woods being traditionally associated with the sideboard. It was not until improved surface coatings became available that the use of pine and other softwoods became popular, because these finishes provided some additional protection against the hard use to which the furniture was subjected.

Because of their size, sideboards generally had two separate sections or carcasses. The top section was constructed with two solid shaped ends between which top and shelf boards were joined. These were then held as a rigid structure by a planked back and a cornice rail. Usually the top rested on the lower section; sometimes it was suspended from, or even built into, the wall. The construction of the base section varied according to type. However, in general, the construction remained close to the traditional frame-and-panel principle, where mortised and tenoned frames formed the structure onto and into which panels were grooved and drawers and doors suspended.

The example illustrated is a fine sideboard from the early eighteenth century and shows these principles clearly. The addition of moldings to the front frame and the arched door detail serves to disguise the severity of line associated with this form of construction. Additionally, the arcaded cornice rail and molding provide proportional balance.

Two-section cornice molding providing proportional balance

H hinge in brass, sometimes referred to as a parliament hinge

Stile extending below the frame to form the foot – a common feature

Horizontal member designed to show at the side – typical of country-made furniture

Scalloped frieze rail suggesting Flemish influence

Planked back and bottom rabbeted and grooved into the carcass

Shelf capped with decorative molding providing safe storage for fragile items

Top carcass side, shaped to provide a graceful line to what would otherwise be a slablike structure. Scratch beading on the leading edge lightens the appearance

Applied bar molding, disguising the otherwise angular appearance of the frame-and-panel structure

Carcass shoe, sometimes permanently fixed to the base section; the top section is doweled into it

Drawer, with fielded front and lining in pine, suspended on rails between front and back framework

Door lock, in steel with brass face escutcheon, the key serving as a handle

Fielded panel, loose fitting in a groove within the framework

Lower carcass of frame and panel construction, providing a reliable and resilient structure

Peak arched door, a feature of early-eighteenth-century country furniture, in this case enhanced by the fielded and molded panel

Static panel decorated to provide symmetry to the front

A

B

C

The cornice rail and molding on the top section often provide a reliable basis for dating.
A. Frog's-back frieze, associated with later sideboards.

B. Fretted arch, found in many furniture periods.
C. Scalloped and pierced frieze, associated with early sideboards.

D

E

F

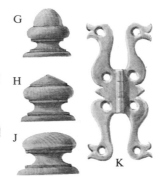

G

H

J

K

To display china, shelves had a plate restraint (**D, E, F**). These examples also show common shelf-edge treatments. Functional fittings, such as the serpent hinge (**K**) and butterfly hinge (**L**), were often of brass; wooden knobs (**G, H, J**) and iron hinges may also be found.

L

The combination of storage facilities offered by sideboards is varied. Most have closed cupboards and drawers in the base section. Some, however,

have an open base shelf, called a potboard, and may include spice drawers, cupboards or clocks in the top section of the dresser.

The writing desk

Towards the end of the seventeenth century written communication began to play a significantly more important role in social and economic life, and in many homes of substance rooms were set aside and furnished especially for this purpose. Thus escritoires (secretaries) and writing tables dominated the libraries and drawing rooms of the period. However, gradually a less formal attitude was adopted and this change was reflected by the innovation of the writing desk.

It is generally held that the slant-front writing desk was directly contrived from the sixteenth-century Bible box as both share the common feature of a sloping fall flap that provides, when opened, a writing surface and, when closed, useful storage behind it and a convenient area from which to read.

Later, the desk sections came to be stacked over secondary carcasses containing drawers. These secondary carcasses were themselves raised on stands or stacked in combinations to bring them to writing height. The lower carcasses were often terraced or curved and contained various combinations of drawers and cupboard arrangements. However, the fashion for multiple supporting carcasses quickly passed and the single supporting carcass came into favor. The main writing desk illustrated is typical of this early eighteenth-century period. It is finished in walnut veneer applied to an oak substrate and the interior of the upper carcass has been fitted for the storage of writing materials. A feature common to this type of desk is a slide beneath the pigeonholes providing access to a well, or void.

As the century progressed the practicality of the writing desk was enhanced by a natural extension of the storage facility by the addition of top carcasses that were often devoted to book storage or fitted to suit a variety of other requirements. This modification consolidated the characteristics of the writing desk and the split, or double, carcass gave way to the single unit.

The important position that writing desks held among contemporary eighteenth-century furniture is evidenced by the tremendous care with which they were made. The desk interiors were intricate and delicately treated with subtle adaptations to conceal secret compartments and other facilities such as candle-slides and pen trays. The exterior workmanship was also of the highest quality.

The convenient storage afforded by the writing desk led to further interesting adaptations, of which the secretaire drawer was perhaps the most novel and popular, as it provided similar facilities but could be more discreetly combined within a wider range of furniture forms.

Split- or double-carcass formation is evidence of the writing desk's ancestry

Walnut veneer applied to an oak ground (the practice of veneering on cheaper softwoods was established later)

Mid molding in walnut to disguise the double-carcass formation

Candle slides incorporated within the drapes

Loper, extended to support the flap

Dummy drawer fronts on the face of the upper carcass (drawers in these double-carcass types are relatively uncommon)

Ogee bracket foot, more commonly associated with later mahogany furniture

Lap dovetail joint in the drawer's corners

Column-fronted drawers operated by a finger pull from within a central cupboard are a common writing desk feature.

A. Drop, or pull, handle (1680s).
B. Solid plate handle (1700s).
C. Pierced, or fretted, handle (1750s).

D. Swan-necked handle (1750s).
E. Octagonal plate or ring handle (1780s).
F. Stamped plate handle (1780s).

Veneered slide for access to a storage well created by the split- or double-carcass formation

Back flap hinge in brass

Hide, or material such as baize, for softening the writing surface

Drawer lining in oak, common to quality pieces of the eighteenth century

Walnut veneer (feather or curl) used in matching pairs

Quadrant molding inset into the drawer front, for masking the drawer's fitting tolerances and for reducing the visibility of the carcass members

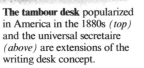

The tambour desk popularized in America in the 1880s *(top)* and the universal secretaire *(above)* are extensions of the writing desk concept.

An American secretary (1750s) shows the writing desk as an item of grandeur rather than one of function.

The writing desk on a stand (1680s) reveals its box-like origin. Note the carrying handles on the sides.

The tallboy

During the late seventeenth century the craft of cabinetmaking developed fully into a separate trade concerned only with furniture making and so became totally distinct from the other woodworking crafts. This was largely due to the wide acceptance of veneering, which had proved so popular that furniture forms were developed to display this art to the full. Carving and paneling disappeared, to be replaced by flat surfaces that showed off the grain pattern and the high polish given to the veneer. Cabinets and chests came to be raised on stands or stacked one on the other to elevate them to a new prominence in the elegant interiors. The mid-eighteenth-century example illustrated is a direct descendant of these pieces, which were once termed clothespresses, and have now become colloquially known as "tallboys".

With the adoption of veneering came some significant constructional changes. Because the frame-and-panel carcasses of earlier years did not provide the necessary flat unbroken surfaces, the slab constructional form was adopted, in which the major components were made from solid sections of wood, joined at the corners by dovetails. From this boxlike structure drawers and doors were suspended. This method was ideal for multiple-carcassed stacking furniture (chest-on-chest) and by the mid-eighteenth century the concept had become commonplace. It is still widely employed as the most economic and practical method of creating storage furniture of substantial size.

The illustrated example demonstrates this principle clearly in that the carcasses and associated sub-frames are constructed to interlock, with the top carcass being slightly narrower and shallower than the base section. This allows it to fit snugly within a cover molding, which also serves as a locating bar. Full advantage has been taken of the sectional and graduated form of the piece by the addition of running moldings used both to disguise joint lines and to provide proportional balance, as can be seen in the pedimented cornice.

The whole top section of this tallboy has been devoted to the storage of folded linen, with sliding trays running within grooves cut directly into the solid sides. However, later examples can be found with hanging space only or with combinations of both sliding trays and hanging space, in which case they are more correctly referred to as wardrobes, or armoires. The base section has the conventional arrangement of drawers suspended on drawer rails and runners.

Tallboys were made in many types of wood, the choice being affected by expanding world trade. However, in most European countries, mahogany and walnut were greatly favored.

Cross references

Designing a storage unit 38
Using a fretsaw and coping saw 68
Chamfering 77, 86, 110
Making a molding 85, 87, 110
Mortise and tenon joints 120
Dovetail joints 130
Housed joint 140
Carcass construction 152, 155
Plinths and bracket feet 153
Making a drawer 156
Making a frame-and-panel door 158
Hinges 159, 258
Relief carving 190
Veneering 192
Surface finishes 202
Handles 261
Locks and escutcheons 262
Honduras mahogany 234, 250
Oak 234, 250
Walnut 231, 232, 233, 246, 248

Butt hinge in brass, secured with steel screws
Three section cornice molding
Solid carcass side in Honduras mahogany dovetailed to a faced softwood top and bottom
Scratch bead used to distract the eye from the protruding hinge knuckles
Carved base molding applied to a softwood sub-frame to form a plinth

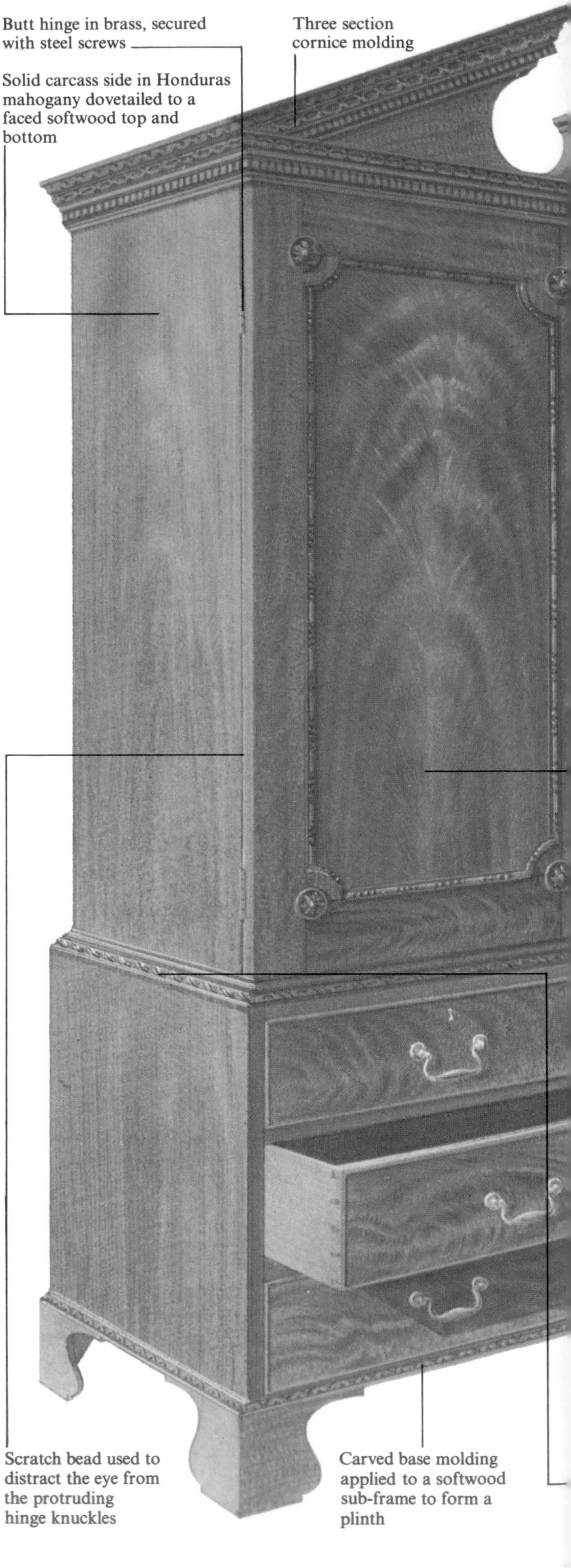

Broken pediment with turned central patera

Detachable cornice sub-frame located by corner blocks glued to the carcass top

Lining tray, in oak with mahogany front, running in grooves cut into the sides

Panel chamfered to reduce its thickness where it meets the frame and secured by beading within the rabbet created by the face molding

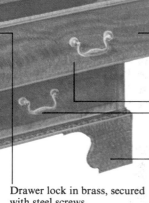

Cock beading of half-round edge section

Inset brass escutcheon

Drawer lining in oak with softwood back

Mahogany curl veneer applied to a softwood ground

Drawer suspended on softwood runners dadoed into the carcass sides, mortised into the front rail and grooved to incorporate a softwood dustboard.

Drawer lock in brass, secured with steel screws

Mahogany curl veneer applied to a solid ground

Mid molding attached to the base carcass, serving as a locating bar for the top carcass

Swan-necked handle, in brass, bolted through the drawer front

Bracket foot, in mahogany with rear facing sections in softwood, secured to the frame with a glue block

Pedimented cornice frames were popular throughout the eighteenth century. The swan-necked pediment (*top*) has a fretted plate above an arched frieze rail and is typical of later examples. Separate pediments and cornices were superseded by shaped and crested head rails (*above*).

The radial, or broken, corner with carved patera was commonly used to decorate framed doors in tallboys of the eighteenth century.

Cock beading was fitted to protect the face veneers of drawer fronts. Its projecting section also disguised the fitting tolerance.

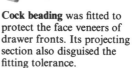

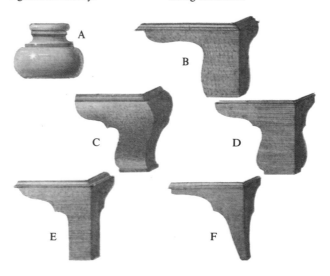

Applied feet were fitted directly or to a molded sub-frame.
A. Bun foot (1700s).
B. Bracket foot (1700s).

C. Double ogee foot (1750s).
D. Single ogee foot (1750s).
E. Bracket foot (1780s).
F. Swept bracket foot (1780s).

The drawleaf table

The drawleaf table first appeared in the late sixteenth century. At the time it represented an innovation contrary to the prevailing techniques and principles of furniture making as it incorporated a relatively sophisticated extension mechanism not found in other contemporary furniture. Carcass furniture in the sixteenth century, for example, was of basic panel-and-frame construction with the addition of only a door and drawers, where required, for specific purposes. Tables were also simple, being either of the dormant or gateleg type. Although it had an extending capacity, the latter table was based on a simple, logical design that was not comparable with the imagination and accuracy required to develop the drawleaf table.

The drawleaf table had a central top under which two sliding extension leaves were positioned. These pulled outwards, either at the sides or, more commonly, at the ends. When extended, the central top dropped into the void left by the leaves and so formed a single continuous surface. To make this movement possible, runners (called lopers) that were attached to the leaves passed through guiding bearers and slots in the main frame. When the table was not extended, the main frame concealed the mechanism.

The main illustration shows a drawleaf table typical of the late sixteenth and early seventeenth centuries and it exhibits the rugged characteristics of the oak furniture of that period. An interesting feature of this table is the cleating, or keying, around the central top and leaves. This was done to minimize the possibility of the wood distorting and so fouling the operation of the table. This feature demonstrates the growing understanding of using timber as a precision material.

Drawleaf tables may be found with a variety of leg arrangements, but traditionally they are of the four-legged type with a stout underframe and stretchers, which are necessary to provide resistance to the forces imposed during extension.

The basic form of the drawleaf table has not changed since the sixteenth century and there have been only some minor mechanical improvements. The principle of the drawleaf, however, has been adapted in a variety of ways. The intricate extending tables of the eighteenth century included circular versions in which segments operated individually to increase the diameter. In the nineteenth century, the principle was incorporated not only in tables but also in writing desks, in order to increase their writing surface. As a result of the improved extension mechanisms now available, the drawleaf is undergoing a revival in popularity — a fact that is a tribute to the foresight of the sixteenth-century furniture-makers.

Cleating applied to the leaves and central top at right angles to the grain of the central panel. Cleating was used in an attempt to limit the effects of warping. In general, however, these cleats were self-defeating and resulted instead in the wood splitting

Timber of quality, for various reasons, has always been prized. The bulbous leg is a fine example of an economic use of wood because, to save excessive waste, applied blocks were glued to the leg in order to increase its diameter where needed on a leg of this type.

The sliding extension leaves of the drawleaf table were held on cantilevered runners passing through the frame and under a centre bearer. The angle of these runners was contrived so as to allow each leaf, when extended, to rise by the thickness of the central top.

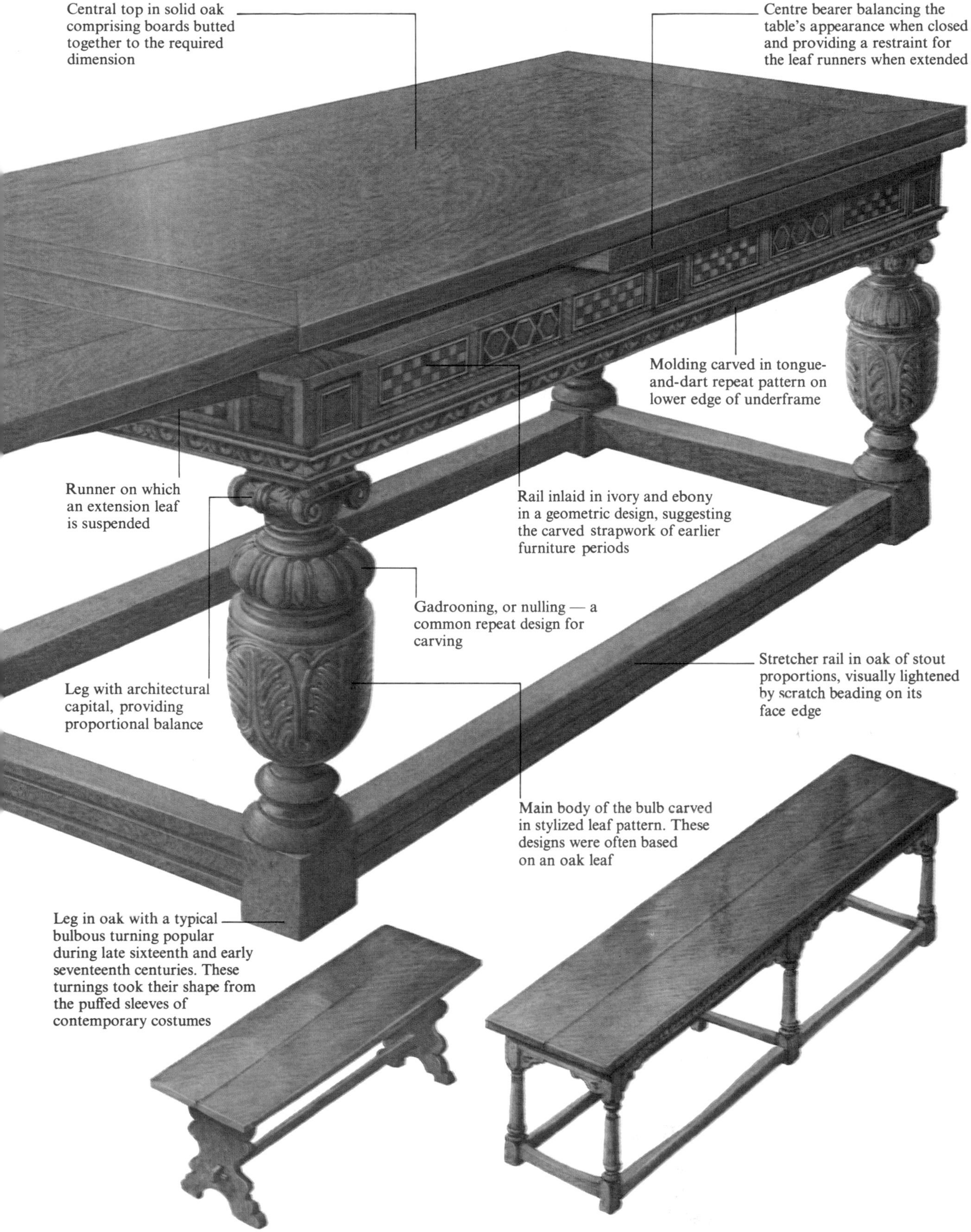

Central top in solid oak comprising boards butted together to the required dimension

Centre bearer balancing the table's appearance when closed and providing a restraint for the leaf runners when extended

Molding carved in tongue-and-dart repeat pattern on lower edge of underframe

Runner on which an extension leaf is suspended

Rail inlaid in ivory and ebony in a geometric design, suggesting the carved strapwork of earlier furniture periods

Gadrooning, or nulling — a common repeat design for carving

Stretcher rail in oak of stout proportions, visually lightened by scratch beading on its face edge

Leg with architectural capital, providing proportional balance

Main body of the bulb carved in stylized leaf pattern. These designs were often based on an oak leaf

Leg in oak with a typical bulbous turning popular during late sixteenth and early seventeenth centuries. These turnings took their shape from the puffed sleeves of contemporary costumes

The sixteenth-century trestle table is an early example of knock-down furniture. The two solid trestles at the ends were joined to a central stretcher by a removable joint, which in most cases was a projected and wedged mortise and tenon, as illustrated (*above*).

The table dormant has acquired the modern name of the refectory table. Such tables were of considerable size and they often incorporated as many as eight or ten legs. The number of legs and stout underframe give these tables a rigidity for which they have become famous.

The gateleg table

The gateleg table has been one of the most popular and enduring table designs ever introduced.

Gatelegs were first made in the last quarter of the sixteenth century, when the Protestant upsurge throughout Europe brought a plainer style of living, centered in smaller, more private apartments. This change was reflected in the furniture of the period and created a demand for smaller, more utilitarian pieces that could be quickly adapted to a variety of uses. Of these pieces the gateleg — or, to use the original name, falling or faldyn — table is a fine example. The term gateleg is of much later origin and is derived from the action and construction of the pivoting sub-frame, which is reminiscent of the conventional gate.

The earliest type of gateleg had a single gate and flap. This type, normally of semicircular form, was designed to stand against a wall with the gated side to the rear. This meant that, to be opened to a full circle, the table had to be moved, which was clearly a disadvantage. A more sensible early adaptation was the triangular type table, which fitted conveniently into a corner and had the gate and flap to the front.

As gateleg tables gained in popularity these single-gated types were superseded by twin-gated ones, which were made in a variety of shapes — ovals, rectangles and polygons being popular. The base frames were often of trestle construction and were joined at the feet by a single planked stretcher. As the gateleg principle developed it was incorporated into other structures — the gated centers of eighteenth-century Hepplewhite tables being a fine example of this.

The table illustrated displays all the rugged solidity of a country-made table of the late seventeenth and early eighteenth century. Most tables such as this would have been constructed from hardwoods that were indigenous to the country of origin; oak, elm, beech and walnut may be found in surviving examples. However, yew and some fruitwoods may also have been used, as well as other hedgerow timber.

The base frame of the example is constructed with baluster-turned legs and bun or bobbin feet. The legs are well joined to the table rails at the top and to the stretcher rails at the base, using drawbore mortise and tenons. This table illustrates well the basic construction of all true gatelegs. The gate should hang from the rails and stretchers, and the underframing should be so proportioned that the supports are equally spaced when the top is opened. These open-frame structures are pleasingly simple, reliable and rigid and lend themselves to adaptation, as may be seen in surviving examples that incorporate drawers, shelves, cupboards and additional gates.

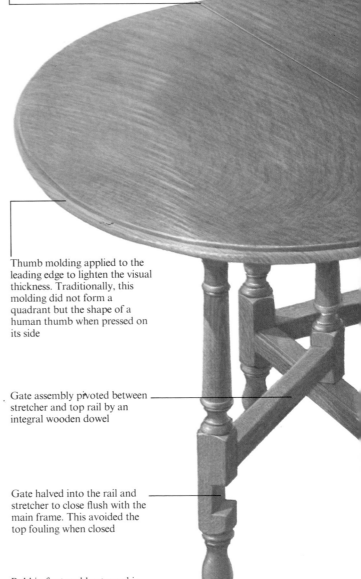

Plain butt joint with steel back flap hinge. This is a common configuration for tables of this period

Thumb molding applied to the leading edge to lighten the visual thickness. Traditionally, this molding did not form a quadrant but the shape of a human thumb when pressed on its side

Gate assembly pivoted between stretcher and top rail by an integral wooden dowel

Gate halved into the rail and stretcher to close flush with the main frame. This avoided the top fouling when closed

Bobbin foot and leg turned in one continuous piece

A   B   C   D

Turnings are ideal for the decoration of gateleg frames. Some typical turnings associated with these tables are shown *(left)*.
A. Gun barrel turning (1760).
B. Spiral twist turning (1640).
C. Bobbin-and-reel turning (1650).
D. Baluster turning used in various periods with numerous decorative features. This turning takes its name from the architectural balustrade.

Top in oak formed from glued and butted boards

Arched end rail suggesting the Flemish influence prevailing at the time

To improve the appearance of folding tables the groove and bead joint was used where the top met each flap. This joint preceded the rule joint.

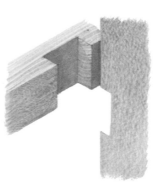

Side rail, sometimes made from softwood to conserve timber: joined to leg with a drawbore mortise and tenon. This joint was used extensively by the early furniture makers to overcome poor adhesives and assembly problems

Where legs and stretchers intersect, halved joints are used even though they weaken the table's structure where it is most vulnerable.

Stretcher rail with scratch beading lightening this otherwise conspicuous component

E

F

G

The semicircular gateleg table (E) operated with a single gate at the rear and the top folded forward when closed. The triangular type **(F),** when closed, fitted neatly into the corner of a room. It had only a single gate, but its top folded downward, covering the front frame when closed. The tilt top table **(G)** had only a single gate, which was pivoted centrally. This enabled the table to fold into a form that was ideal for storage.

The Pembroke table

Pembroke tables take their name from the ninth Earl of Pembroke, who in the mid-eighteenth century is said to have designed a new style of table. The precise definition of a Pembroke table, however, is difficult to establish as the term is now used to describe a variety of tables that in many cases have contrasting appearances, although some features can be identified as being common to all: they are mostly of moderate size, light in structure and contain a drawer or drawers. A more specific generic feature of this family of tables is perhaps the top arrangement of a central section flanked by two rising flaps that are held in the open position by fly support brackets incorporated within the underframing. This design differs very distinctly from drop-leaf and gateleg tables, where the flaps are supported by hinged or pivoted legs.

The most usual form of Pembroke table is that shown in the main illustration. This style of four-legged table was popularized in the late eighteenth century and the example shown emphasizes the delicacy of treatment common to these tables. The carcass contains a single drawer, so a dummy drawer has been designed on the rear frieze rail to give the table a symmetrical appearance. The top is oval in shape with rule joints, which give a pleasing appearance to the hinge detail. In common with most Pembrokes, the legs terminate in casters and are of sufficient section to conceal the side rails and fly support mechanism, which consists of two traversing brackets stoutly pivoted by knuckle joints. These mechanisms occur in opposing pairs or singly, depending on the size and quality of the table. In general all Pembroke tables in the late eighteenth century displayed a scrupulous attention to detail that cabinetmakers in succeeding periods have been at pains to perpetuate.

During the peak of their popularity, in the last quarter of the eighteenth century, Pembroke tables provided the leading designers with the ideal medium on which to display the highly stylized decorative treatments popularized by them. Hepplewhite and Sheraton in England and Phyfe in America all produced designs for Pembrokes that succinctly focused their own personal styles. The English craftsman utilized subtle blends of mahoganies and satinwoods in conjunction with classical motifs of marquetry, painted decorations and bandings, while Phyfe exaggerated the graceful lines of the Pembrokes using indigenous timbers such as walnut and mulberry.

Pembroke tables are now in common use in both Europe and America, where numerous versions of the original concept, such as the pedestal breakfast table and the sofa table, may be found.

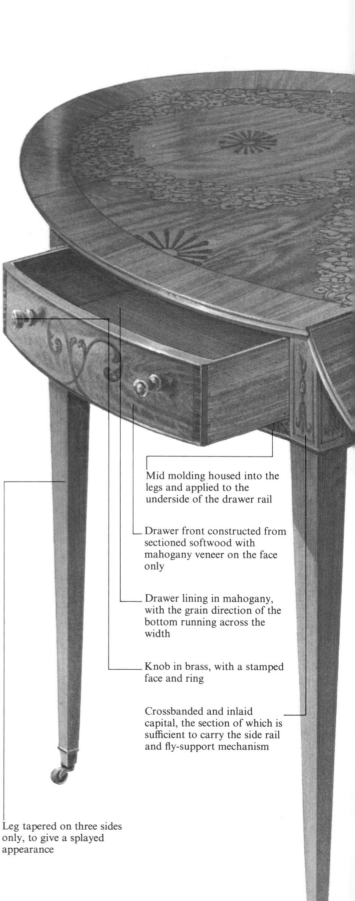

Mid molding housed into the legs and applied to the underside of the drawer rail

Drawer front constructed from sectioned softwood with mahogany veneer on the face only

Drawer lining in mahogany, with the grain direction of the bottom running across the width

Knob in brass, with a stamped face and ring

Crossbanded and inlaid capital, the section of which is sufficient to carry the side rail and fly-support mechanism

Leg tapered on three sides only, to give a splayed appearance

Caster in brass, secured with steel screws

Dummy drawer to the rear, giving a symmetrical appearance

Rule joint between central top and flap

Floral decoration and motifs executed using a combination of marquetry background, and painted foliage with highlights

Astrigal molding worked on the leading edges to lighten the appearance of the top

Crossbanded borders in mahogany with the grain running at right angles to the perimeter

Top in solid mahogany veneered on the face with Honduras mahogany, applied in book-matched pairs on the center top and continued on the flaps

The rule joint is used to conceal the otherwise unsightly gap that would be left when a flap is lowered. The geometrics of this joint require the use of a specialized hinge with one flap approximately one-third longer than the other to clear the hollow in the cove of the joint.

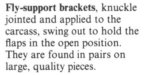

Fly-support brackets, knuckle jointed and applied to the carcass, swing out to hold the flaps in the open position. They are found in pairs on large, quality pieces.

A. Threaded caster with an astrigal cap.
B. Round socket, or cup, caster for turned legs.

C. Square socket, either flush fitting or flanged (*above*).
D. Lion's paw toe caster found on the pedestal-type table.

The breakfast table of the eighteenth century used the same constructional principles as the Pembroke but had an enclosed lower carcass.

The sofa table is not strictly a Pembroke but has similar ancestry. True Pembrokes would have the flap along the side rather than the ends.

The Windsor chair

The success of the Windsor chair has been attributed to its immense practicality and its simplicity of design.

Since its introduction in the early part of the eighteenth century the Windsor chair has undergone numerous minor changes, determined by prevailing fashions. However, its basic form has remained unchanged. The term Windsor chair is generally applied to all chairs that display the simple characteristics of peg, or stick, construction used in conjunction with a solid saddled seat. Various combinations of back and leg arrangements may be found.

It is generally thought that this type of construction had been used by chairmakers for some time prior to the evolution of the recognizable Windsor. This construction technique provided a degree of flexibility during manufacture as the wedge-fit used for the slats and spindles made assembly similar to the procedures used in the craft of coopering. Tension was provided by compressive force, and stress by the bracing action of the back stretcher sections. This provided the chair with some unique qualities, in that it provided additional strength when forces were imposed and a high tolerance to material deformation.

The materials traditionally associated with chairs of this type have always been those most amenable to the constructional techniques used. In Europe, for example, ash, elm and yew were commonly utilized for the steam-bent sections, whereas for the turned and sculptured members denser woods such as beech, oak and hickory were used, together with a wide variety of fruitwood and hedgerow timbers. For this reason it was common to find an individual chair made in three or more different woods and this added to its charm.

Windsor chairs are found in various countries, each country producing its own distinctive styles and types. Individual localities have also become associated with particular adaptations and styling; even the craft terminology varied among the diverse clusters of local craftsmen working close to the standing forests from which their woods were taken. The low-backed chairs of Philadelphia, where the making of these chairs was centered from the beginning of the 1700s, and those of Suffolk are fine examples of this.

In general it is possible to subdivide the Windsor's form into three main categories: the tall, gently curving comb back; the distinctive hoop, or bent, back; and the low-backed tub. Each type has an infinite number of variations in its own developmental history.

Among contemporary adaptations, some retain the peg construction but most rely on improved adhesives for rigidity, which together with loose upholstery tend to debase the original elegance of the Windsor chair.

Four main types of Windsor chair are shown; within each, a wide variety of turning, splat and rail shapes exists.

A. Mendlesham Windsor (1800s).
B. Gothic Windsor (1750s).
C. Smoker's bow (1850s).
D. American fan back (1780s).

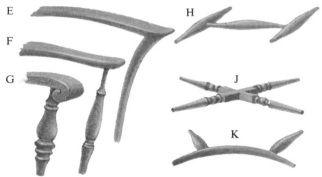

E. Bentwood, or swept, arm and upright (1750s).
F. Turned spindle imitating the leg shape (1730 onwards).
G. Stout baluster arm (1850s).

H. Simple plain-turned H form.
J. Cross- or X-frame stretcher associated with circular seats.
K. The classic crinoline, or cowhorn, stretcher.

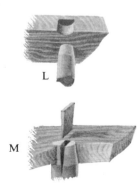

The turned sections are reamed with a slight taper and compression fitted (L) or are through wedged into a bored hole (M).

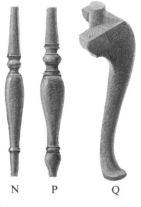

Front legs are traditionally the turned baluster type (N, P). Both these and the cabriole leg (Q) are used with plain legs at the back of the chair.

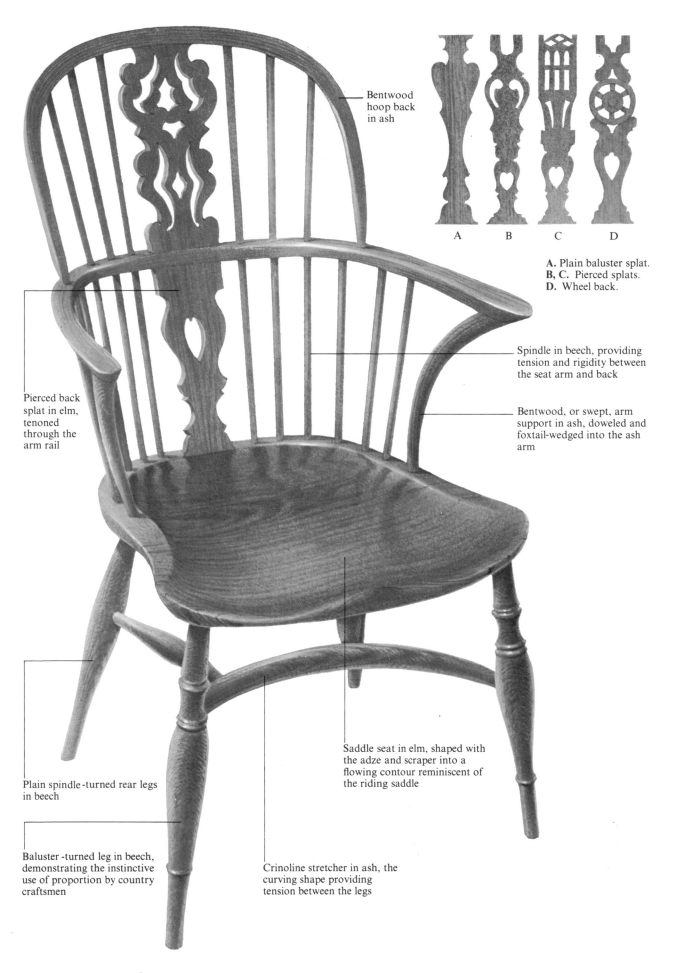

Bentwood
hoop back
in ash

A. Plain baluster splat.
B, C. Pierced splats.
D. Wheel back.

A   B   C   D

Spindle in beech, providing
tension and rigidity between
the seat arm and back

Bentwood, or swept, arm
support in ash, doweled and
foxtail-wedged into the ash
arm

Pierced back
splat in elm,
tenoned
through the
arm rail

Saddle seat in elm, shaped with
the adze and scraper into a
flowing contour reminiscent of
the riding saddle

Plain spindle-turned rear legs
in beech

Baluster-turned leg in beech,
demonstrating the instinctive
use of proportion by country
craftsmen

Crinoline stretcher in ash, the
curving shape providing
tension between the legs

The walnut chair

At the end of the seventeenth century European chairs were noted for their tall, slender appearance and graceful proportions. However, the new century brought change and a style of chair emerged that was based on the contemporary proportions, yet possessing greater continuity of line. This new style, attributed to Daniel Marot, a French designer and furniture maker, represented such a fundamental change in design that its effects influenced other furniture makers well into the mid-eighteenth century.

In contrast to the previous angular style, the new eighteenth-century chairs were given gently curving backs that undulated in profile and had a central solid support or splat, which replaced the caned panels previously in fashion. Perhaps more important was the innovation of the cabriole leg, the graceful lines of which were in complete harmony with the new taste and which proved so popular that adaptations were to be found in the majority of succeeding furniture periods. It is thought that the cabriole leg was an adaptation of the S-scroll leg and that it represented a revival of the classical use of natural shapes and forms. This is borne out by the fact that in its early use the cabriole leg often terminated in a cloven foot or hoof. As it developed to a more sturdy form, the cabriole leg provided sufficient stability to enable furniture makers to dispense with supporting stretcher rails to the underframes.

The chair shown in the main illustration is typical of chairs of the early eighteenth century and demonstrates the subtle modification by the English and Dutch craftsmen of the period of Marot's more florid style. This translation successfully blended the elegance of the original French concept with a sturdy practicality. This chair has the rising shape of its crest rail accentuated by C scrolls and hipped rear uprights, around which a vase-shaped splat and scroll-over arms add to the curvilinear design. An interesting feature is its cabriole legs, which terminate in knurl, or whorl, feet. The cabriole leg proved immensely popular and on fine-quality chairs it was used in conjunction with curved seat rails and a carved ornamentation such as shells, husks and acanthus leaves added to the knee and wings.

Early-eighteenth-century chairs were almost exclusively made in walnut and often bore decorative veneers on their flat upright surfaces. Walnut burl was laid on the splats and seat rails, and crossbanded veneer was laid around the flat faces of the back.

The international origins of this style, humored by solid craftsmanship, give chairs of this period a restrained elegance that is second to none.

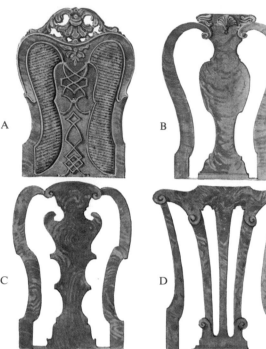

A

B

C

D

These backs are all of early-eighteenth-century origin. The French concept was modified by English and Dutch craftsmen. **A.** Florid style of an early design by Marot.

B. Anglo-Dutch variation with a baluster splat.
C. Hipped back of late origin with a vase-shaped splat.
D. Early Georgian adaptation with a pierced splat.

Seats in the early eighteenth century were heavily shaped, often having only one straight side. They were of the drop-in type, in which a beech frame was lap halved together and

then profiled to fit within a rabbet cut into the seat rails. The illustration (*above*) of the most popular frame shape at that time shows this treatment clearly.

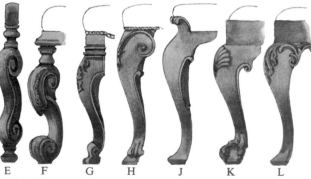

E   F   G   H   J   K   L

The development of the cabriole leg is illustrated (*above*). **E, F.** Scrolls suggesting the later form of cabriole. **G.** Early cabriole. **H.** Early cabriole with hoof foot.

J. Fully developed cabriole with hipped section and pad foot.
K. Fully developed cabriole with claw-and-ball foot.
L. Fully developed cabriole with pad foot.

Crest rail with shell-form carving. Because of their double curving shape, these rails are sometimes known as cupid's bows, of which they are reminiscent if held vertically

Carved patera applied over the joint line between the splat and crest rail

Solid central splat of baluster form, faced with walnut burl veneer. The floral marquetry is evidence of Dutch influence

Carved ear suggesting a C scroll formed between it and the patera on the crest rail

Scroll-over, or shepherd's crook, arm, the curving shape of which contributes greatly to the chair's elegance

Rear leg cut and shaped from a single section of wood

Back frame, with face veneered in walnut burl

Splat shoe, providing both strength and tolerance during fitting

Drop-in seat upholstered onto a beech frame

Rear leg terminating in a pad foot. Notice the changing section as the leg rises

Curving front rail, with face veneered in walnut, shaped to blend in with the legs

Cabriole leg with acanthus leaves at the knee and hip. Notice the knurl, or whorl, foot and C scrolls bridging the leg and wings

The balloon-back chair

More than any other item of furniture, the balloon-back chair has increased in popularity through the fashionable interest in antiques. This is perhaps due to the chair's unique decorative quality and the fact that it possesses an elegance that enables it to look equally good whether placed on its own, in pairs, or among furniture from other periods.

Although the balloon-back chair reached its peak of popularity in Europe and America during the mid-nineteenth century, the term "balloon back" was not coined until the last quarter of the century and it is thought to be of American origin. The chair's style is almost exclusively English and its development may be clearly plotted through the dining- and drawing-room chairs of the early nineteenth century.

The balloon back has its origins in the classical Grecian-style chairs of the late Regency period. These chairs had a very square appearance with a wide crest rail, extending over the upright back legs, and a horizontal rail positioned halfway between the crest rail and seat. During the 1830s these chairs, particularly those made for drawing-room use, had become elaborately covered with scrolls. While the overall construction remained the same, the crest rails had developed a distinctive downward line, created by the use of scrolls at either end.

The transition of the fully integrated crest rail and rear legs to form a continuous curve seems to have been made in the late 1830s. These balloon backs had a distinctly flattened crest rail that closely resembled the proportions of their scrolled predecessors. As the century progressed the upward curve of the crest rail became more exaggerated and the rear legs developed an inward curve or waist. The only reminders of their classical origins were the tapering padded seats and the straight, turned front legs. The final modification took place around 1850, when gently curving cabriole legs (in this case sometimes referred to as French legs) were introduced, to provide balance to the exaggerated shape of the back. These legs were normally fielded with a carved hip and were used in conjunction with serpentine-shaped seat rails. The pleasing proportional balance this created is well shown in the main illustration.

By 1860 the balloon back had become a standard pattern with several adaptations to suit a variety of uses. Dining chairs remained relatively plain and were made in quality woods such as rosewood, mahogany and walnut. Drawing-room and bedroom balloon backs, however, had lively curved embellishments to mid- and crest rails and were made in such timbers as walnut, maple and birch.

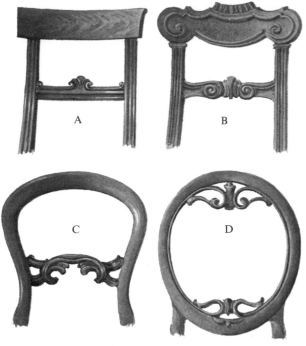

The early balloon back of the 1840s was based on the Grecian and scroll-top chairs, which started a line of development, ending in the variations of the 1860s.
A. Grecian style chair (1830).
B. Scroll-top chair (1835).
C. Early balloon back (1840).
D. Double C scroll (1865).

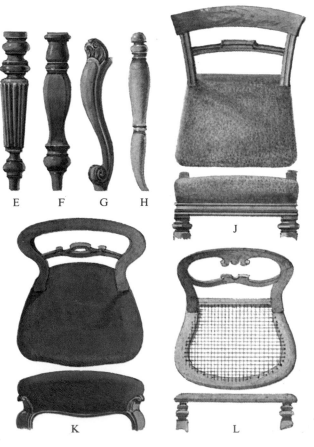

Early counterparts of the balloon back had angular tapering seats (J) with fluted and turned legs (E, F). True balloon backs had a serpentine-shaped seat (K) and French legs (G). Later variations adopted cane seats (L) with turned and splayed (or horned) legs (H).

Crest rail in walnut doweled into the rear legs, a vulnerable feature on chairs of this type

Carved mid rail in walnut tenoned between the rear legs and often so heavily decorated as to make it weak

Upholstered seat containing coil springs, the exaggerated rising shape of which was fashionable during the 1850s

Seat rail in beech profiled to accept tacks; tenoned into the leg

Serpentine-shaped front rail with an applied walnut face-beading on the lower edge, providing continuity of line with the leg

Gimp, or braided strip, applied to conceal the tack heads

Rear leg in walnut cut from a single section of wood; the inward curve at the hip is carved to accentuate the ballooning effect

Rear leg rectangular in section at the base of the seat for ease of joining, gradually tapering to an oval section at the foot

Front leg in walnut, terminating in modestly proportioned scrolled foot

French, or cabriole, leg carved at the hip and knee, greatly contributing to the elegance of the balloon back

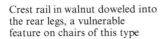

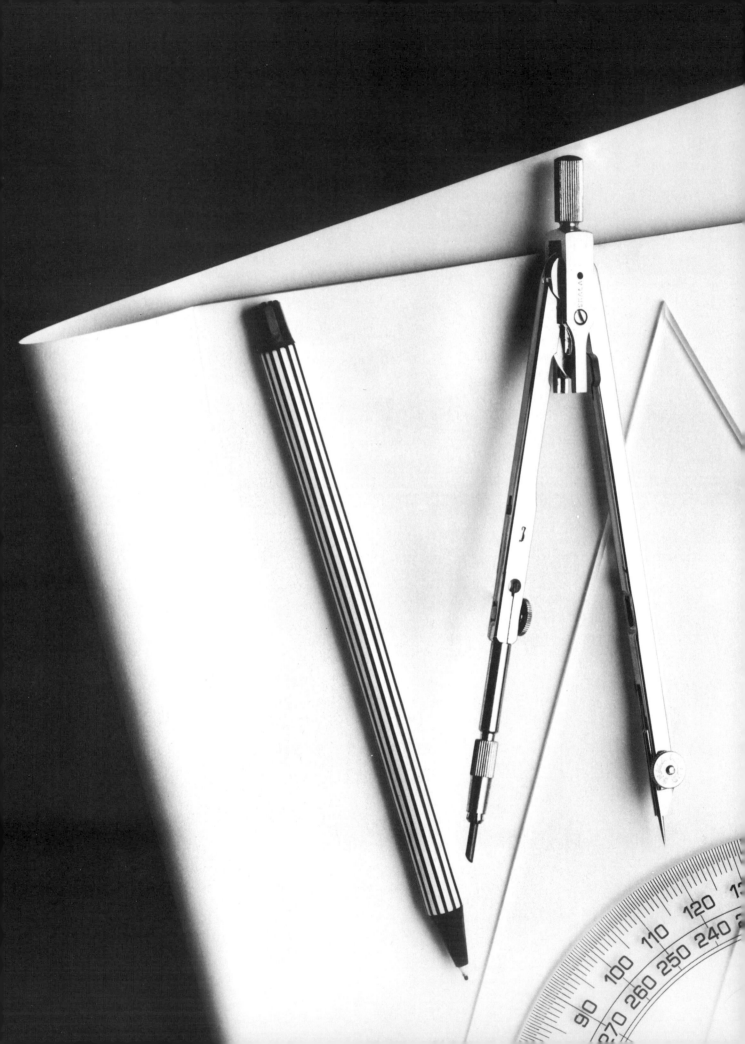

Design

An analysis of all the problems that need to be
resolved before starting any project

Design tools

Before attempting to make any article from wood, such as a piece of furniture, the main dimensions and method of construction need to be established. For very simple pieces it is possible to work from a free-hand sketch, but usually an accurate drawing either of the whole item, or details of its critical dimensions, is essential. An effective drawing can be made with very little equipment. The most basic equipment required to make a working drawing is a flat surface, such as a drawing board, to work on and straightedges, such as a T square and triangle, to draw against.

The drawing board should have a straight edge to guide the T square and a semihard surface that will accept the point of a compass. At its simplest the drawing board is a sheet of cardboard on a table.

The T square is a straightedge with a stock firmly fixed at right angles across one end. On some T squares the straightedge is tapered on one side. It is better, however, to buy a T square that is parallel-sided so that the worker can use both edges, making it easier to draw near to the bottom edge of a board. A second advantage of a parallel-sided T square is that it can be used by both left- and right-handed people. A beveled edge will improve drawing accuracy as the worker can see the drawing edge and therefore position the T square with greater precision.

Standard triangles – 30°/60° and 45° – are the most important single instruments and it is therefore worth buying the best available. A large adjustable triangle is an advantage when drawing the long lines and angles that are common in furniture drawings. As with all other plastic instruments, buy tinted transparent plastic in preference to plain as it can be seen more easily when covered by scraps of paper and drawings. Clean T squares, triangles, rules and templates regularly; most can be washed in warm soapy water.

The board and straightedge can be bought as a single unit. A "parallel motion" straightedge is fixed to the board and operated on pulley wheels with either tensioned wire or cord guides. A "drafting machine" has a head with two rules set at right angles to each other, which can be adjusted to any angle on the surface of the board.

For best results both pencils and pens must be properly maintained. Avoid dropping any type of pencil as the lead breaks very easily — within the casing as well as at the tip. Sharpen a wood-cased pencil at a shallow, even angle to give a gradual taper and therefore a slower thickening of line as it wears. When using a pen with a tubular nib and plunger it is essential to use the ink recommended by the manufacturer and to wash the pen out after use. Ink flow is controlled by capillary action. If the ink is too thin it will blot; if it is too thick it will starve the nib.

Avoid a coarse-textured cartridge paper when using a pencil as it will wear the point quicker; this will cause carbon dust to collect on the surface, making it easier to smudge the drawing. Buy a thicker paper if drawing in ink or if intending to make many revisions. Printed grid papers can save constant measuring. They are available printed with different sized squares and isometric grids and can either be drawn onto directly or be used as a backing to a tracing paper overlay.

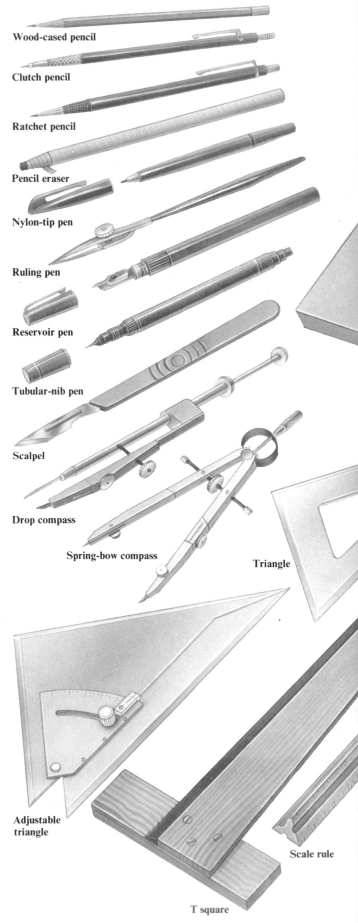

Wood-cased pencil

Clutch pencil

Ratchet pencil

Pencil eraser

Nylon-tip pen

Ruling pen

Reservoir pen

Tubular-nib pen

Scalpel

Drop compass

Spring-bow compass

Triangle

Adjustable triangle

Scale rule

T square

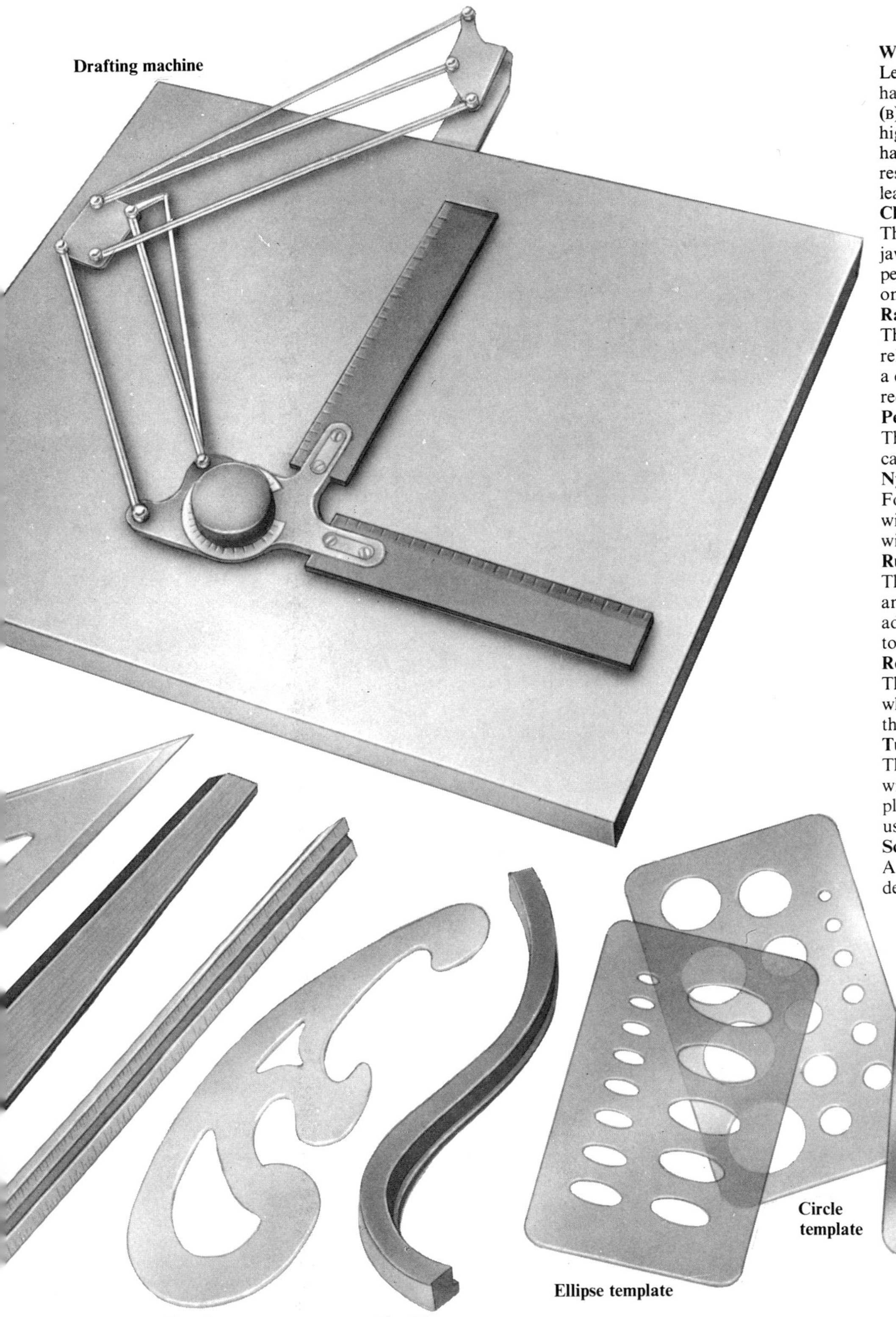

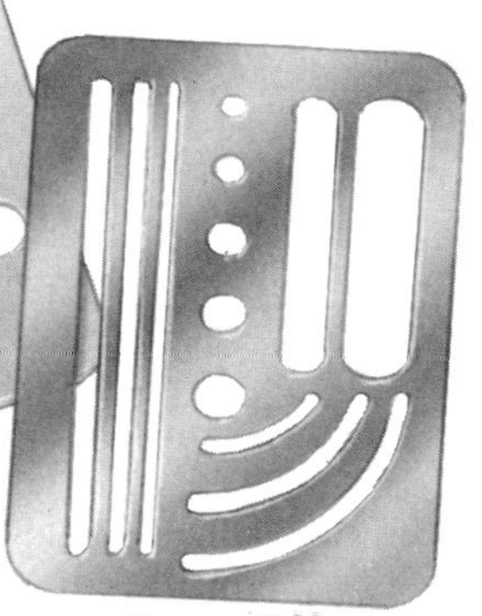

Drafting machine

Wood-cased pencil
Leads are graded as to hardness (H) and blackness (B). Grades are numbered, the highest numbers being the hardest and blackest respectively. The blacker the lead the quicker it will wear.

Clutch pencil
The lead is gripped by a set of jaws, released by pressing the pencil's top. Sharpen the lead on fine garnet paper.

Ratchet pencil
The lead is very fine and is released from the top through a controlled feed. It does not require sharpening.

Pencil eraser
The type with a coiled-paper casing is the most convenient.

Nylon-tip pen
For rough pen work. The type with a metal casing at the tip will not leave ink on the rule.

Ruling pen
The nib is filled directly from an ink dispenser and can be adjusted, by turning the screw, to a wide range of thicknesses.

Reservoir pen
The pen has an ink reservoir to which nibs for varying line thicknesses can be attached.

Tubular-nib pen
The pen has an ink container with a range of tubular-nib and plunger attachments. It can be used at any angle.

Scalpel
A very sharp knife with a detachable blade.

Circle template

Ellipse template

Eraser shield

French curve

Flexible curve

Drafting machine
A pair of parallel arms clamped to the edge of the board allows the drawing edges to be positioned anywhere.

Drop compass
The pin forms a vertical axis, making it easy to draw very small circles.

Spring-bow compass
Used for most circle drawing.

Use trammel heads for very large circles (*see* page 55).

Triangle
An adjustable triangle with wide arms for greater rigidity is useful but expensive. All edges should be smooth and beveled to allow for greater accuracy.

T square
Choose a square with parallel sides and beveled edges.

Scale rule
Used to measure in proportion to produce scaled-down drawings. Choose one with thin scale lines and thin edges for accurate measuring.

French curve
Set of templates used for drawing free curved shapes.

Flexible curve
Used for drawing large flowing shapes. The curve can be bent to follow any shape. It has a lead core with a plastic sleeve.

Ellipse and circle templates
Especially useful for drawing small ellipses and circles.

Eraser shield
Used to protect areas of the drawing around a line or detail to be erased. Made from thin flexible steel.

Basic design techniques

There are basic drawing tenets, methods and conventions that help to simplify and speed up the drawing process and to avoid the problems of misreading the final result. Drawing methods relate to an understanding of basic geometrical shapes, proportion and scaled drawings. In order to produce a drawing of manageable size it is necessary to reduce, in exact proportions, the intended dimensions of the object. To do this, a scale rule is ideal. Choose a scale to suit the size of the object being drawn and the amount of detail needed to be shown. Full-size drawings of such details as moldings are often drawn alongside scaled drawings of the whole subject. The usual scale for individual pieces of furniture is 1:5 (metric) or 1:4 (U.S.), and for room layouts 1:20 (metric) and 1:24 (U.S.). The scale 1:5 means that 1 centimeter on the scaled drawing represents 5 centimeters on the object.

Of the various methods of laying out a drawing the most universal is known as third-angle projection; the individual face-on views usually shown are one or more of the following: the front, the side and the top. By representing these views pictorially, however, with the three surfaces adjoining each other, the projection can be more easily read and understood. The easiest pictorial projection to draw is an isometric one.

In drawings, there are conventional ways to present and record dimensions, diameters, radii, sections and types of material, so that they can be readily interpreted both by the designer and the intended user of the finished object. For clarity, dimensions are added not to the outline itself but between the extension lines, which are extensions of the object's outlines. Write the figures centrally above the dimension line, omitting the unit of measurement, which can be specified in a title box describing the drawing. All dimensions should be so placed that they are readable from the bottom right-hand corner of the drawing.

Of the many methods and skills that will be acquired with practice, the following should prove useful. When running a pencil along a straightedge, rotate the pencil slowly to give an even wear and an even thickness of line. One of the main problems when drawing in ink with a pen that is not equipped with a tubular nib is the possibility of the ink being drawn under the straightedge by capillary action; this invariably results in the ink smudging when the straightedge is removed. To avoid this, use bevel-edged rules and triangles with the bevel downwards, so that the edge is not resting on the work. If the straightedge is not bevel edged, angle the pen away slightly to prevent the ink from flowing back under the edge. To avoid smudging when using a T square on a drawing board, secure two pieces of string so they are held flat and taut across the board, top to bottom. Rest the T square on the string.

When drawing up large areas of work, cover parts already completed with tracing paper to avoid any smudging. An ink line drawn on tracing paper can be removed easily by scratching with a scalpel; the surface may then need to be lightly burnished with the back of a teaspoon to avoid subsequent lines spreading. Finally, good lighting is essential for accurate drawing. Apart from a general light, an adjustable local light will be needed to highlight the work area.

Basic geometrical shapes

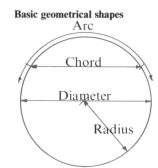

Circle parts.

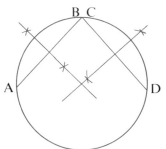

Finding the center of a circle.

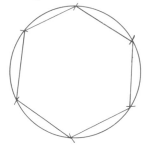

Drawing a hexagon.

Drawing an octagon.

Basic geometrical shapes

The circle is the key to a number of geometrical shapes. To find the exact center of a circle, draw a chord AB. Set a compass to over half AB. From A scribe arcs inside and outside the circle. Repeat from B, intersecting the arcs. Join up the points of intersection. Draw a chord CD. Repeat this process. The lines intersect at the center. To draw a hexagon, first draw a circle with a radius equal to the side length of the required hexagon. With the compass at this setting, step around the circumference six times. Join up the points. To draw an octagon within a square, draw the diagonals of the square. Set a compass to half the length of a diagonal. With the compass point on a corner intersect the adjacent sides. Repeat from each corner and join up the points of intersection. To draw an equilateral triangle, first draw the line AB to the required length. Set a compass to the line length and describe two arcs, one from A and one from B; join up the lines from the point of intersection C. To draw an ellipse, first decide and draw up the major and minor axes, AB and CD, which bisect each other at right angles. Set a compass to AX. From C intersect AB giving points N_1 and N_2. Insert pins at N_1, N_2 and B. Tie string tightly from N_1 to N_2, via B. Remove pin B. Keeping the string taut, draw the ellipse.

Drawing an equilateral triangle.

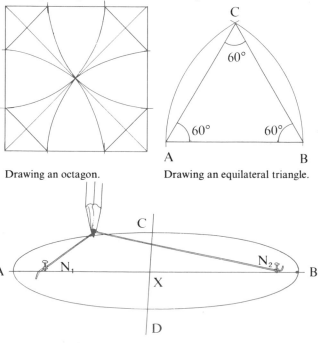

Drawing an ellipse.

Isometric projection

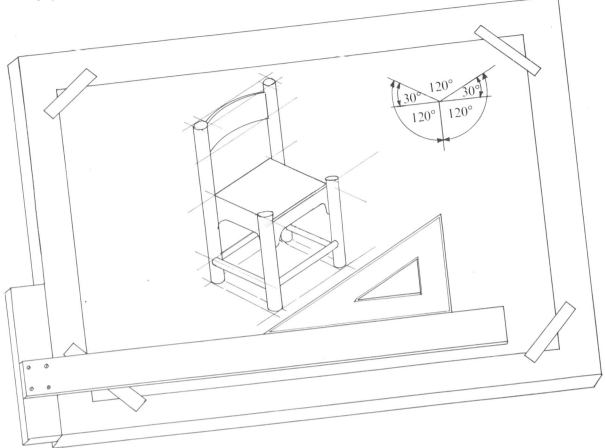

Third-angle projection

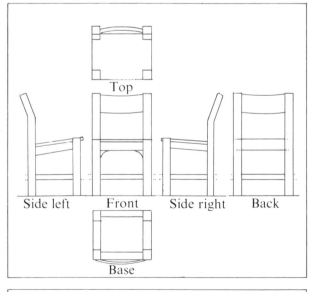

Top

Side left   Front   Side right   Back

Base

Drawing symbols

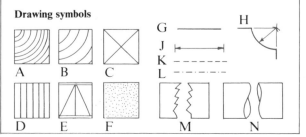

A   B   C

D   E   F

G
J
K
L

H

M   N

Third-angle projection
One of the most
straightforward ways of
representing a three-
dimensional object is to draw
the individual views from the
front, side and top. The front
view (front elevation) is drawn
first; the view from the right
(side elevation) is drawn on the
right of it and the view from
the left is drawn on the left; the
top view (plan) is drawn above
the front view. By following
these conventions the drawing
will be universally understood.
All views should be drawn to
the same scale in proportion to
the intended dimensions of the
object. As many as six views
can be drawn; for simple
objects, however, front and
side elevations should convey
all the necessary information
to describe the object.

Drawing symbols
For clarification, symbols can
distinguish the types of
material to be used, such as
finished hardwood (A), finished
softwood (B), nominal-sized
timber (C), multi-ply (D),
blockboard (E), chipboard (F).
Lines of different thickness are
used to clarify parts of the
drawing. Use the thinnest lines

Isometric projection
Of the many pictorial methods
of presenting a drawing, the
isometric projection is one of
the most common and most
straightforward.
Conventionally, all vertical
lines are drawn vertically and
all other lines are drawn to the
side of each vertical at 30° to
the horizontal base line. Use a
30°/60° triangle with a T
square. The three isometric
axes around which all drawings
are made are shown on the
drawing (*above*). Because of
this rigid geometrical method,
which does not follow the laws
of perspective, the picture will
appear distorted. Being
mechanically correct, however,
it can be used for taking off
dimensions. This makes the
projection particularly useful
for complicated structures.

for initial guide lines and the
thickest (G) for the main
outline. Make a thin line to
indicate diameters and radii
(H) and for dimension lines,
drawn between extension lines
(J). Use thin, short dashes for
hidden details (K), a chain line
as a center line (L), jagged lines
for rectangles (M) and wavy
lines for cylinders (N).

Planning a project

A methodical approach to design, following the inception of an idea, is likely to ensure the best solution to the problems of intended use and construction. This design approach can be broken down into basic stages: the brief, or analysis, of the idea; sketch drawings; scaled drawings; and models and rods (full-size drawings). A cutting list is then made of the required materials.

The brief identifies the intended use and function of the object within its proposed context. It can set limits on the materials and finishes and how they are to be used. It can set visual guidelines; thus, for compatibility with existing furniture or room setting, the intended proportions, positioning and color can be compared. The brief need only be a short statement clarifying the needs for the object. Even where ideas may seem very clear and the furniture simple, a brief is always useful as a written record, providing a check on progress throughout the project. In the example illustrated the brief would describe the following requirements: a table with fixed top to seat two people; to be used mainly with one side against a wall with seating positions either at the ends or on one side; stable hardwood supporting the structure; veneered top with a durable finish harmonizing with the chairs.

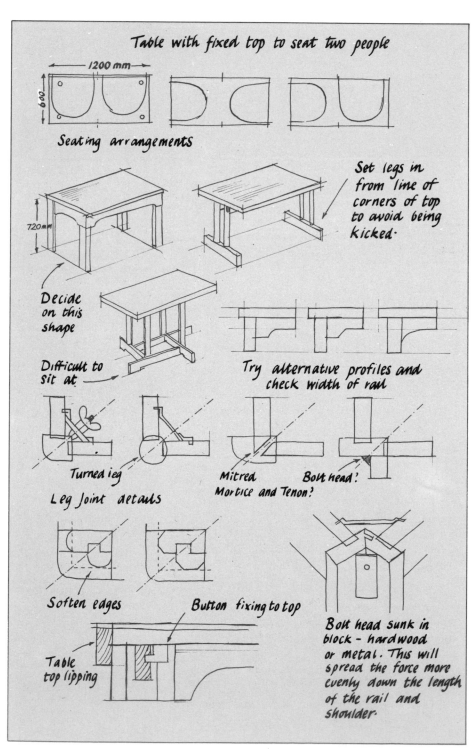

Table with fixed top to seat two people

1200 mm
600

Seating arrangements

720 mm

Set legs in from line of corners of top to avoid being kicked.

Decide on this shape

Difficult to sit at

Try alternative profiles and check width of rail

Turned leg

Leg joint details

Mitred Mortice and Tenon?

Bolt head?

Soften edges

Button fixing to top

Table top lipping

Bolt head sunk in block - hardwood or metal. This will spread the force more evenly down the length of the rail and shoulder.

Sketch drawings
The designer's sketchbook translates the written brief into a series of visual solutions. Among the general considerations incorporated will be the material sizes and thicknesses, with appropriate allowance made for joints. It can also include a more precise breakdown of materials required, such as fixtures to be used and special tools. From the various sketches and ideas, one will be selected and developed. Check that the sketch chosen conforms in all aspects with the outline on the original brief. In the example illustrated the following are a few of the considerations that have been made: the size of the top; its height and rail width to give sufficient knee clearance; the leg structure and the method of connecting the rails with joints, knock-down fittings or dowels; the lower shape of the rail; and, finally, how to fix the top.

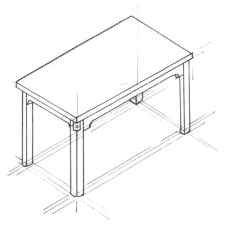

Scaled drawings

The scaled drawing is the basis for confirming the decisions taken at the sketch stage and for establishing accurate dimensions and proportions. The most straightforward drawing uses third-angle projection to show, respectively, the front, side and top views. A scale of 1:5 (metric) or 1:4 (U.S.) will usually result in an overall drawing of manageable size. An isometric drawing, showing the three views combined, is a useful aid to a fuller appreciation of the design. When designing furniture to fit a particular room, it may also prove helpful to apply the same principles of scaled drawing to the room itself. Cardboard cutouts of each item of furniture can be moved around a scaled plan of the room to establish the position and suitability of the new piece. The example illustrates an isometric projection.

Models and rods

Full-size drawings against which the work can be laid for checking are known as rods. The drawing is usually a cross section, either vertical or horizontal, to show the joints more clearly; it can be laid out on lining paper or a sheet of three-ply or thin board. It is especially useful when defining a particularly intricate part of the design, such as the position of the pivot on a folding chair, and for checking the work at progressive stages. For a more complete evaluation of the overall appearance of the object, a scaled model, made from stiff cardboard or balsa wood, is helpful. Details of balsa models can be taken direct from the scaled drawings; when using stiff cardboard, take the dimensions from the scaled drawings and redraw the object, making as many surfaces as possible continuous with their adjacent sides for cutting and folding into a three-dimensional form. The example illustrates a rod for the joint in the table leg, drawn using third-angle projection and also as an isometric drawing.

The cutting list

Using the scaled drawings or rods, the cutting list can be prepared for ordering the wood — either in nominal or finished sizes. Nominal is the size that the boards are cut on the band saw or a circular saw. When planed, the board's thickness is reduced by at least 3 mm/$\frac{1}{8}$ in. Prepared wood is indicated on the cutting list either as finished sizes or s4s (surfaced on all four sides). List the names of the parts, the number of pieces of the same size; then, in order, the length, width and thickness, allowing 13 mm/$\frac{1}{2}$ in of waste in length and 6 mm/$\frac{1}{4}$ in in width. Specify the wood required and recheck the total number of parts. The example illustrates the cutting list for the table.

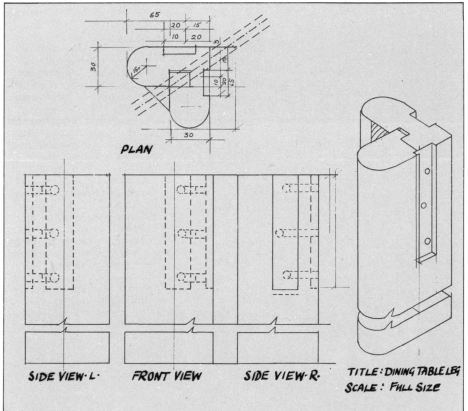

PLAN

SIDE VIEW·L·   FRONT VIEW   SIDE VIEW·R·

TITLE: DINING TABLE LEG
SCALE: FULL SIZE

Cutting List	All finished sizes · Units in m.m.				
ITEM	Nº OFF	MATERIAL	LENGTH	WIDTH	THICKNESS
Leg Members	4	Pine	735	70	30
Leg Members	4	Pine	735	50	30
Rails	2	Pine	1,052	100	20
Rails	2	Pine	440	100	20
Top	1	Pine	1,212	606	23
Buttons for top	14	Pine	45	35	20
Leg inserts	4	Mahogany	102	20	20
Pellets to cover bolt heads	4	Mahogany	150	20	20
Dowels	24	Beech	40	11 (dia)	
Corner leg braces	4	Large complete with 6 mm/$\frac{1}{4}$" mushroom-headed bolts, washers and wing nuts.				

Storage

Ideally, well-designed storage should be capable of coping with the changing demands of an individual or a family, it should be easily accessible, make economical use of space and protect stored items against dust, mildew and extremes of temperature. The main restraints on any design are the size of the items to be stored, the available space for storage, the height to which the items can be lifted and the distance that can be reached to retrieve them. These ideal requirements and restraints should also be considered in the context of the area needed for circulation around other objects or work surfaces and the daily routines in each room.

Storage items can be categorized into those used most frequently, such as food, shirts and toys, and those used only occasionally or seasonally, such as lawn-mowers, winter overcoats and suitcases. Those items used frequently should be easily accessible and those used infrequently should be well protected.

Storage units can be categorized into those that are free standing and those that are fixed. Apart from individual pieces, free standing storage can be considered in modular form; unit-size carcasses can be fitted out individually to suit particular requirements and be arranged, rearranged and added to at will. The modular system can also be applied to fixed storage. Manufactured boards fixed from the floor to the ceiling can be used to take adjustable shelves and units. Fixed storage can be built in or be attached to a wall. Whatever the type of storage, fixtures and fittings can influence the design (*see* pages 256–65).

Key dimensions

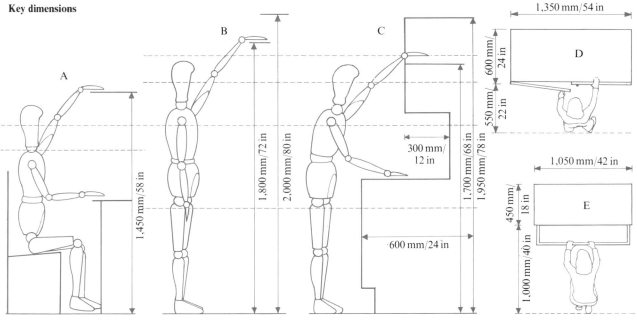

The central zone for storing items most frequently used or heavier items requiring two hands to lift with minimal effort is between 700 mm/28 in and 1,300 mm/52 in. Longer-

term storage or that for lighter-weight items not requiring direct visual access should be located above or below this zone. The average eye level for a standing adult is from

1,550 mm/62 in to 1,750 mm/70 in, and for a sitting adult, 1,100 mm/44 in to 1,250 mm/50 in. Easy vertical reach when sitting should be considered (**A**), as should both comfortable

reach and full stretch when standing (**B**) and when leaning over a work top (**C**). Allow sufficient area when using a wardrobe (**D**) and a chest of drawers (**E**).

Free-standing modular storage

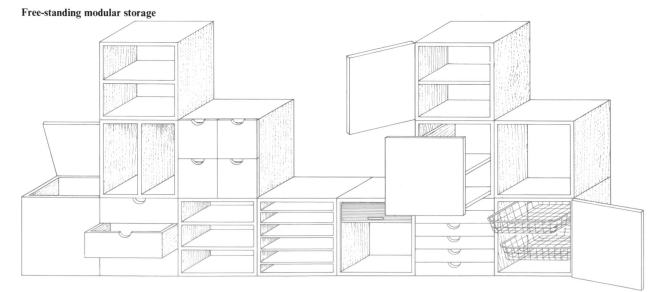

Drawer fronts

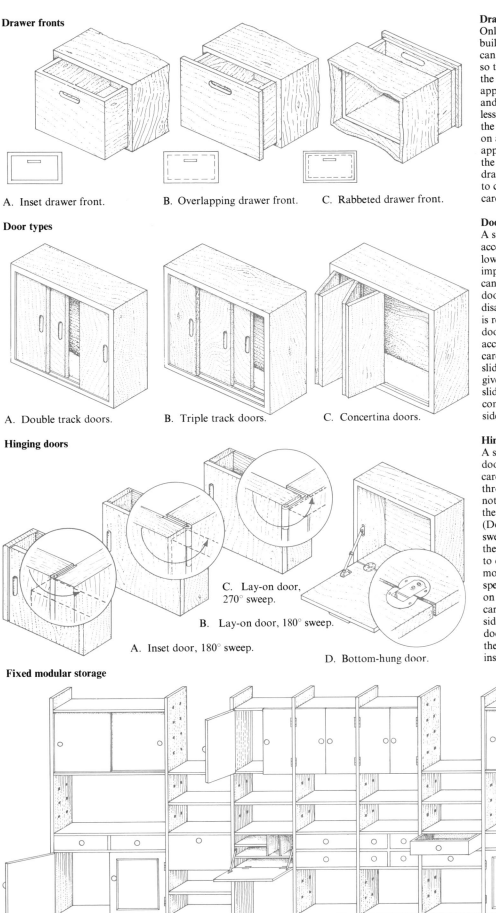

A. Inset drawer front.

B. Overlapping drawer front.

C. Rabbeted drawer front.

Door types

A. Double track doors.

B. Triple track doors.

C. Concertina doors.

Hinging doors

C. Lay-on door, 270° sweep.

B. Lay-on door, 180° sweep.

A. Inset door, 180° sweep.

D. Bottom-hung door.

Fixed modular storage

Drawer fronts

Only when a carcass is well built from well-seasoned timber can each drawer be made so that it is visually framed by the carcass without any gaps appearing between the carcass and the drawer front (**A**). If a less rigid carcass is used or if the carcass is to be positioned on an uneven floor, gaps can appear between the carcass and the drawer. To avoid this, the drawer front can be extended to cover whole or part of the carcass (**B**), or be rabbeted (**C**).

Door types

A side-hung door provides best access to a cupboard, but at low level such a door can be impractical and at high level it can be dangerous. Sliding doors do not have these disadvantages; however, access is restricted. Three sliding doors on a double track give access to only one-third of the carcass at one time (**A**); three sliding doors on a triple track give access to two-thirds (**B**). A sliding, pivoting door (**C**) combines the advantages of the side-hung and sliding doors.

Hinging doors

A simple butt hinge fixed to a door inset flush into the carcass allows the door to open through 180° (**A**). If the door is not inset flush into the carcass the sweep will be restricted. (Dotted lines show maximum sweeps.) A door fully covering the carcass edges can be hinged to open through 180° (**B**). By moving the pivot point specialized hinges enable lay-on doors to open when two carcasses are butted side by side (**C**). For bottom-hung doors, the hinge should allow the flap to open flush with the inside of the carcass (**D**).

Tables

Of the many variations in table design, there are four basic types from which all others developed. Each type reflects the distinct requirements of the designer faced with the following key problems. Is the table for regular or infrequent use? How much space is available? How many will use the table at any one time?

The first type is the fixed-top table, such as refectory and coffee tables; it is the simplest to design and construct, whether for regular or infrequent use. The second is the collapsible table, such as trestle and card tables, which resolve the problem of infrequent use. The third and fourth types are the extending tables, the first having visible flaps, such as on gateleg and Pembroke tables, the second having hidden leaves, such as on drawleaf and swivel-top tables.

Whatever the type of table, the considerations of height are applicable to all. The height from the floor ranges from 675 mm/27 in to 725 mm/29 in for tables related to seat heights. The lower height applies to machinist's tables, such as those used by typists, the higher to dining tables. The height of tables used while standing, as in the kitchen or workshop, is about 900 mm/36 in. For occasional tables the range can be between 300 mm/12 in and 550 mm/22 in.

The key dimensions

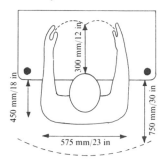

The key dimensions, when deciding the size of the table top and the position of the supporting structure, are a width of 575 mm/23 in per person, allowing for a reasonable amount of elbow room, and a clearance of 300 mm/12 in from the edge towards the center of the table for the sitter's legs. When one is seated at a table, the back of the chair is usually about 400 mm/16 in to 450 mm/18 in from the edge of the table; allow, therefore, a minimum of 750 mm/30 in from the edge of the table for circulation. A square table will take four people comfortably. A rectangular table provides greater seating capacity than a square one of the same area. The minimum comfortable width for a rectangular table is 750 mm/30 in. The minimum diameter for a round table — the most sociable shape — is 1,000 mm/40 in to seat four people. An oval table combines the best qualities of the round and rectangular tables. Leaves or flaps can be added to any of these tables. In each case consider the position of the supporting structure and the problem of the sitter's position relative to the table legs. (Table-leg positions are indicated by black dots.)

Fixed-top and collapsible tables

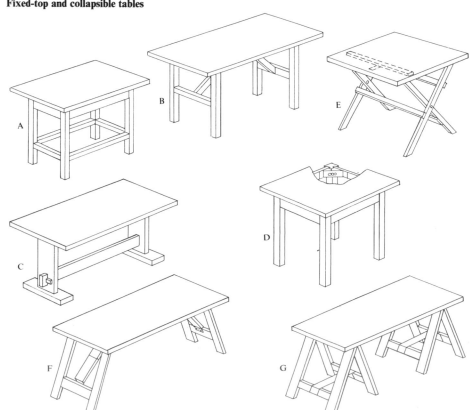

Fixed-top and collapsible tables
A fixed top can be almost any size, and it can be supported on free-standing rigid structures or folding structures. The rigid structure can be very simple, consisting of a free-standing leg assembly glued and screwed to a top (**A**) or legs braced to a simple batten support (**B**). The refectory table (**C**) is a long, narrow solid-top table. Large tables can be difficult to transport, so it is worth considering a detachable top or legs; a hanger bolt and wing nut is the simplest way of fixing removable legs (**D**). A common form of folding table is the games table (**E**). Four irregularly spaced rails act as stops when the table is folded, and a turn catch locks the tabletop against a rail in the upright and folded positions. Different leg assembly is preferable for long tables (**F**). The trestle table is the classic folding table; the two trestles are kept upright with a web strap and are located between two rails fixed the length of the top (**G**). The design is ideally suited to seasonal use.

Visible-flap tables

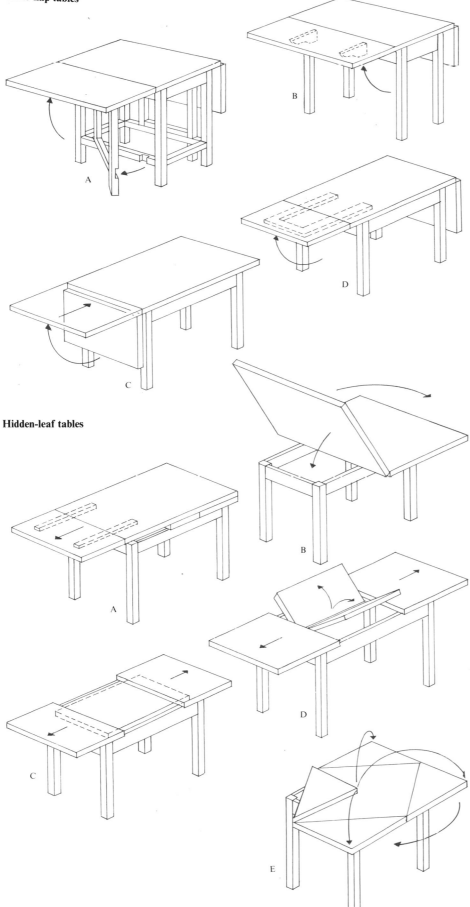

Visible-flap tables
The main disadvantage of visible-flap tables is the restriction on seating when the flap is not in use. However, the folding-leg frame, such as in the gateleg table, offers the greatest extension of area of any type of table. The most usual visible-flap table is the double gateleg (**A**). A whole range of tops can be used with one or more gates pivoted on the main frame. Instead of gates, the Pembroke table (**B**) has support brackets pivoted at the sides. These can either comprise one long bracket or two short ones, depending on the size of the flap. The single-flap table has a top that slides in grooves; a fall flap hinged to the edge of the table can be raised and the whole surface slid across so the leg frame supports both the flap and the top (**C**). The drop-leaf table has a fixed top housing a slider frame for each flap; the flap is raised and the slider is pulled out to support it (**D**).

Hidden-leaf tables

Hidden-leaf tables
The drawleaf table (**A**) consists of two leaves and a center bearer supporting a loose top; all are of identical thickness. Runners screwed to the leaves slide between guiding bearers and slots in the main frame. The swivel-top table (**B**) has a double top hinged along one edge, which is pivoted, allowing it to move through 90°. The most satisfactory shape, when swiveled and opened out, is a square. An alternative to end leaves is to use a center leaf. The simplest type has a sliding top divided across its width. When both halves are pulled apart, a loose center leaf, stored below the top or away from the table, is fitted centrally onto the frame (**C**). In a similar version, the center leaf is in two halves hinged together and pivoted at the sides to cross rails; the complete leaf can be folded up and swung down within the table frame (**D**). The envelope-top table (**E**) is a variation of the swivel top. The table has a square lower top pivoted to the frame, allowing it to move through 45°, and an upper top made of four triangular sections which is hinged to the lower top.

Chairs

Traditional chair design and construction demand a high degree of specialist skill, the main problem being that one component rarely meets at a right angle to the next. Modern designs, however, employ straight-sectioned timber, where possible, and timber that has been bent by steaming or laminating for curved work.

Before starting a design there are three key points to consider — stability, durability and comfort. Will the chair fall over when a sitter starts rocking it? Will it stand many years of use and misuse? Are chair arms necessary? Will the type of upholstery be suited to the chair's use? Finally, the visual appearance has to be considered. Will the chair suit the room setting?

Of all these considerations comfort is the one on which chairs are primarily judged. However, comfort is not measured or guaranteed by the amount of soft upholstery. The essentials to comfort are the angles and positions of the individual components and the support given by the completed frame. If they are correct the body muscles should not have to work significantly to maintain a comfortable posture.

A front seat height of 400 mm/16 in to 450 mm/18 in is suggested so the sitter's feet may rest on the floor, with a seat depth not exceeding 460 mm/18½ in if the back is straight across its width, or 420 mm/16¾ in if the back is curved. This will ensure that the front edge of the seat does not exert pressure at the back of the knees, resulting in blood starvation in the legs. The seat should slope back slightly, between 5° and 8° from the horizontal or by about 20 mm/¾ in in height between the front and back legs. This slope will help to hold the natural tilt of the pelvis and lumbar curve of the spine. About 200 mm/8 in above the seat the chair back should be shaped to follow the lumbar curve and then be extended, for a high back, at a 20° to 25° angle to the vertical to support the shoulders. The front edge of the armrests should be about 200 mm/8 in above the seat and set back by a minimum of 60 mm/2½ in to ensure that the knuckles are not crushed when the chair is pulled up to the table's edge. The amount of upholstery should be kept to a minimum.

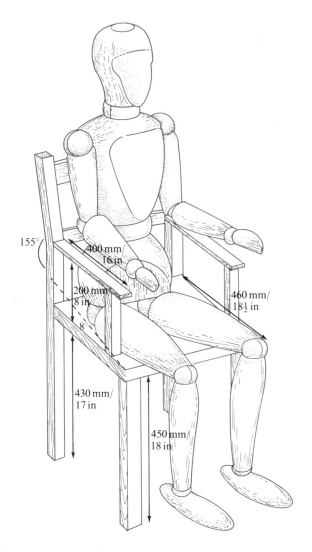

The chair dimensions illustrated are the maximum suitable for the "average-sized" person; they should therefore be considered carefully in relation to the worker's specific needs. In particular, when designing dining chairs, ensure that they relate to the height and structure of the table.

Rigid frame

Top-stacking frame

Side-stacking frame

Stacking chairs

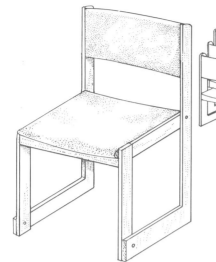

Stacking chairs

The design of most stacking chairs is derived from two basic types — those that stack from the front and those that stack from the top. A front-stacking chair has part of the structure offset to allow one frame to pass over the next. Usually the side frames are constructed in two L shapes, one forming a back leg and floor rail, the other the front leg and seat rail; the two are offset at the join. A top-stacking chair has legs fitted to the outside edges of the seat frame, again slightly offset to give a tolerance for stacking. Both types can be stacked five or six chairs deep before the stack becomes unstable. With the addition of arms, the chairs will be stronger and more stable when stacked.

Side-stacking principle

Top-stacking principle

Folding chairs

Folding chairs

The design of most folding chairs is derived from two basic types — those that fold from the front to the back and those that fold from side to side. The deck chair and park chair are typical examples of the first and the "director's" chair is typical of the second. These two basic types can however be combined. The key feature of any design is the positions of the pivot points. One way of assessing the correct positions is to use strips of thin cardboard pinned at each pivot point. The pivots themselves, whether standard hinges or bolts, will determine the stability and durability of the chair. Make a prototype of the pivots early on in the design process.

Front-to-back folding principle

Side-to-side folding principle

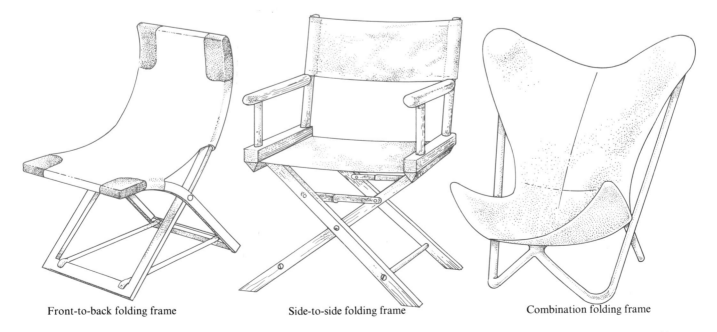

Front-to-back folding frame

Side-to-side folding frame

Combination folding frame

Beds

As well as being the most used piece of furniture in a house, a bed can also dominate any planning and layout of a room. Only when the bed is disguised or folded away can the room take on an independent character. The variations on bed design are thus for the most part centered on the problems of space saving and storage. The three main types of rigid space savers are the stacking bed, the nest of beds and the bunk bed. The bunk bed can be developed by converting the lower bed into a table or a storage unit.

The interrelationship of the bed base with the mattress is the key to comfort. The sleeping surface must be horizontal with no sag; it should allow the spine to maintain its natural shape. This is best achieved by increasing the firmness of the mattress from a soft top surface down to a more resistant bottom.

The design of a bed will be limited by the standard sizes available if a proprietary mattress is to be used. It is therefore worth considering making a mattress, using rubber foam or layers of polyurethane foam of varying densities and original-shape recovery rates.

The ideal support

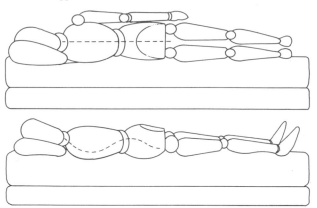

The human spine naturally forms a straight line when viewed from the front and back, and an S shape when viewed from the side. The ideal mattress should hold the spine as close as possible to this shape. A very firm mattress gives better support than an oversoft one.

Circulation and storage area

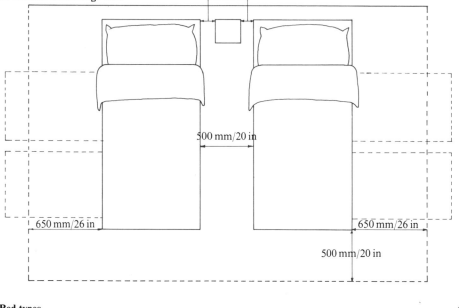

100 mm/4 in

500 mm/20 in

650 mm/26 in

650 mm/26 in

500 mm/20 in

Circulation and storage area
The minimum circulation space around a bed when bedmaking should be 650 mm/26 in at the sides and 500 mm/20 in at the foot and between twin beds. Allow more space if there is to be a storage area underneath. The dotted lines represent drawers pulled out to their fullest extent. A bed should be easy to move for room cleaning. To ensure that it is unlikely to move accidentally, fit the base with glides and casters, for free movement, or glides with wheels, for movement in one direction.

Bed types
Stacking beds (**A**) are useful when a room has to double as a bedroom and play- or sitting room. Up to four beds can be stacked together with either just the mattress in position on each bed or with each bed fully made up. Each supporting leg on bunk beds (**B**) can be divided into two to allow the bunks to be separated into two single beds. A double bed (**C**) can be built from two single beds, connected at the base by a hook. The mattresses are zipped together. This construction makes it easier to move the bed from room to room and gives the opportunity of varying the firmness of the mattresses according to requirements. A nest of beds (**D**) is an alternative to stacking beds.

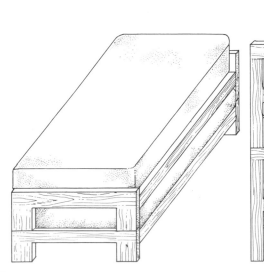

A. Stacking beds

B. Bunk beds

Bed bases

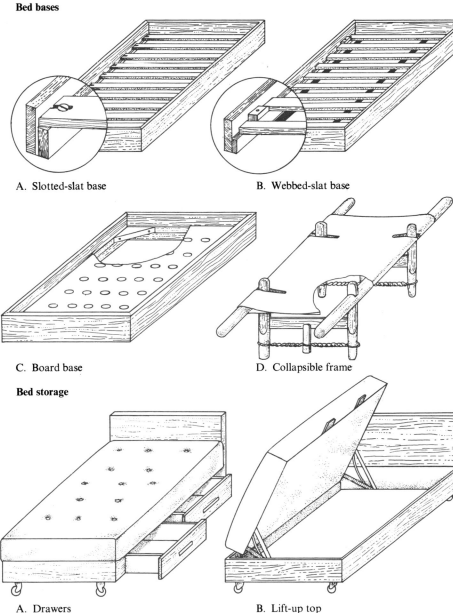

A. Slotted-slat base

B. Webbed-slat base

C. Board base

D. Collapsible frame

Bed bases

The dimensions of the bed base will be restricted if a proprietary mattress is to be used. Examples of standard sizes are 900 mm/36 in, 1,500 mm/60 in and 2,000 mm/ 80 in in width, and 1,900 mm/ 76 in and 2,000 mm/80 in in length. Ideally the bed should be 200 mm/8 in longer than the user's standing height. The height of the bed will be a compromise between a comfortable height for sitting (up to 450 mm/18 in) and one for easy bedmaking (600 mm/24 in). The height can affect the appearance of the room — a low bed making a room look much bigger. The base can consist of solid wood or five-ply slats, either slotted and screwed (**A**) or attached to webbing (**B**) to allow the slats to flex; a manufactured board drilled over its whole area to ventilate the mattress (**C**) or a collapsible frame with a tourniquet action for tensioning the canvas base (**D**).

Bed storage

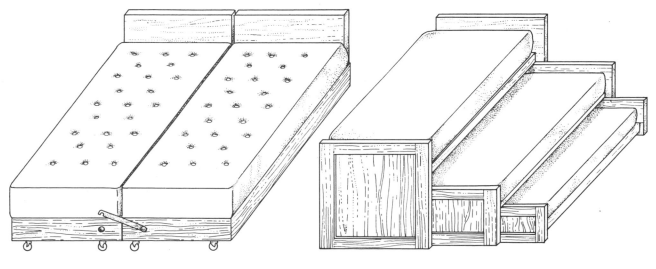

A. Drawers

B. Lift-up top

Bed storage

The base frame can be designed so that it is high enough to accommodate storage units underneath. Drawers can be suspended from or fitted into the base (**A**). Alternatively, a detached storage box on casters is simpler to make; however this has the disadvantage that the box will not move with the bed. A bed with a lift-up top (**B**) can have its full base area used for storage. This design, which incorporates spring-loaded hinges, is usually only suitable with lightweight foam mattresses.

C. Linked single beds

D. Nest of beds

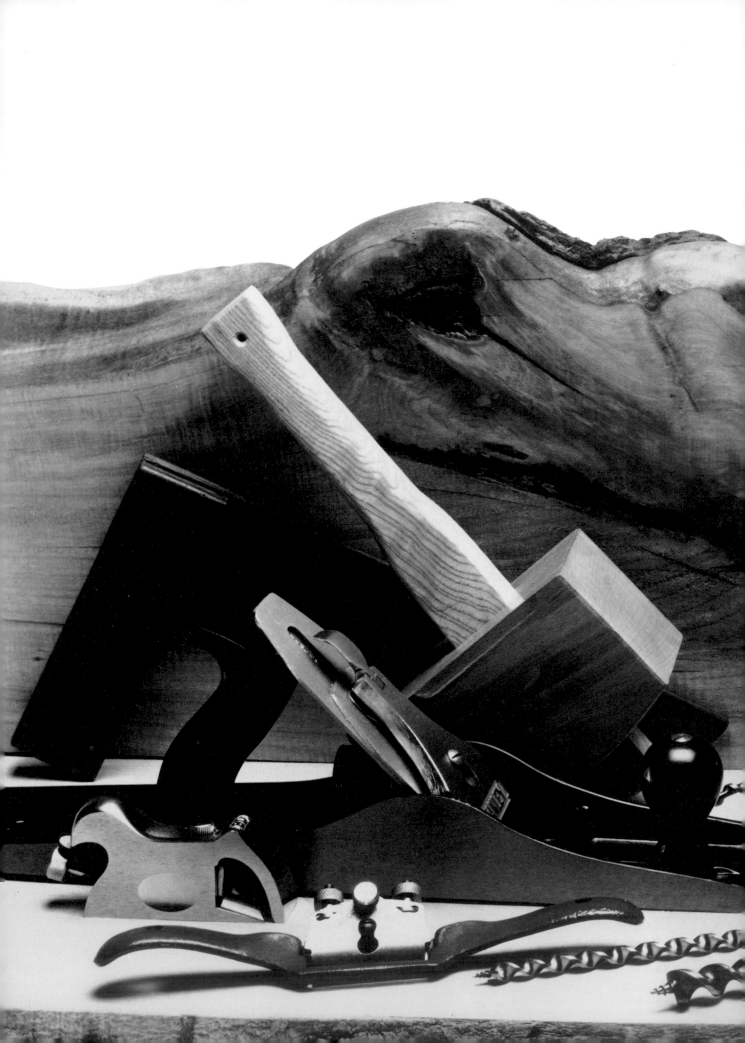

Tools

A survey of all the hand and machine tools
required to produce quality work

The tool kit

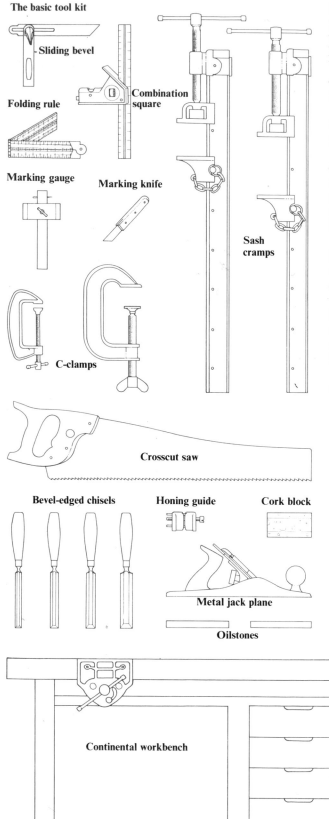

The basic tool kit

Sliding bevel

Folding rule

Combination square

Marking gauge

Marking knife

Sash cramps

C-clamps

Crosscut saw

Coping saw

Tenon saw

Bevel-edged chisels

Honing guide

Cork block

Punches

Cross-peen hammer

Mallet

Metal jack plane

Oilstones

Screwdrivers

Bradawl

Pincers

Suggesting a basic kit must inevitably involve personal preferences largely dependent on the nature of the work attempted; whatever the tool eventually selected, however, always buy the best quality at the outset. Working with inferior tools is an unnecessary handicap. Equally, the best results cannot be achieved if the tool is misused or not sharpened or maintained.

The first consideration should be a sturdy and adaptable workbench. The continental-type bench is preferable, because it can hold the wood at both ends.

Of those tools for measuring and marking, the combination square is far more versatile than the traditional 100 mm/4 in try square. Of the gauges a marking gauge is the most essential. A marking knife is preferable to a pencil, as it makes a finer, clearer line for the saw to follow.

Many ingenious holding devices are available, but initially two C-clamps, one large and one small, are sufficient for general work. Two medium-length sash clamps are also indispensable for clamping frames.

Of the handsaws, a crosscut saw is a sensible first buy. It cuts both with and across the grain. A tenon saw makes most joining cuts, and the coping saw is useful not only for its normal work of cutting curves but also for removing the waste when dovetailing.

A range of best-quality bevel-edged chisels is also essential. Rectangular-section chisels are marginally

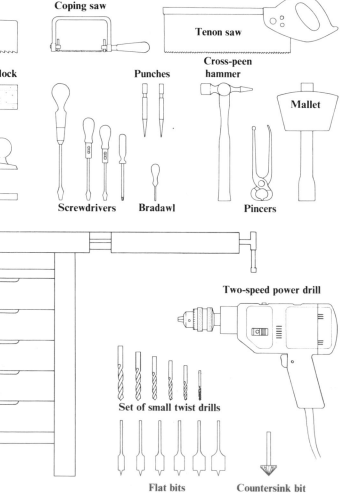

Continental workbench

Two-speed power drill

Set of small twist drills

Flat bits

Countersink bit

stronger, but clumsier. A chisel beveled on three sides gives a clear view of the area being worked; suggested sizes are 3 mm/$\frac{1}{8}$ in, 6 mm/$\frac{1}{4}$ in, 13 mm/$\frac{1}{2}$ in, 19 mm/$\frac{3}{4}$ in and 25 mm/1 in.

Most planes are designed for specific operations, but a good general choice is a metal jack plane. It will do the work of a jointer plane, such as planing edge joints, and in most cases that of a smoothing plane, such as cleaning up wide surfaces. For smoothing with abrasives a cork block is essential to ensure an even surface. Two oilstones, one medium and one fine, are needed to keep blades in prime condition, together with a honing guide for maintaining the correct bevel.

A bradawl with the blade integrated with the handle and a nail-punch are essentials. Of the screwdrivers choose a long cabinet screwdriver for heavy-gauge screws, a large and a small ratchet screwdriver for general use and a narrow rigid screwdriver for repairing items such as locks.

A cross-peen hammer and a pair of pincers cope with most nailing work. Despite the introduction of so-called unshatterable chisel handles, a wooden mallet will cause less damage than a hammer. A two-speed power drill is the best first-time buy for the basic kit. Select a set of small twist drills up to 6 mm/$\frac{1}{4}$ in, a countersink bit with a round shank and a set of six flat bits. A second drill will be needed for accessories.

The follow-up tool kit

The basic tool kit as described will be quite adequate for general work. The additional tools (*illustrated below*) are probably the most useful for extending the range of the original kit. It is important, however, to realize that they are mostly specialist tools. It is therefore essential to consider whether the work justifies the additional cost.

If the intention is to make traditional joints, a mortise gauge will be necessary to set out the double lines of the mortise and tenon joint, as well as two mortise chisels (3 mm/$\frac{1}{8}$ in and 6 mm/$\frac{1}{4}$ in) to lever out the waste. For dovetails a cutting gauge makes a very fine line for accurate working. Use a dovetail saw for very fine joinery work. For large frames a long try square is very effective for checking squareness. Two pairs of clamp heads will provide two extra sash clamps of variable length. A ripsaw and panel saw, or a good power jigsaw, will also justify their purchase.

A metal smoothing plane is a good addition to the jack plane. The remaining planes illustrated are all important in their own right, but study their specialized uses and only buy when the need arises. Buy two spokeshaves: one with a flat sole for forming convex curves; the other with a curved sole for concave curves. When finishing hardwood surfaces a cabinet scraper is invaluable if used correctly.

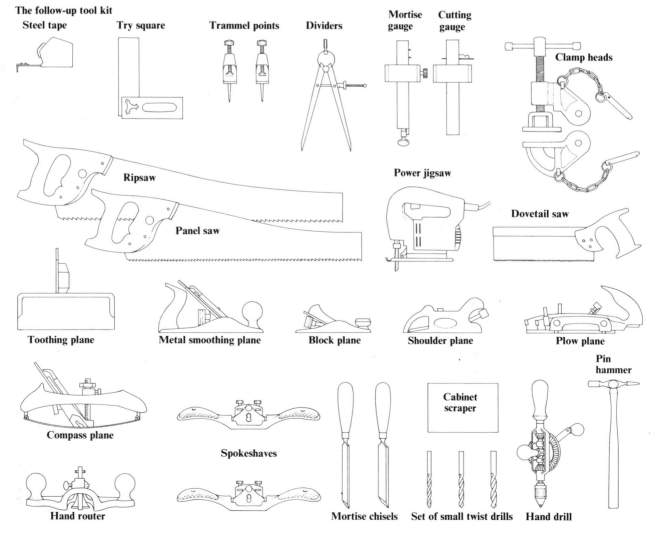

The follow-up tool kit

Steel tape · Try square · Trammel points · Dividers · Mortise gauge · Cutting gauge · Clamp heads · Ripsaw · Panel saw · Power jigsaw · Dovetail saw · Toothing plane · Metal smoothing plane · Block plane · Shoulder plane · Plow plane · Compass plane · Spokeshaves · Cabinet scraper · Pin hammer · Hand router · Mortise chisels · Set of small twist drills · Hand drill

The workshop

The workshop should ideally be a space set permanently apart from the rest of the house. A closed door will discourage outside interference and will allow the woodworker to leave work in progress.

The workshop must be warm, well lit and have good ventilation. Windows and a corrugated vinyl transparent roof will provide natural light. Where windows cannot be opened, a ventilator must be installed to provide good air circulation. Beware of naked flames in the workshop and always keep a fire extinguisher close at hand. A basic first aid kit should be readily available in case of injury.

Thoughtful and methodical planning of the work area will continually benefit efficient working, especially if the workshop is kept tidy and the tools are ready to hand.

The most important item in the workshop is the workbench, around which other fixtures and equipment should be arranged. Drawers are useful for keeping pencils, a rule, a steel tape and small odds and ends. Cupboards will store tools, which should be racked separately to protect them. Each tool should be considered separately when deciding the best place for storage and the method of storage, such as clips and magnetic strips. Return each tool to its proper place after a working session. Planes should be laid sole downwards on a piece of lightly-oiled felt. Chisels and saws should be fitted with plastic sheaths, wherever possible, to protect their blades. Hand tools should be rubbed over with turpentine and grease occasionally to clean off resin and to prevent rust.

A portable workbench can be folded and neatly hung against the wall from strong hooks. A plastic waste bin should be kept in the workshop to hold wood shavings, used abrasive paper and other litter.

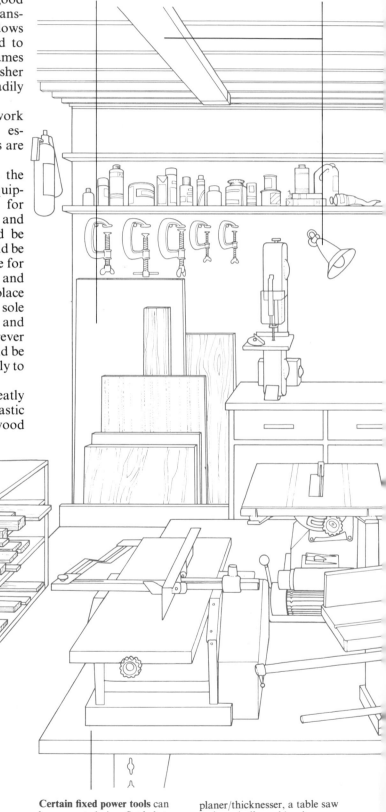

Store manufactured boards upright on edge against a dry wall and lodge them securely with a wooden batten fixed to the workshop floor.

Overhead fluorescent tubes give shadowless illumination, and an angled spotlight over the work area gives an effective concentrated light.

Lay short ends of wood horizontally for easy identification. Stack longer pieces vertically between mild steel rods.

Certain fixed power tools can be connected via a flat belt to a common motor that is built into a specially designed table. The system illustrated shows a planer/thicknesser, a table saw and a spindle molder; however, any combination of power tools can be included on a table of this type.

Keep all toxic materials, such as thinners and lacquers, in only small quantities. Place all adhesives well away from direct heat.

Store nails, screws, pins, etc., in small separate containers, which must be kept dry. Label them clearly and arrange them methodically for easy access.

Tools placed in cupboards, drawers and under cover are protected from dust and rust. A bag of silica gel will further discourage rust.

Many hand tools can be racked with spring clips, rubber webbing, magnetic strips, hooks, projecting dowel rods or drilled blocks.

Fit power outlets above the work surfaces close to the power tools so their cords can be short and they will not trail on the workshop floor.

Stout shelving provides extra storage space and is especially useful for portable power tools that are too heavy to hang on the walls.

Cover concrete floors with vinyl sheeting for warmth. A rubber mat or slatted board will provide additional comfort and warmth while working.

Stack sawhorses one upon the other to leave as much floor space as possible for assembly work and general movement.

The workbench

The main purpose of the workbench is to support the wood being worked. To do this it must be sturdy, especially to withstand diagonal stresses. The top should be absolutely flat and level, made from thick hardwood and supported by four solid legs with a bracing structure. The whole bench must stand at a comfortable height for the individual worker.

The workbench should also accommodate wood of varying proportions and shapes; clamping devices and bench dogs are therefore built into it. Work-top mortises are either part of the commercially available bench, or they can be made. They take metal or nylon bench dogs, which can be adjusted to the level required. Mortises can also be used to take a holdfast. Those cut into a vertical batten of solid wood, known as a deadman, also take a holdfast, used for supporting awkward work such as doors. Metal spring-loaded bench stops can also be sunk into the work top.

The woodworker's bench vise, made of cast-iron with wooden jaw facings, is the main clamping device. It can be fitted by the worker. It is usually bolted to the underneath of the bench, close to a leg for maximum stability. The rear jaw is fitted 10 mm/$\frac{3}{8}$ in down from the bench top and is sunk flush into the apron. Some benches have an integral sliding tail vise. Mortises are worked into the tail vise and the front of the bench. By using the tail vise in conjunction with bench dogs the wood can be gripped at both ends. The tail vise can also hold a deadman. Whatever the vise, it will not withstand heavy hammering.

Jigs act as guides for accurate planing and sawing; they are simple to construct. Among the most useful and easily made is the bench hook, which fits either in a vise or against the bench apron. The shooting board is more practical than the bench hook for planing thin boards and end grain.

A portable workbench can be used in addition to or instead of a permanent workbench. Many accessories are available, to make it extremely versatile. A pair of sawhorses is invaluable when sawing large pieces.

Woodworker's bench vise
Opening capacity 113–375 mm 4$\frac{1}{2}$–15 in. For holding the work, usually horizontally. Place the work centrally where possible. When the work is placed in one side only, support the other with waste wood of the same thickness to minimize strain.

Mortise

Portable workbench
This small rigid bench has adjustable legs and a folding structure for neat storage. The two boards can expand to give a larger working surface or can contract to act as vise jaws for holding both parallel or tapered work. The mortises hold the bench dogs.

Sawhorses
For supporting lengths of wood for sawing and marking out. The narrowly splayed out legs and frame may be built from softwood but the top must be hardwood. With a pair, one top should be wider than the other for neat stacking. The components can be assembled with brackets.

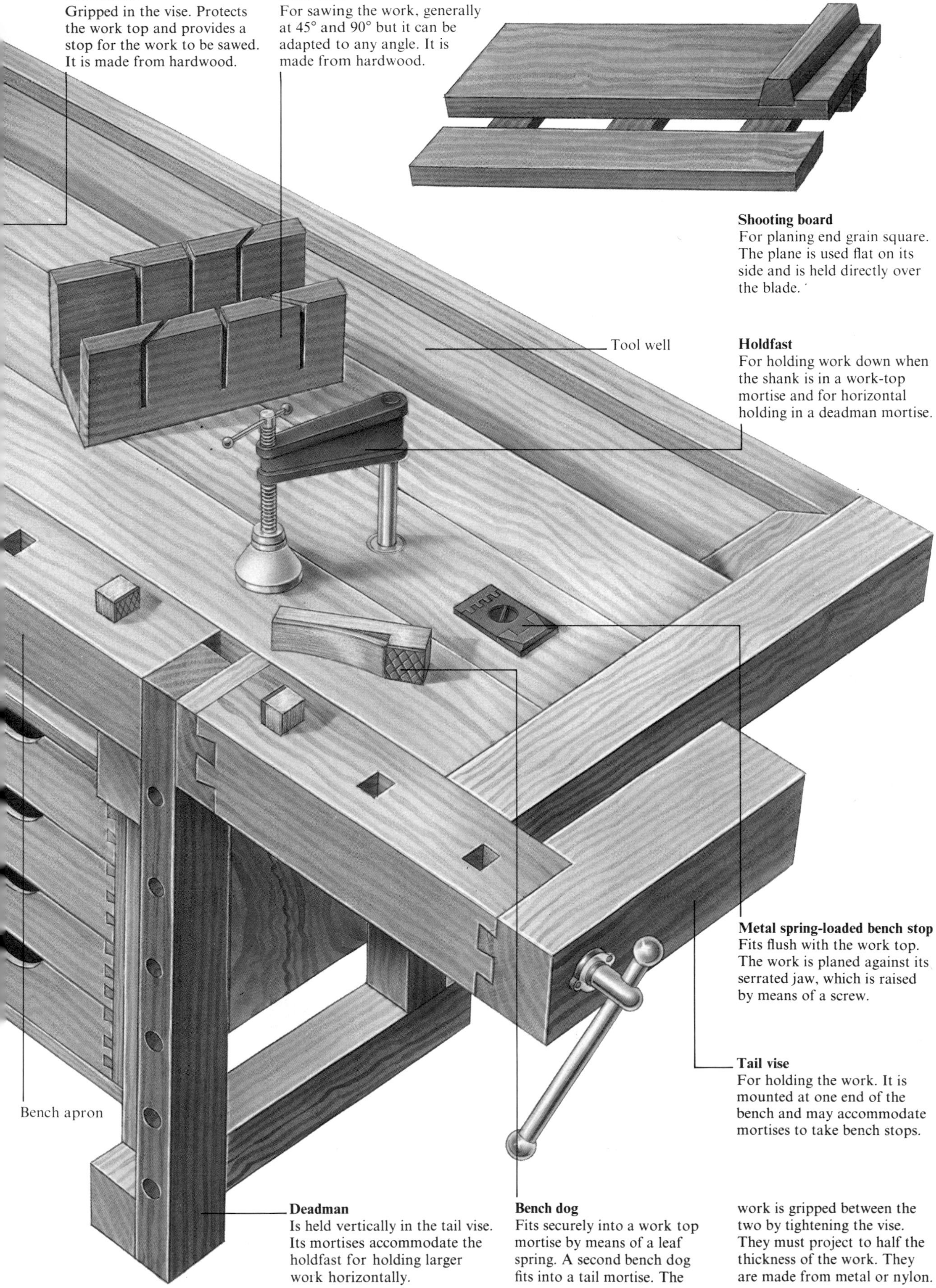

Bench hook
Gripped in the vise. Protects the work top and provides a stop for the work to be sawed. It is made from hardwood.

Miter box
For sawing the work, generally at 45° and 90° but it can be adapted to any angle. It is made from hardwood.

Shooting board
For planing end grain square. The plane is used flat on its side and is held directly over the blade.

Tool well

Holdfast
For holding work down when the shank is in a work-top mortise and for horizontal holding in a deadman mortise.

Bench apron

Metal spring-loaded bench stop
Fits flush with the work top. The work is planed against its serrated jaw, which is raised by means of a screw.

Tail vise
For holding the work. It is mounted at one end of the bench and may accommodate mortises to take bench stops.

Deadman
Is held vertically in the tail vise. Its mortises accommodate the holdfast for holding larger work horizontally.

Bench dog
Fits securely into a work top mortise by means of a leaf spring. A second bench dog fits into a tail mortise. The

work is gripped between the two by tightening the vise. They must project to half the thickness of the work. They are made from metal or nylon.

Measuring and marking tools 1

The most ancient units of measurement are those based on the human foot, outstretched arm and hand. Then, around 2500 BC, the Egyptians used calibrated wooden or stone rods based on the cubit (a forearm's length) and later the Romans used a folding bronze foot rule.

Measuring and marking tools must be treated with special care: if edges, blades and surfaces, hinges, locking screws and rivets are damaged the tools lose their accuracy and are therefore of no further value. Always mark cutting lines, joints and anything else requiring accuracy with a sharp knife or a scribing point, rather than with a pencil.

The try square is essential for preparing and checking a true edge. On the best try squares the stock end of the steel blade is L shaped.

Of those tools for measuring or marking flat surfaces, a spirit level with a long body is more accurate than a short one. A chalk line and reel can double as a plumb bob for testing vertical surfaces, provided the reel is sufficiently weighty.

Of those tools for marking parallel lines, the cutting gauge, with its cutting edge, is ideal for marking across the grain; a marking gauge when used across the grain will merely scratch the wood.

Retractable tape
Length 900–5,000 mm/
36–192 in. Flexible steel or
fiberglass tape in a case.
Folding rule
Length 300–1,825 mm/
12–72 in. For measuring
and laying out.
Steel rule
Length 150–900 mm/6–36 in.
Made of stainless steel. Can be
used like a straightedge.
Straightedge
Length 300–1,825 mm/12–72 in.
For testing straightness and
flatness. A strip of parallel-
sided steel with one beveled
edge for cutting and scribing.
Spirit level
Length 75–1,825 mm/3–72 in.
Use central vial to check level
of a surface and the two vials
at each end to check for
plumb. The bubble in each vial
should appear centrally.

Sliding bevel
Blade length 190–323 mm/
7½–13 in. For marking or
checking any angle. The slotted
blade is set against a protractor
or the work.
Miter square
Blade length 150–300 mm/
6–12 in. Set at 45° for miters.
Try square
Blade length 150–300 mm/
6–12 in. For checking
straightness and squareness of
adjacent surfaces.
Combination square
Blade length 300 mm/12 in.
Combines the functions of the
rule, try square, miter square
and spirit level. The square
head slides along the steel rule.

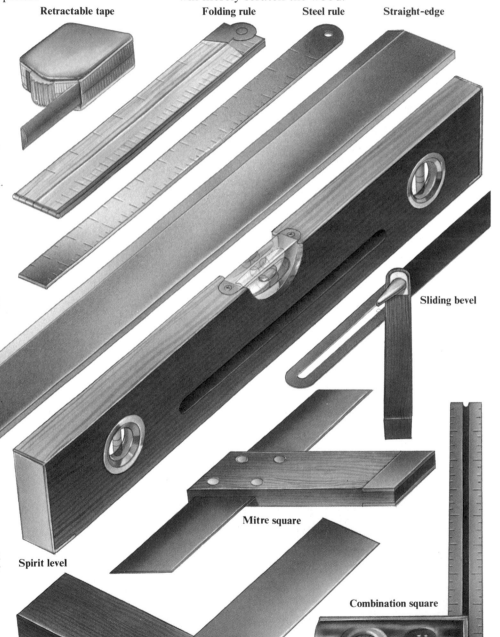

Retractable tape **Folding rule** **Steel rule** **Straight-edge**

Sliding bevel

Mitre square

Spirit level

Combination square

Try square

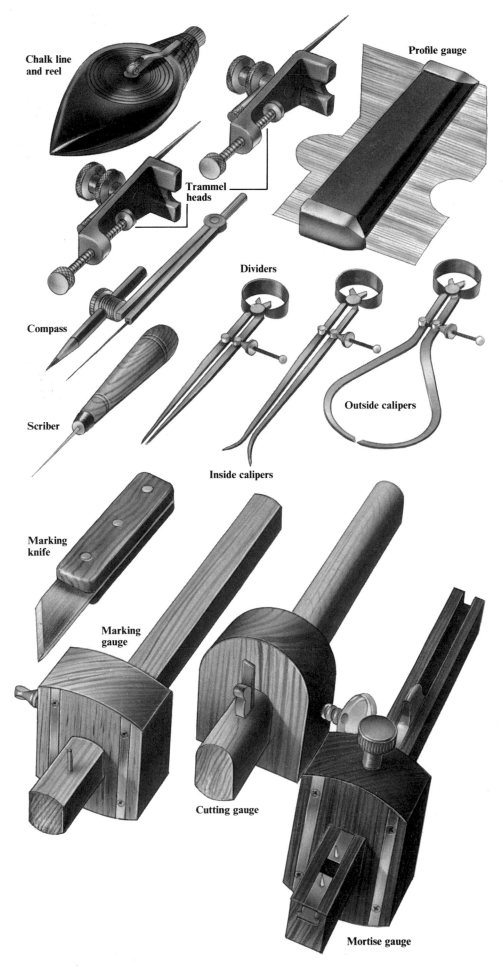

Chalk line and reel

Trammel heads

Profile gauge

Dividers

Outside calipers

Compass

Scriber

Inside calipers

Marking knife

Marking gauge

Cutting gauge

Mortise gauge

Chalk line and reel
Length 5,000–30,000 mm/ 196–1,200 in. For marking long straight lines by "snapping" the chalked cord against a surface.

Trammel heads
Length 100 mm/4 in. Two steel points that, when fixed on a wooden beam, can scribe a large circle.

Profile gauge
Length 150 mm/6 in. For tracing and checking work. When pressed against a shape the steel needles slide back to reproduce the outline.

Compass
Capacity 100–200 mm/4–8 in. For marking a small circle.

Scriber
Blade 100 mm/4 in. For marking with and across the grain.

Dividers
Capacity 100–200 mm/4–8 in. For scribing a circle, stepping off and measuring.

Inside calipers
Capacity 100–200 mm/4–8 in. For measuring the inside diameter of a pipe, tube and bowl.

Outside calipers
Capacity 100–200 mm/4–8 in. For measuring the outside diameter of a cylinder.

Marking knife
Blade 50 mm/2 in. The blade is ground to a beveled skew for marking against a rule or straightedge.

Marking gauge
Length 160–240 mm/6¼–9½ in. For marking a line parallel to an edge, along the grain or on end grain. The stock, which acts as a fence and is locked by the thumbscrew, is often faced with brass to protect against wear. The spur at the end of the stem marks the line.

Cutting gauge
Length 160–240 mm/6¼–9½ in. For marking across the grain parallel to an edge. The flat blade in the stem, which is secured by a wedge, can be removed and sharpened. Also cuts veneers and cardboard.

Mortise gauge
Length 925 mm/9 in. For marking both sides of a mortise and tenon. Two spurs, the inner one adjustable, protrude from the sliding stem to mark parallel lines.

Measuring and marking tools 2

Before marking out can begin on any piece of wood, it must have datum lines: one flat side and one edge square to it. Choose the better looking side to be the face side. Plane it flat and level. With the straightedge, check across the face for flatness — also along its width, its length and from corners to diagonally opposite corners. When the face is true, pencil a loop — the face mark — in the center, extending the line to the better of the two side edges. Plane this face edge square with the face side and check it along its length with the straightedge. To square up the remaining surfaces, set the marking gauge to the required width and, from the face edge, mark both sides. Plane to the gauge marks. Set the gauge to the required thickness and gauge both edges from the face side. Plane to the gauge marks. Accurate measuring, marking out and checking are essential. Tools must be carefully set.

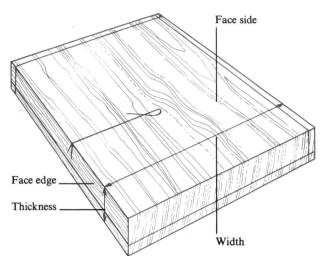

Face side

Face edge

Thickness

Width

Setting and using gauges

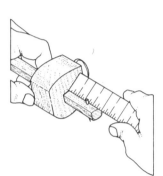

1. Set stock using a rule.

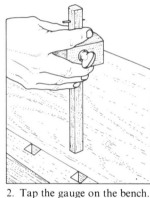

2. Tap the gauge on the bench.

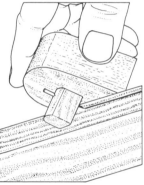

3. Hold the stock firmly.

Setting and using gauges
Loosen the thumbscrew. Set the distance between stock and spur by pushing the stock away with a rule or pushing it towards the spur with your thumb (1). Tighten the thumbscrew and check the setting. If it is incorrect do not loosen the thumbscrew but tap either the stock or the spur end of the stem on the bench (2). Recheck and, if correct, tighten the thumbscrew. When using any gauge, keep the stock tight against the work (3).

Drawing lines and outlines

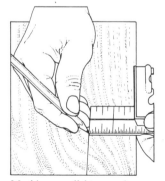

Marking parallel to an edge.

Finger gauging.

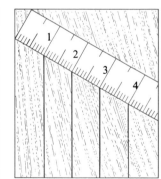

Making equal subdivisions.

Marking parallel to an edge
Set a combination square to the required distance. Then hold the head tightly against the edge and, with a pencil against the end of the rule, run it along the board.
Finger gauging
To draw a line close to an edge, use the middle finger as a gauge for the pencil.
Making equal subdivisions
To divide up a given width into equal spaces, angle a rule between the two parallel sides until the readings on the rule are easily divisible by the number of spaces required.
Scribing
To shape a board to fit against an irregular surface first rest it at right angles to the surface; then draw a compass along with the point running against the surface.
Using a profile gauge
Use the gauge to duplicate moldings or turnings. Push the needles fully out on one side. Hold the gauge square to the shape and press against it. The gauge will then show the profile of the shape and the negative profile.

Scribing.

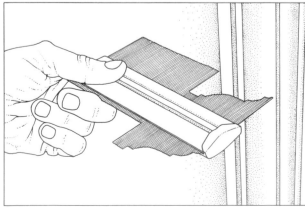

Using a profile gauge.

Measuring and checking verticals, horizontals and planes

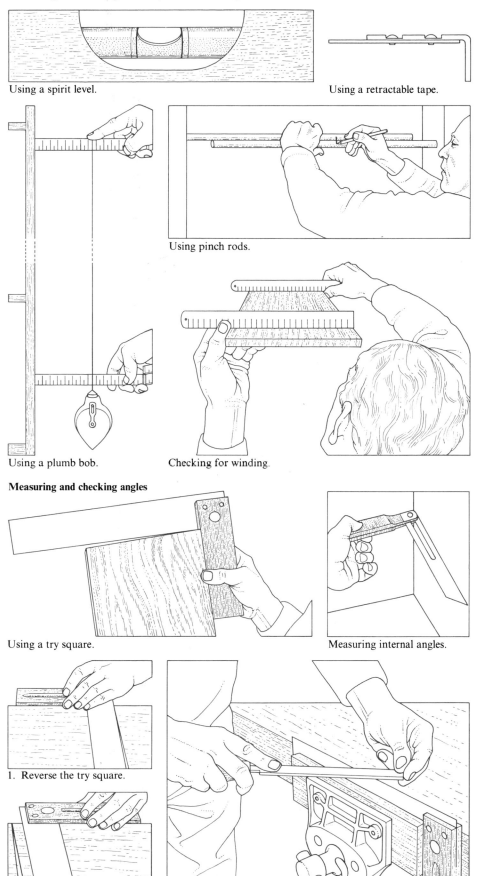

Using a spirit level.

Using a retractable tape.

Using pinch rods.

Using a plumb bob.

Checking for winding.

Measuring and checking angles

Using a try square.

Measuring internal angles.

1. Reverse the try square.

2. Draw a second line.

3. File blade for correction.

Using a spirit level
The air bubble will be centered between the datum lines when the surface is level or plumb.

Using a retractable tape
Hook the end of the tape over the work to take an external measurement. For an internal measurement butt the metal hook against the work.

Using a plumb bob
To check if a surface is plumb, hold the string against a rule at the top of the surface and a set distance, such as 75 mm/3 in, away from it. When the bob stops swinging, ask an assistant to check with a second rule that the measurement from the surface to the line is the same at the bottom as at the top.

Using pinch rods
Use two overlapping battens to take a measurement. Put coinciding marks on both; then transfer the measurement to the wood to be cut.

Checking for winding
To check whether a board is twisted in its length, known as being "in winding," place a steel rule at each end and sight across the top edges to see if they are parallel. Two parallel battens can be used instead.

Using a try square
When checking for squareness, always hold the stock of the try square tight against the work otherwise the tool is being used as a straightedge and not as a square.

Measuring internal angles
Walls are rarely square to each other, so always measure and set out the first angle with a sliding bevel.

Checking a try square
With prolonged use or if it is dropped, a try square can become inaccurate. To check it, plane a perfectly straight edge on a board. With the stock of the try square tight against the edge, draw a line along the blade. Reverse the square and check whether the blade coincides with the line (**1**). If not, draw another line along the blade (**2**). The blade must be corrected so that it bisects these lines. File along both edges of the blade, working diagonally with a smooth file. (**3**). Check the try square again. When correct, drawfile to remove the initial file marks. Smooth the edges with an oilstone.

57

Clamps 1

Clamps are used for holding materials during working and while glue dries. Originally all clamps were made of wood, but these have mostly been superseded by metal and tough plastics. The hand-screw and cam clamp, however, still have wooden jaws.

The C-clamp is the most common general-purpose clamp. It has many variations: miniature clamps; deep-throat or long-reach clamps; and spin-grip clamps, which can be tightened quickly with one hand. The edge clamp, toggle clamp, quick-release clamp and cam clamp have similar uses to the C-clamp.

The problem of clamping miters and frames has been solved in many ingenious ways. The lightweight miter clamp, which is in fact a set of four corner clamps, is useful for picture frames, while the heavier metal frame clamp has greater capacity. The versatile web clamp can be adapted as a frame clamp by using four right-angled wooden or metal corners.

A number of bar clamps are needed when any large flat surface such as a table top is being glued. Clamp heads fixed to a batten can be made up to a clamp of any length required.

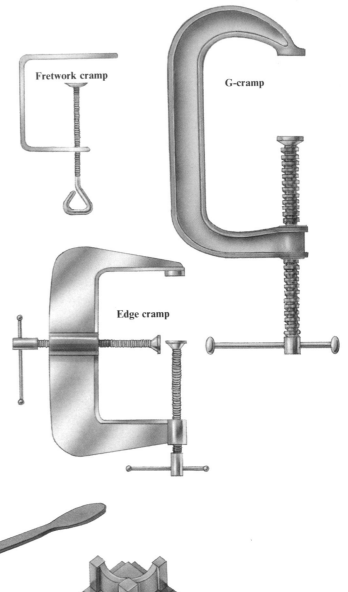

Fretwork cramp

G-cramp

Edge cramp

Fretwork clamp
Capacity 25–100 mm/1–4 in. A small lightweight steel clamp.
Edge clamp
Capacity 38–75 mm/1½–3 in. Depth 50–112 mm/2–4½ in. For pressure in two directions.
C-clamp
Capacity 50–300 mm/2–12 in. Most varieties have a swivel shoe on the screw end.
Quick-release clamp
Capacity 100–600 mm/4–24 in. Lower jaw slides on the steel bar and is locked and released by a single turn of the handle.

Toggle clamp
Holding capacity 360 kg/800 lb. Strong steel clamp with trigger release and an optional sliding lower jaw. The upper jaw pivots up to 84° backwards.
Cam clamp
Capacity 200–575 mm/8–23 in. The lower beechwood jaw locks on the steel bar by cam action.
Hand-screw
Capacity 50–300 mm/2–12 in. Steel screws pivot in wooden jaws, which can exert parallel pressure or can be angled for concentrated pressure.

Hand vice

Web cramp

Frame cramp

Joiners' dog

Mitre cramp

Hand vise
Capacity 19 mm/¾ in. For gripping very small work.
Web clamp
Length 3,500–6,000 mm/ 144–240 in. Canvas or nylon webbing tensioned by ratchet mechanism for clamping large, irregular and round shapes.
Joiners' dog
Width 25–75 mm/1–3 in.

Tapered steel staples driven into end grain for holding glued butt joints.
Frame clamp
Capacity 50–100 mm/2–4 in. Holds right-angled joints securely during assembly.
Miter clamp
Capacity 32 mm/1¼ in. Four corners with strong spring-loaded jaws hold the work.

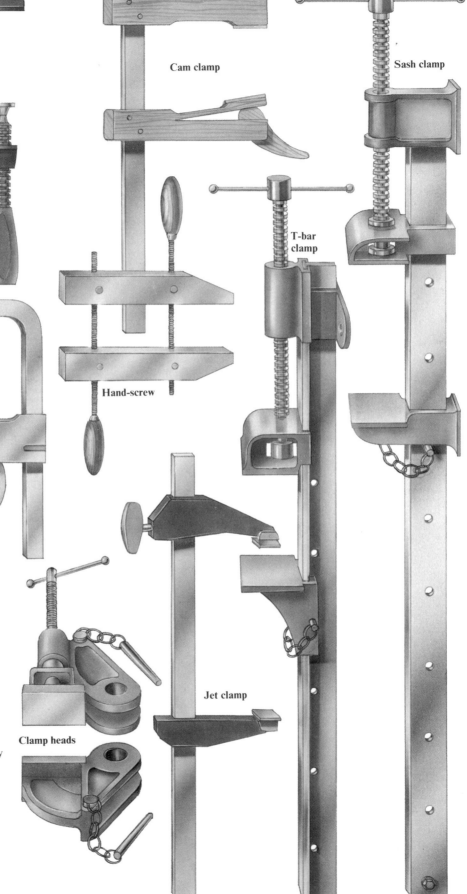

Quick-release clamp

Cam clamp

Sash clamp

T-bar clamp

Toggle clamp

Hand-screw

Jet clamp

Clamp heads
Can be fitted to a board of any length over 38 mm/1½ in wide and 25 mm/1 in thick. The heads are locked with pegs placed in holes in the board.

Jet clamp
Both jaws can be moved along a steel bar of standard section and of any length. The clamp heads are locked by a spring-loaded wedge. Both heads can be taken off and reversed to apply outward pressure for jacking work apart.

T-bar clamp
Capacity 760–1,985 mm/ 30–78 in. A heavy-duty joiners' clamp. Any bar clamp over 1,525 mm/60 in long will usually be of T-section steel, which resists bending. A separate lengthening bar can be added.

Clamp heads

Sash clamp
Capacity 450–1,370 mm/ 18–54 in. The movable tail slide is secured with a pin through the steel bar. Final adjustment is made by the screw on the fixed head. Two clamps can be bolted end to end to cover a wider span.

Clamps 2

Wood can be bruised if it is gripped too hard between the metal jaws of C-clamps and sash clamps. Clamps with wooden or plastic-capped jaws, however, are less damaging. Always place a piece of waste wood between the clamp and the work to spread the pressure. For odd shapes and angles, scraps of carpet can be used instead of waste wood. When gluing up assemblies make sure that the bar or spine of the clamp is aligned with the work; if it is not, the weight of the clamp may pull the work out of square. To prevent clamps from pulling the work out of alignment, always alternate clamps on either side of the work when more than two are being used. For example, when clamping the sides to the top and bottom of a carcass, sash clamps should be placed alternately front and back.

Most cabinetwork requires a number of sash clamps to be used in assembly. Improvised clamps can be made from battens and waste wood.

C-clamp

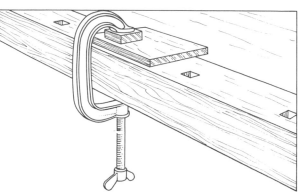

Always use waste wood to protect the work from the metal jaws. The wing nut is usually positioned towards the ground, leaving the working area free. A swivel shoe on the screw end locates easily on slanting surfaces. The leverage exerted by the screw thread creates great pressure.

Hand-screw

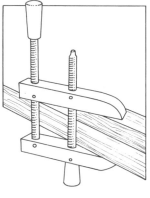

To open and close, hold a handle in each hand and revolve both handles. Tighten the inner handle; then apply final pressure with the outer handle, making the jaws parallel or angled. As long as the wooden jaws are kept clean no waste wood is needed to protect the work.

'Cam clamp

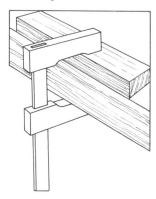

Move the lower beechwood head up the steel bar until it presses against the work. Lock with the lever, which is worked by cam action.

Jet clamp

Slide the lower jaw up the steel bar. Lock it by turning the handle clockwise. Release the clamp with a single turn of the clamp handle.

Quick-release clamp

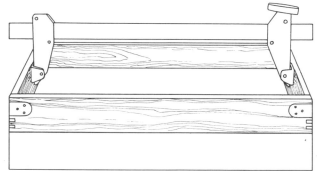

Move both heads into position along the steel bar. The spring-loaded wedges in the heads will automatically lock themselves. Fix in position by tightening the knob on one of the heads. To apply an outward pressure for jacking work apart, remove the heads from the steel bar and reverse them.

Joiners' dog

Drive into end grain to draw the ends of boards together. The tapered steel points leave marks, so the work will need filling or covering.

Hand vise

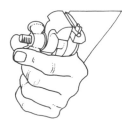

Grip a hand vise in the bench vise or hold it in the hand. Use when working on metal fittings that are too small to hold by other means.

Web clamp

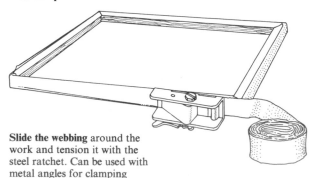

Slide the webbing around the work and tension it with the steel ratchet. Can be used with metal angles for clamping around corners.

Sash clamp

Adjust the clamp initially by sliding the tail shoe along the steel bar and securing it with a pin through one of the holes in the bar. To move the tail shoe,

tap it along the bar with a mallet. (It is made of cast iron and a hammer may shatter it.) Make final adjustments with the tightening screw at the

other end of the bar. Use for clamping all large constructions together during assembly, while the glue dries, checking the pressure is even.

Ensure enough clamps of the right length are available before beginning assembly. Use T-bar clamps in the same way as sash clamps.

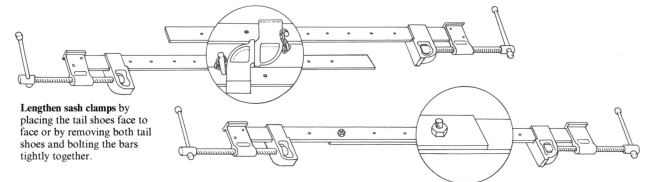

Lengthen sash clamps by placing the tail shoes face to face or by removing both tail shoes and bolting the bars tightly together.

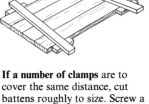

The work must be adapted to fit the clamps when even clamps specially devised to hold awkward shapes cannot be used. Glue clamping blocks to

the work with brown paper between. Then clamp the work. When dry, remove the work. Knock off the blocks; then soak or smooth off the paper.

If a number of clamps are to cover the same distance, cut battens roughly to size. Screw a pivoting cleat to each end. Put the battens in place. Tap them

so they slant, pivoting the cleats and making the battens fit tightly. Cleats rigidly screwed to each end of a batten can be used with folding

wedges tapped in to supply pressure. Bore holes in a batten to take a movable tail shoe, locked by a bolt and wing nut for rough adjustment.

At the other end, fix a cleat with one sloping side matching the slope of a wedge. Make fine adjustments by tapping in the wedge.

Miter clamp   **Clamp heads**

   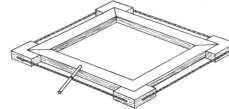   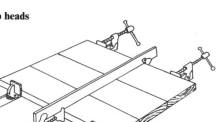

Place the work against the corner fence, where two sprung jaws will hold them while the glue dries on a miter joint or while pins are hammered in.

Make a miter clamp with blocks of wood recessed at the inner corners. Apply pressure with strong cord resting in grooves on the outside. Twist the cord around a stick, rest the stick against the frame to lock the windlass.

Use clamp heads in pairs on a 25 mm/1 in section wooden batten. Bore holes in the batten at regular intervals. Place the clamp heads and batten over the work and lock in position with pegs placed in the holes.

Hand- and backsaws 1

Saws are of three main types, categorized according to the job they do: handsaws for rough cutting and conversion; backsaws for joints and fine work; and special saws for curves and shapes. Handsaws have flexible tempered steel blades; a well-tempered blade when bent from tip to handle should spring back into a perfectly straight line. Backsaws have a brass or steel strip folded over the top of the blade to give rigidity and add weight along the line of the cutting edge.

The basic design of the handsaw was established in the mid-seventeenth century. The English type, incorporating an oval handhole and angled grip, became the standard pattern almost everywhere for both handsaws and backsaws. An alternative to the traditional beechwood or hardwood handle is one made of polypropylene; this is usually molded onto a Teflon-coated blade, which reduces friction.

A saw's teeth are specifically designed to cut either along the grain, as with the ripsaw, or across the grain, as with the crosscut saw. A crosscut saw will also cut along the grain, but less efficiently than the ripsaw. The greater the number of teeth in a given length of blade, the finer and slower the cut. Some woods can be naturally tough; in such cases use a saw with smaller teeth. The number of teeth is expressed as so many points per 25 mm/1 in. Saws also have their teeth set to prevent the blade binding in the kerf or cut.

Handsaw parts

Taper ground blade for kerf clearance

Skew back for good balance

Traditional beechwood handle set low for efficient cutting

Backsaw parts

Back edge reinforced with brass or steel for rigidity

Traditional beechwood handle set high for correct balance

Straight parallel blade

Ripsaw
Blade 600–650 mm/ 24–26 in. 4½ points to 25 mm/1 in. For boards 13 mm/½ in and upwards, when cutting with the grain.

  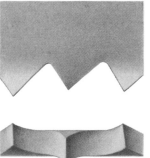 

Ripsaw teeth are set as shown above, with their tips bent over in alternate directions to prevent binding.

Each tooth is filed to form a chisel edge. This enables it to cut the fibers cleanly along the grain of the wood.

The teeth on a crosscut saw are sloped on their leading edge as well as being filed at an angle and set.

Filed in this way, each tooth is sharply pointed to act as a knife to sever the fibers of the wood.

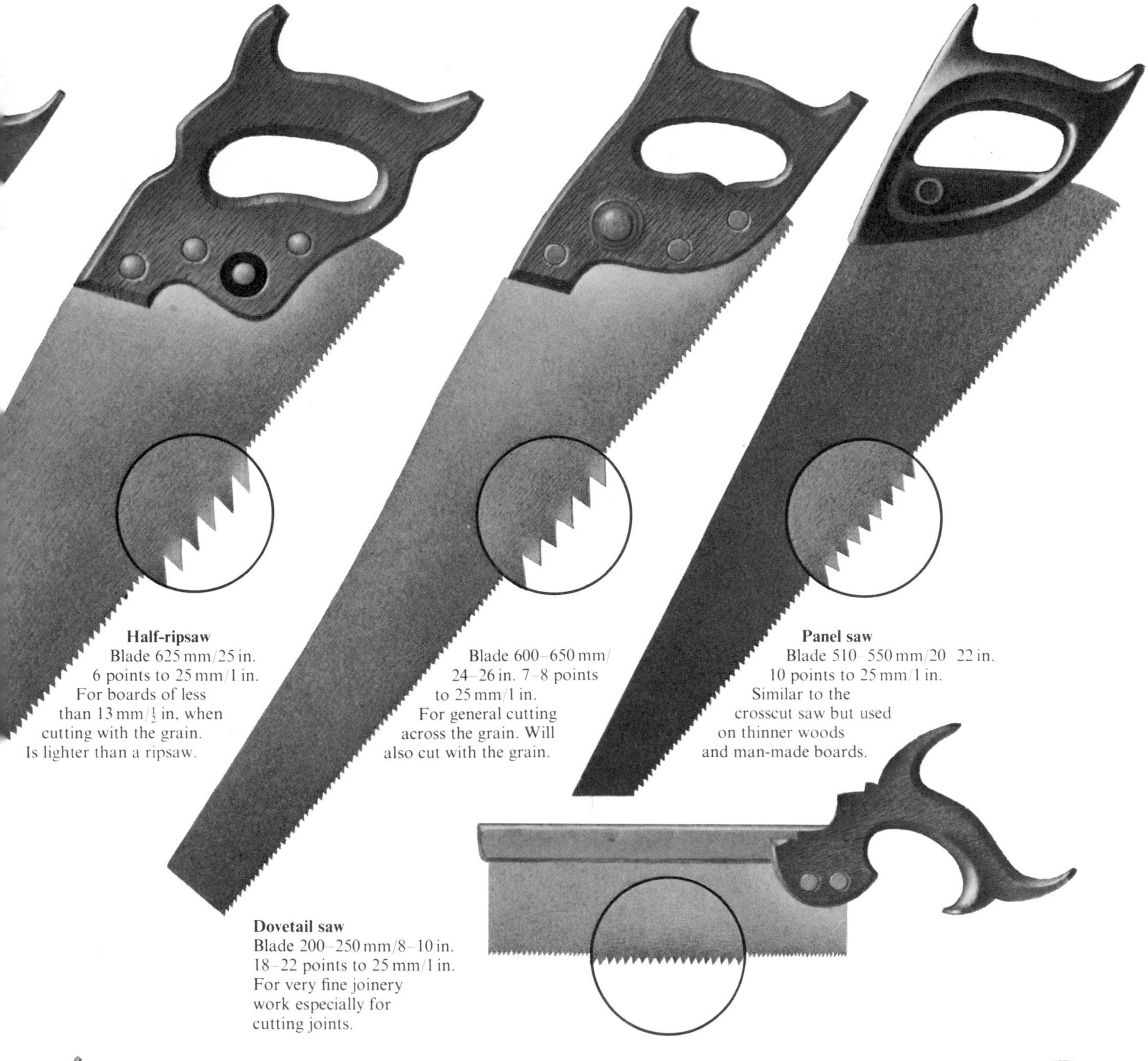

Half-ripsaw
Blade 625 mm/25 in.
6 points to 25 mm/1 in.
For boards of less
than 13 mm/½ in. when
cutting with the grain.
Is lighter than a ripsaw.

Blade 600–650 mm/
24–26 in. 7–8 points
to 25 mm/1 in.
For general cutting
across the grain. Will
also cut with the grain.

Panel saw
Blade 510–550 mm/20–22 in.
10 points to 25 mm/1 in.
Similar to the
crosscut saw but used
on thinner woods
and man-made boards.

Dovetail saw
Blade 200–250 mm/8–10 in.
18–22 points to 25 mm/1 in.
For very fine joinery
work especially for
cutting joints.

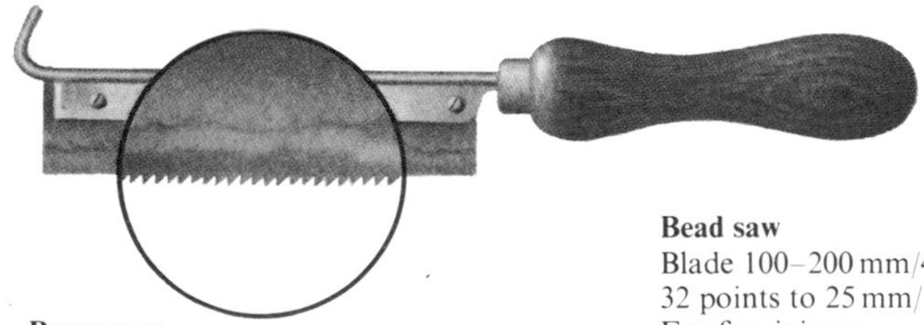

Razor saw
Blade 125 mm/5 in.
Points too numerous
to count. For extremely
delicate work. Replace
blades when blunt.

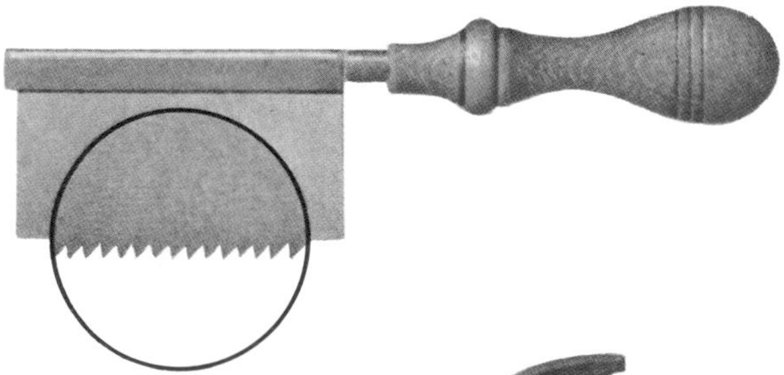

Bead saw
Blade 100–200 mm/4–8 in.
32 points to 25 mm/1 in.
For fine joinery
work, leaving a
smooth finish.

Tenon saw
Blade 300–350 mm/12–14 in.
12–14 points to 25 mm/1 in.
For general bench
work especially when
cutting joints.

Hand- and backsaws 2

Sawing is a skillful process; mastery is achieved only with constant practice, but attention to the following points will make the work easier. A blunt tool is more difficult to control than a sharp one. A sharp saw needs little pushing; unnecessary force makes it more difficult to follow a line and can cause the saw to buckle. The saw should be well-balanced and should cut smoothly, virtually under its own weight. For efficient cutting, saw with the blade in line with your arm and shoulder; start the cut low down and bring the saw up to the correct cutting angle. Then move the saw up and down in long steady movements.

Saw vertically wherever possible, even if this means that the wood must be gripped out of upright in the vise to bring the cutting line to the required angle. If the saw binds in the wood, rub three drops of oil on the blade. (Avoid oils that will stain the wood). In extreme cases, drive a narrow wooden wedge into the saw-cut to hold it open. Wipe any resin residue off the blade, using kerosene; then wipe with oil.

If saw teeth are overset, that is alternate teeth are pushed too far over, it will be more difficult to work the saw. This can be corrected by side dressing. (This and other methods of maintaining saws are covered on pages 72–3.) To bring a wandering saw back to the cutting line, twist the blade towards the line, working with the narrow front end of the saw.

Ripping

To saw on only a single stool overhang the board a bit at a time, steadying it with the knee. Saw halfway in from each end of the board.

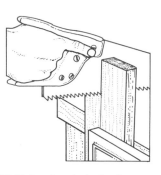

Hold short lengths in the vise. Secure the wood so that only a little bit projects at a time; otherwise the wood will vibrate, making the cut inaccurate.

If a level higher than the stool is more convenient for ripping, fasten the wood to the bench and hold the saw upright, using both hands.

Place the board across two stools and start sawing at the overhang. Then lift the saw and board back and continue sawing between the stools.

Using a ripsaw

Ripsaw teeth are specifically designed for cutting along the grain. If used for crosscutting the ripsaw may jump and cause an accident. Should ripping prove difficult, try changing to a half-ripsaw or in extreme cases to a crosscut or even panel saw.

Short boards can be ripped in the vise or on a single stool. Longer work needs supporting on two stools. An angle of 60° between the saw and the work is comfortable when ripping on stools, but in a vise a ripsaw should be held horizontally to the work.

Controlling a handsaw

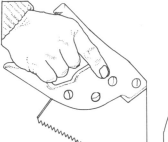 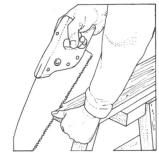

To control a handsaw with accuracy needs experience. A forefinger held along the handle aids control.

The saw must be guided to the cutting line. Steady the blade against the thumb. Start with a few backward cuts.

Using a crosscut saw

To cut a groove across the grain, the wood fibers must be cut before the waste is removed or they will splinter. This is the principle of a crosscut saw: the knifelike points of the teeth sever the fibers.

The small teeth of a crosscut saw make a clean cut with minimal splintering. It will also rip effectively, though it is slower than the ripsaw because of its smaller teeth. A panel saw is a smaller version of a crosscut saw. It too can be used for ripping, although it cuts even slower than a crosscut saw. When crosscutting the most suitable angle to hold the saw blade is at 45° to the work.

Using a backsaw

Backsaws are designed for fine accurate work. Their blades are thinner and teeth finer than handsaws. The weight of the steel or brass back enables the saw to cut under its own weight so that force need not be used to make the cut.

Both tenon saws and dovetail saws can cut along or across the grain and for bench work both are used with a bench hook. The tenon saw is excellent for cutting three-ply giving a cleaner cut than crosscut saws. The back will not interfere with deep sawing if the saw is held at a shallow angle to the work. The dovetail saw is for similar but finer work to the tenon saw.

Crosscutting

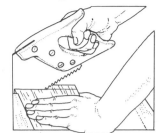
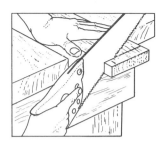

To saw off a short end hold the board firmly against the bench well and start cutting. Stop just before the end.

To complete the cut change the direction of the saw to prevent the offcut from falling and splintering the board.

Alternatively place the board in the well of the bench and support the offcut with the free hand while sawing.

To saw off a long end stand at the end of the bench, rest the board along the well and support the offcut while sawing.

Controlling a backsaw

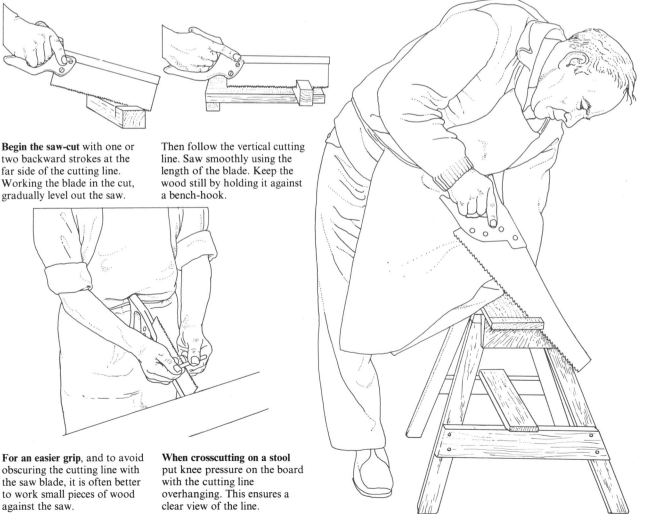

Begin the saw-cut with one or two backward strokes at the far side of the cutting line. Working the blade in the cut, gradually level out the saw.

Then follow the vertical cutting line. Saw smoothly using the length of the blade. Keep the wood still by holding it against a bench-hook.

For an easier grip, and to avoid obscuring the cutting line with the saw blade, it is often better to work small pieces of wood against the saw.

When crosscutting on a stool put knee pressure on the board with the cutting line overhanging. This ensures a clear view of the line.

Saws for shaping 1

For a saw to cut around a curve the blade has to be narrow or it will bind in the wood. To cut a tight curve, the blade must be very narrow; however, the narrower the blade the weaker it becomes and this inevitably leads to compromise in the design of saws. The solution that has been used for centuries is to hold the blade in tension in some kind of frame. Frame saws with the blade in the center of the frame were used by the Romans, and the cantilever principle of the modern bow saw was known and in use as far back as the thirteenth century. The blade in the bow saw is held between two uprights, or cheeks, that pivot on either side of a central stretcher rail. The blade is tensioned by pulling the tops of the cheeks towards each other by means of a twisted cord and toggle or, in some bow saws, by a threaded rod tensioned by wing nuts. Coarse- and fine-toothed blades can be used in the bow saw and the blade can be swiveled to allow the saw to cut parallel to the edge of the board.

Thick boards, 16 mm/$\frac{5}{8}$ in and upwards, must be cut by a heavily framed saw such as the bow saw; however, it will not be possible to cut sharp curves. A coping saw with its narrow blade can tackle curves of very small radii, but its use is confined to thin woods and the softer metals. Since rigidity of the coping saw and fret saw is determined by the quality of the steel in the frame, do not buy anything less than the best.

The two frameless saws illustrated — the compass saw and the keyhole saw — have their advantages and disadvantages. Because they are frameless, they can be used in places where other saws would be restricted by their frames. Although the compass saw blade tapers, allowing it to start a fine cut, its overall width is still restricting. On a shallow curve, however, it will cut almost as quickly as a handsaw on a straight line. The compass saw shown is one of a nest of three saws fitting on a single handle with each additional blade reduced in size. The keyhole saw blade is narrow but, because it has been thickened to counteract the tendency to buckle, it can become hot and hard work to use.

Continental bow saw
Blade 300–675 mm/12–27 in.
Depth of cut 150–200 mm/6–8 in.
9 points to 25 mm/1 in.
Two operators can use the
saw for crosscutting.

Compass saw
Blade 125–275 mm/5–11 in.
9 points to 25 mm/1 in.
Three interchangeable blades can
be fitted according to the
curvature of the cut.

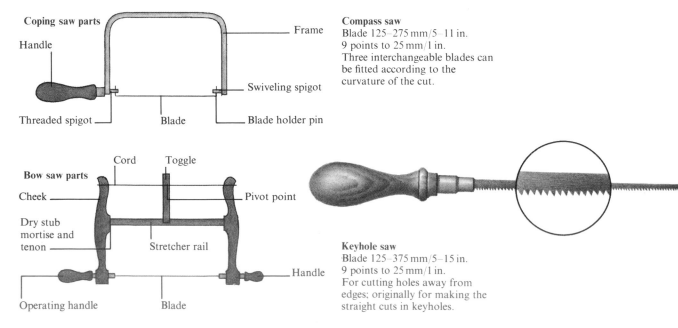

Coping saw parts

Handle

Frame

Swiveling spigot

Threaded spigot

Blade

Blade holder pin

Bow saw parts

Cord   Toggle

Cheek

Pivot point

Dry stub
mortise and
tenon

Stretcher rail

Handle

Operating handle   Blade

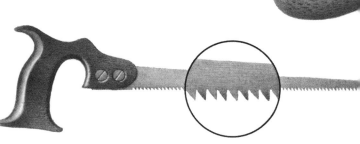

Keyhole saw
Blade 125–375 mm/5–15 in.
9 points to 25 mm/1 in.
For cutting holes away from
edges; originally for making the
straight cuts in keyholes.

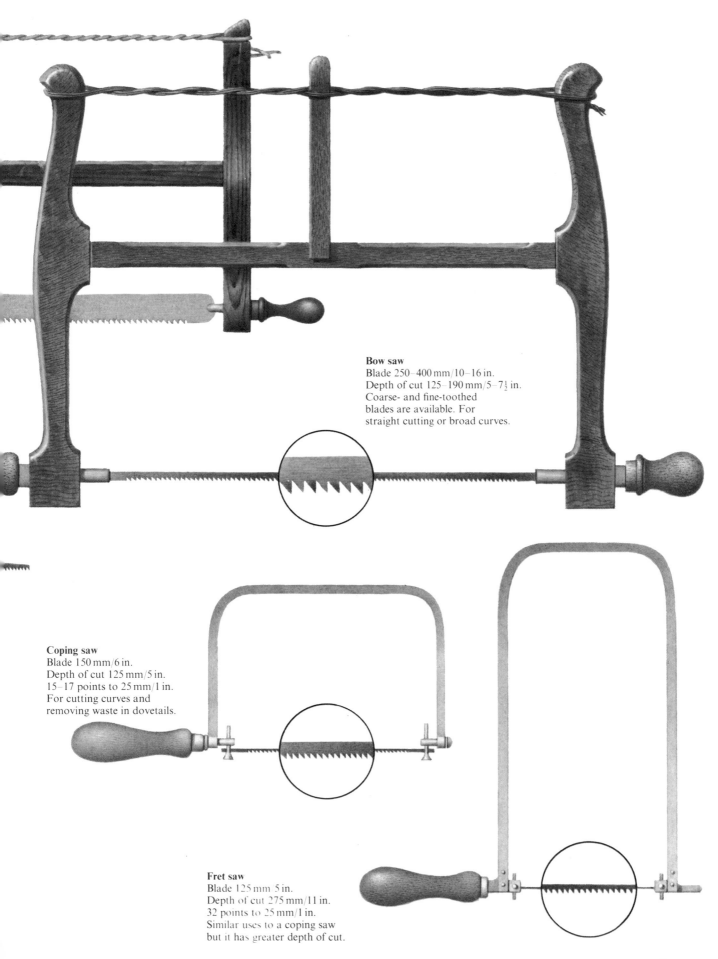

Bow saw
Blade 250–400 mm/10–16 in.
Depth of cut 125–190 mm/5–7½ in.
Coarse- and fine-toothed
blades are available. For
straight cutting or broad curves.

Coping saw
Blade 150 mm/6 in.
Depth of cut 125 mm/5 in.
15–17 points to 25 mm/1 in.
For cutting curves and
removing waste in dovetails.

Fret saw
Blade 125 mm/5 in.
Depth of cut 275 mm/11 in.
32 points to 25 mm/1 in.
Similar uses to a coping saw
but it has greater depth of cut.

Saws for shaping 2

When using saws for cutting curves, face the work directly and stand with the feet apart. The saw should be introduced at right angles to the work and the blade held in the horizontal position. The bow saw is operated by its handle. Both bow saw and coping saw are used with two hands: one hand wrapped tightly around the other with the fingers astride the blade.

The worker always has three options while sawing shapes: to change the direction of the blade; to switch saws; or to readjust the position of the wood in the vise. Use these options freely, depending on ease of working and how much support the vise is providing. Altering the direction of the blade is quickly done on most frame saws. Place a hand on each side of the blade; otherwise the blade will twist and perhaps break. Be prepared to switch to a saw with a wider blade when the opportunity arises; the narrower the blade, the more difficult it is to cut a straight line.

Frequently change the position of the wood so that the point of sawing remains near the vise jaws. If too high the wood will vibrate and the blade may break.

With the bow saw, cutting is on the push stroke; this comes naturally to most woodworkers. With the coping saw, where the blade can be fitted with the teeth facing towards or away from the handle, cutting on the push stroke tends to compress the frame and loosen the blade, whereas cutting on the pull stroke conserves the blade; however, the cutting line may be obscured because the fibers are raised.

Correct stance

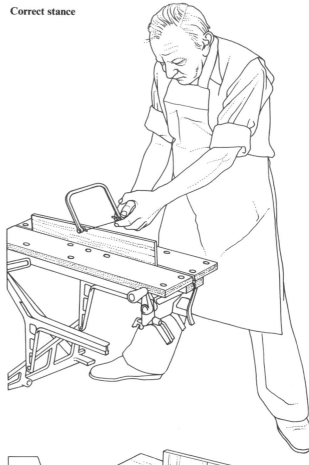

Coping saw

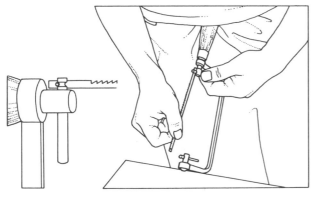

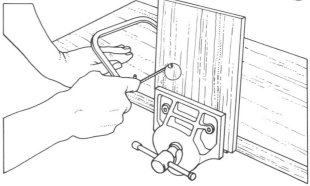

Spigot pins are split and grooved so that when a blade is slid into the split the pin will slot into the groove.

To change a blade compress the frame against the bench to allow the blade pins to pass the spigots.

For an interior cut release one end of the blade. Pass it through a drilled hole, refit the blade and start sawing.

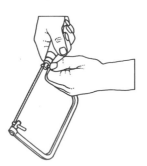

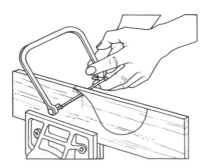

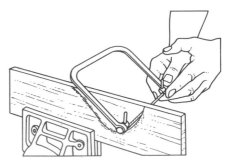

When tightening the handle, after replacing the blade, hold the spigot to prevent the blade twisting and breaking.

To prevent the blade slipping out of line when both hands are on the saw handle, start the cut inside the cutting line.

Finish a cut by adjusting the blade angle in an upward direction. Alternatively, start a fresh cut at the other end.

Bow saw

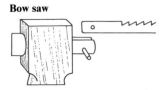

To fit a bow saw blade pass the holding pins through the handle extension into the blade and out the other side.

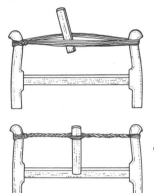

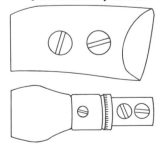

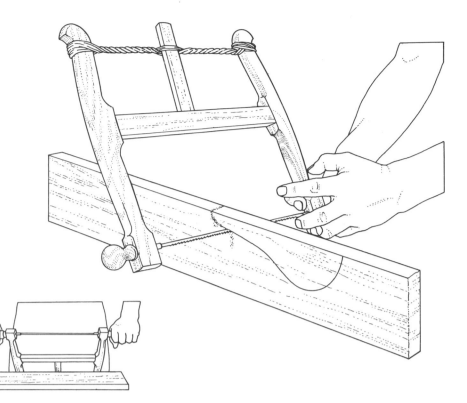

To tension the blade turn the toggle and cord. When the blade is taut tuck the toggle against the stretcher rail.

To adjust the angle of a bow saw blade hold the frame steady against the bench while turning both handles.

To cut curves in thick boards use a bow saw's sturdy frame fitted with a toughened blade. Keep the blade horizontal.

Fret saw

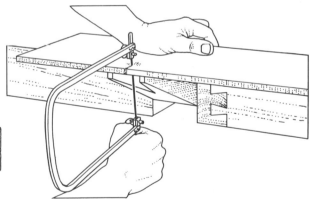

A fret saw blade is held, with the aid of wing nuts, by a friction grip. The frame is compressed for blade tension.

A V-shaped block supports the work when fret sawing. Clamp the block level with the bench top and cut within the V.

The fret saw is usually worked vertically, cutting on the down stroke. The blade teeth face towards the handle.

Compass saw and Keyhole saw

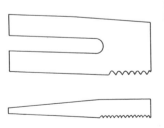

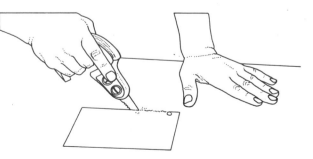

A compass saw blade is slotted for easy changeover. It passes around two screws in the handle, which are then secured.

A keyhole saw handle is slotted throughout its length. The blade, held by the two screws, can project to any length.

To start an interior cut with a keyhole saw or compass saw, first bore a hole to accommodate the blade tip.

Power handsaws

Sawing is a common task when working with wood and may be tiring to do by hand. Power saws speed up this work and do it more accurately than a handsaw.

A circular saw makes straight accurate cuts with and across the grain. The size of a circular saw is determined by its blade diameter. A size of 125 mm/5 in to 190 mm/7$\frac{1}{2}$ in is sufficient for most purposes. The maximum depth of cut is approximately one-third the blade diameter. The blade should not be exposed during working: all reputable makes of saw have a fixed upper blade guard and a lower spring-loaded guard, which retracts automatically.

Jigsaws and saber saws are extremely versatile tools that will cut up to 63 mm/2$\frac{1}{2}$ in in softwood and 25 mm/1 in in hardwood and particle boards. Their major task is to cut curves but they can also make straight cuts, although they cannot match a circular saw. Many different blades can be fitted. Variable speed models are available; use a slower speed for curves and harder materials.

When using these saws, support the work securely so it cannot move and ensure that the blade will be unobstructed beneath the work. Prevent vibration by making the cut close to the workbench. Start a cut by resting the sole on the workpiece, with the blade clear of the surface and in line with the cutting line. Switch on the saw and allow it to develop full speed before advancing it through the work. Do not push or the motor will overload; jigsaw blades may break. Never force or twist the blade: it may cause kickback. When the cut is complete, turn off the saw and continue to hold it until the blade stops; in partial or blind cuts, let the blade stop before removing it from the cut. For accurate straight cuts, always run the sole against a guide fence, either the rip fence supplied with the tool or a home-made T-guide.

Blunt circular saw blades are best resharpened by a professional; jigsaw blades should be discarded.

As with all power tools, never make any adjustments to these saws without first switching off and unplugging the tool. Check the power cord regularly for cuts and wear, and take care in use that the cord is not beneath the work, where it might accidentally be cut.

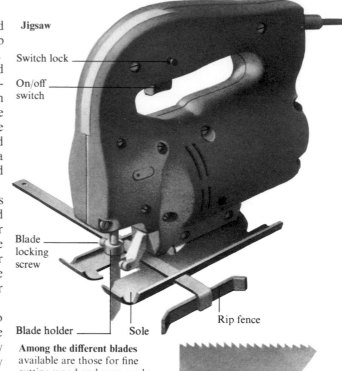

Jigsaw

Switch lock

On/off switch

Blade locking screw

Blade holder

Sole

Rip fence

A

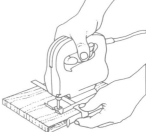

B

Among the different blades available are those for fine cutting wood and man-made boards (**A**) and those just for plywoods and particle boards. Others will rough cut wood only (**B**) while some will also cut soft metals.

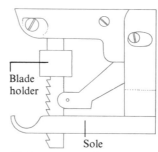

Blade holder

Sole

Blades are bought for specific manufacturers' saws. The teeth cut on the upward stroke, so place the face side of the work downwards.

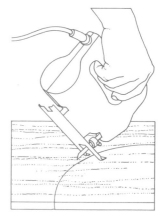

Before ripping lubricate the fence with candlewax. Set the fence to the required distance from the blade and press it against the side of the work.

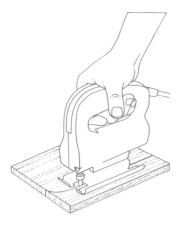

Cut a curve freehand slightly to the waste side of the cutting line. Using a firm grip support the board fully. Do not force the saw.

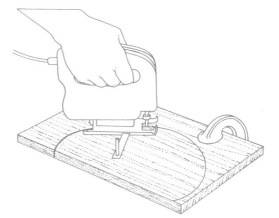

To cut a circle drive a nail through the center of the work, invert the fence and slip it over the nail. Then start the cut, using the nail as a pivot.

To begin a plunge cut tilt the saw on its front end with the blade well clear of the work. Switch on and slowly lower the blade through the wood.

Circular saw

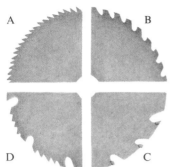

A B
D C

A variety of blades is available for different kinds of work. The most useful are those for crosscutting **(A)**, ripping **(B)**, and combination cuts **(C)**. The planer blade **(D)** gives a smooth finish. A carbide-tipped blade is much harder than an ordinary steel blade.

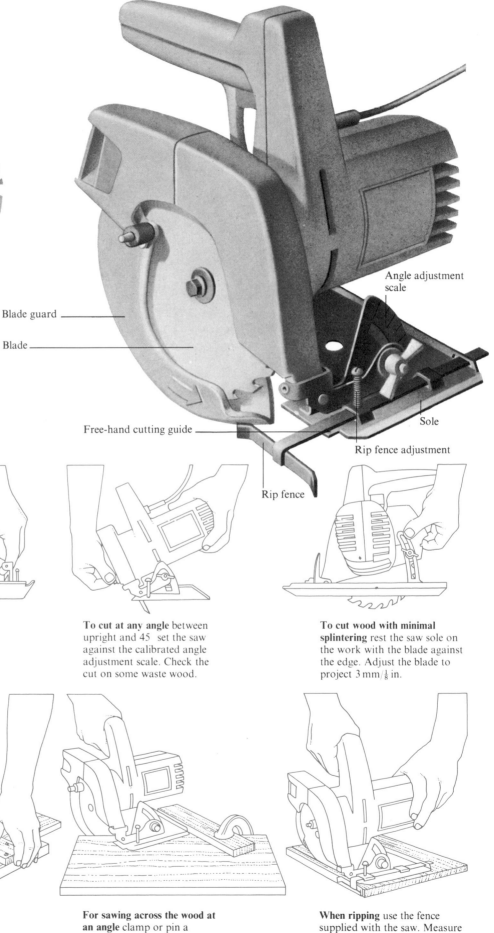

Angle adjustment scale

Blade guard

Blade

Sole

Rip fence adjustment

Free-hand cutting guide

Rip fence

To fit a blade lower the bottom blade guard. Withdraw the bolt and washer. With the new blade over the spindle hole replace the bolt and tighten.

To cut at any angle between upright and 45 set the saw against the calibrated angle adjustment scale. Check the cut on some waste wood.

To cut wood with minimal splintering rest the saw sole on the work with the blade against the edge. Adjust the blade to project 3 mm/⅛ in.

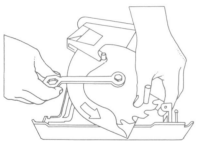

Stock

Before crosscutting make a T-guide with an overlong stock. Cut through the stock and the work. The stock can then be used as an exact T-guide.

For sawing across the wood at an angle clamp or pin a temporary fence to the work. The fence must be thin enough to clear the motor.

When ripping use the fence supplied with the saw. Measure from inside the fence to the blade allowing for the set of the saw teeth.

Sharpening and maintaining saws

A sharp saw needs no forcing, but once the teeth become dulled the saw will cut slowly and inaccurately. If it is forced, it may wander from the cutting line, jam in the kerf or buckle. The complete process of renewing the edge on saw teeth is called refitting and it involves three quite separate stages — topping, setting and sharpening. All hand- and backsaw blades can be refitted. The blades of frame saws should be discarded when blunt. Circular blades for power saws should be professionally sharpened.

During each stage of refitting the saw must be held securely along the length of the blade. It can be held in the vise with battens on either side. A special saw-sharpening vise is available which clamps to the bench. The disadvantage with these methods is that the light may not be good enough, the best possible light being required for accurate refitting. A purpose-made saw clamp is better. Make the first attempt at refitting on a new saw, because the angles of the teeth will be relatively easy to follow. If it is difficult to keep a saw to the cutting line, it is likely that the teeth on one side are higher than those on the other. Topping is the remedy.

In topping, the teeth are filed to the same height with a smooth, flat file, which must be held horizontally; make a topping clamp to ensure this. If the file is not held horizontally, the teeth on one side are liable to be filed more than on the other. Each tooth touched by the file will show a bright white spot of new metal. If, after two or three strokes, some teeth are still not showing a white spot, the teeth are too uneven and the saw should be sent to an expert for reshaping.

After topping, the teeth must be set. This involves bending the tips of the teeth alternately to right and left of the blade. The result is that the kerf formed during cutting is wider than the thickness of the blade, giving the blade clearance and reducing friction. The width of the kerf should be no more than one and a half times the thickness of the blade. The teeth should be set to no more than half their depth and to the same side to which they were originally bent. Bending them the

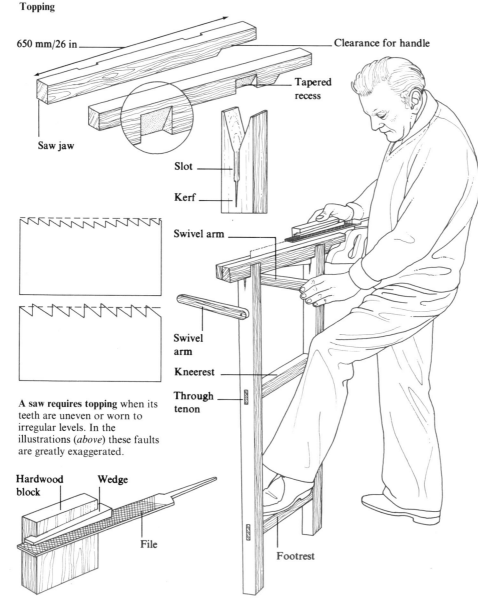

Topping

650 mm/26 in

Clearance for handle

Tapered recess

Saw jaw

Slot

Kerf

Swivel arm

Swivel arm

Kneerest

Through tenon

Footrest

A saw requires topping when its teeth are uneven or worn to irregular levels. In the illustrations (*above*) these faults are greatly exaggerated.

Hardwood block

Wedge

File

Topping clamp.

Topping a saw in a saw clamp.

Topping
The tops of the teeth are filed with a smooth, flat file to level them. To ensure the file is held flat, make a topping clamp from a grooved hardwood block. Taper the top of the groove. Hold the file in the groove with a wedge. Some woodworkers prefer a handsaw edge to be slightly rounded in its length to produce a more effective cut for the energy used. This can be done in stages at each topping.

A saw clamp holds the saw firmly along the length of its blade and can be set up anywhere in a good light by propping its swiveling arms against a wall. Cut the saw jaws long enough to support the longest handsaw blades. Shape the jaws at one end to provide clearance for the saw handle. Cut tapered recesses in the outside edges of the jaws to fit into the uprights of the saw clamp. Line the recesses with abrasive paper to increase friction. Make the frame from two uprights joined by two rails through tenoned at convenient heights to act as a footrest and a kneerest. Attach the swiveling arms to the uprights with screws and washers. Cut the uprights to accommodate the saw jaws. Then cut a slot to provide a clearance for backsaw backs. Make a kerf below the slot for handsaw blades.

To top a saw, clamp the saw blade in the saw clamp, teeth projecting well above the saw jaws. Hold the topping clamp against the saw blade so that the file is flat on the teeth. Run it along until a bright white spot of new metal is visible on every tooth.

other way may break them. The setting must be uniform on each side; for the saw to work true, the teeth on each side must do the same amount of work.

The easiest way to set a saw is using a proprietary saw set. Professional saw doctors use the thin cross-peen of a hammer to top the teeth, on an anvil. This method is quick but demands much experience. Over-setting is a mistake amongst amateurs; this makes the saw inefficient as more effort is needed to cut the wide kerf and more wood than is necessary is wasted as sawdust. If the set is excessive, it can be remedied by side dressing. Lay the saw flat on the bench and run the edge of a lubricated oilstone lightly along the teeth for two or three strokes on each side. Subsequent sharpening without setting brings the saw back to normal.

In sharpening, the cutting edge or point is renewed on each tooth. A saw can be sharpened two or three times before it needs resetting, but it should be topped lightly before each sharpening. This has the advantage of ensuring that the teeth remain level throughout and

the white spots provide guides for sharpening. The aim in sharpening is to file away the bright white spot left on each tooth by topping. Tapered triangular files are used for sharpening. The correct length of file to be used depends on the number of points per 25 mm/1 in on the saw. The file must fit correctly in the gullet — the space between the saw teeth. Always discard files when they start losing their cutting edge. Once dulled, control of the file is lost and consequently saw-tooth angles will be inconsistent. The teeth of rip and crosscut saws are differently shaped, and so must be sharpened accordingly. The teeth of dovetail and tenon saws are more difficult to sharpen as they are so small. Light is all-important, and this emphasizes the value of a portable saw clamp.

When a saw is in very bad condition, an extra shaping stage can be introduced after topping. Shaping would only be necessary if the saw had been incorrectly sharpened, repeatedly. It should then be put into the hands of an expert for recutting.

Setting

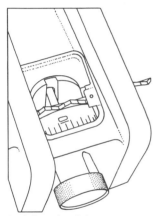

1. Adjust the dial.

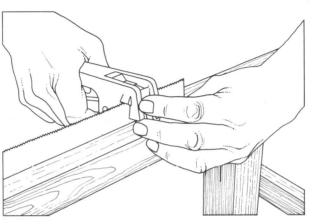

2. Place the saw set over one tooth.

Sharpening

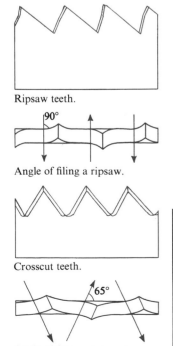

Ripsaw teeth.

90°

Angle of filing a ripsaw.

Crosscut teeth.

65°

Angle of filing a crosscut saw.

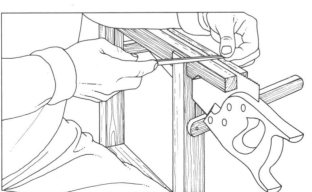

Sharpening a ripsaw.

Choosing the correct file for sharpening		
Saw, type of	**Points per 25 mm/1 in**	**File length**
Dovetail	18–22	150 mm/6 in
Tenon	12–14	175 mm/7 in
Panel	10–12	200 mm/8 in
Crosscut	7–8	225 mm/9 in
Rip	4½–6	250 mm/10 in

Setting

The easiest way to set saw teeth is with a pliers-type saw set. Adjust the dial according to the number of points per 25 mm/1 in (**1**). This controls the amount and depth of set. Place the saw set over one tooth, lining up the plunger squarely with the center of the tooth, and squeeze the handles (**2**). The plunger presses the tooth against the anvil, the bevel of which was set when the dial was adjusted. Set each alternate tooth in turn along one side of the saw, then turn the saw around and set the intermediate teeth to the other side of the blade.

Sharpening

Lower the saw in the saw jaws so that the teeth and no more than 3 mm/⅛ in of the blade are showing. Select a file of the correct length. Fit a handle to the file. Holding the file at each end, begin filing at the toe end of the saw. Work on every saw tooth that bends away from the worker. To sharpen ripsaw teeth, seat the file snugly in the gullet, holding it horizontally and at right angles to the saw. File only on the forward stroke, applying even pressure to adjoining teeth. Remove half the white spot on adjoining teeth. Then turn the saw around and work back again. Sharpen crosscut, panel and tenon saw teeth in a similar way, but hold the file at an angle of about 65° to the blade and lower the file handle slightly. Sharpen the teeth on a dovetail saw at right angles like ripsaw teeth. After sharpening, side dress the saw very lightly to remove any wire edge, using the edge of an oilstone.

Chisels and gouges 1

A chisel has a long narrow blade that is used to cut deep narrow recesses for joints, such as mortises, and shallow recesses for butt hinges; it is also for paring and chamfering. A gouge has a curved blade and is used to cut and smooth grooves and hollows. Cabinetmakers' gouges have blades with parallel sides; carvers' gouges can be parallel or fishtail.

The cutting edge of chisels and gouges is across the end of the blade, the width of which designates the size of the tool. Handles can be made of wood or plastic.

Chisels and gouges are classified by the way in which the handle is fastened to the blade: either by a tang or a socket. A tang is a tapered end inserted into the handle.

A socket chisel has the end of the blade formed into a conical socket, which fits over the handle. A socket chisel is generally used on heavier work than a tang chisel. A heavy-duty chisel can have a shock-absorbent leather washer between the bolster of the blade and the handle to absorb the blows of the mallet.

Chisel blades can be rectangular in section or beveled to give clearance when cleaning out corners or working in undercut or acute angles, such as dovetail housings. Bevel-edged chisels are lighter and less robust than rectangular-sectioned chisels. Mortise chisels have strong but narrow blades for chopping and levering out waste.

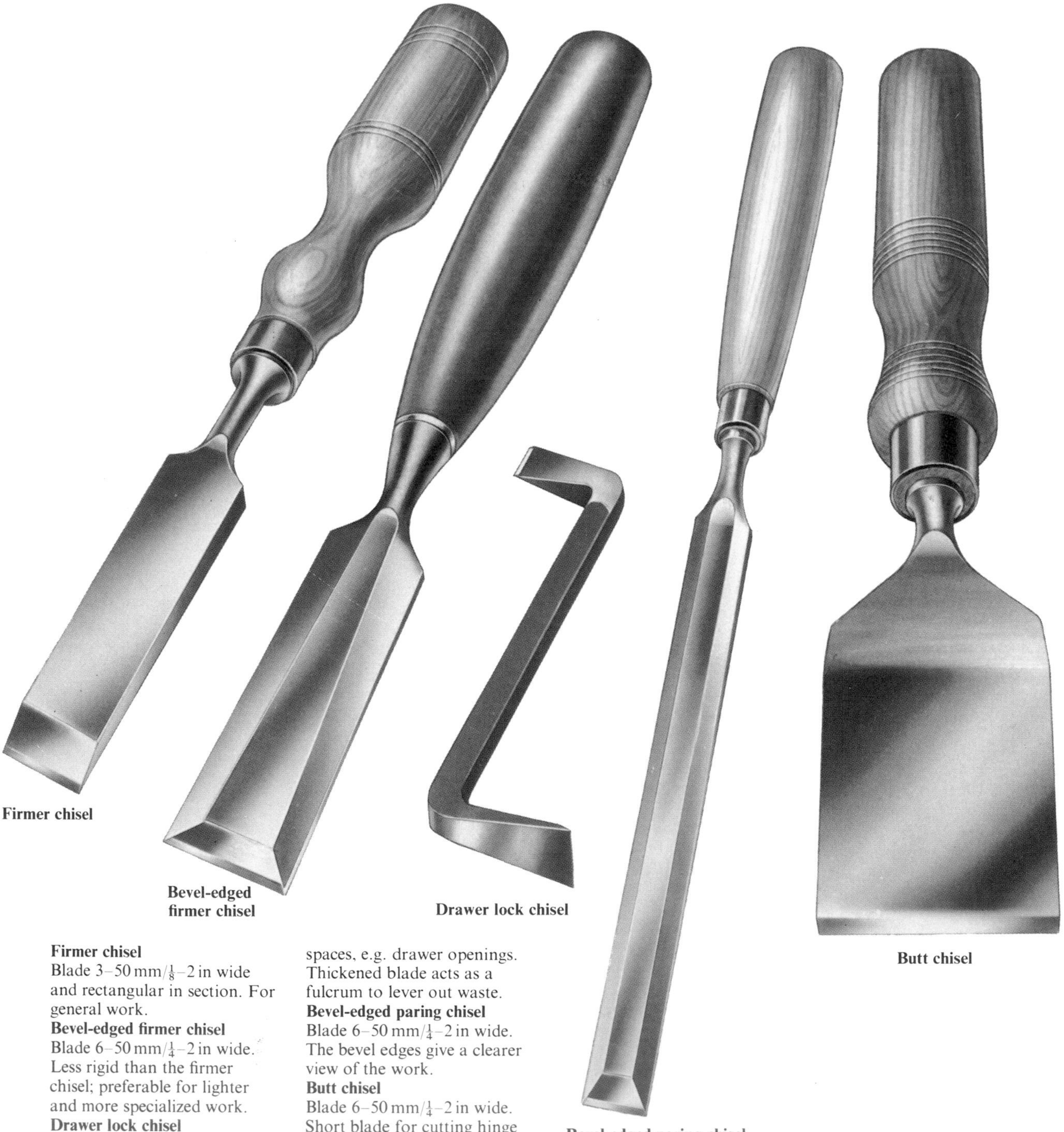

Firmer chisel

Bevel-edged firmer chisel

Drawer lock chisel

Butt chisel

Bevel-edged paring chisel

Firmer chisel
Blade 3–50 mm/$\frac{1}{8}$–2 in wide and rectangular in section. For general work.
Bevel-edged firmer chisel
Blade 6–50 mm/$\frac{1}{4}$–2 in wide. Less rigid than the firmer chisel; preferable for lighter and more specialized work.
Drawer lock chisel
Blade 10–16 mm/$\frac{3}{8}$–$\frac{5}{8}$ in wide. Use with a hammer in close

spaces, e.g. drawer openings. Thickened blade acts as a fulcrum to lever out waste.
Bevel-edged paring chisel
Blade 6–50 mm/$\frac{1}{4}$–2 in wide. The bevel edges give a clearer view of the work.
Butt chisel
Blade 6–50 mm/$\frac{1}{4}$–2 in wide. Short blade for cutting hinge housings. Preferred by many workers as a general chisel.

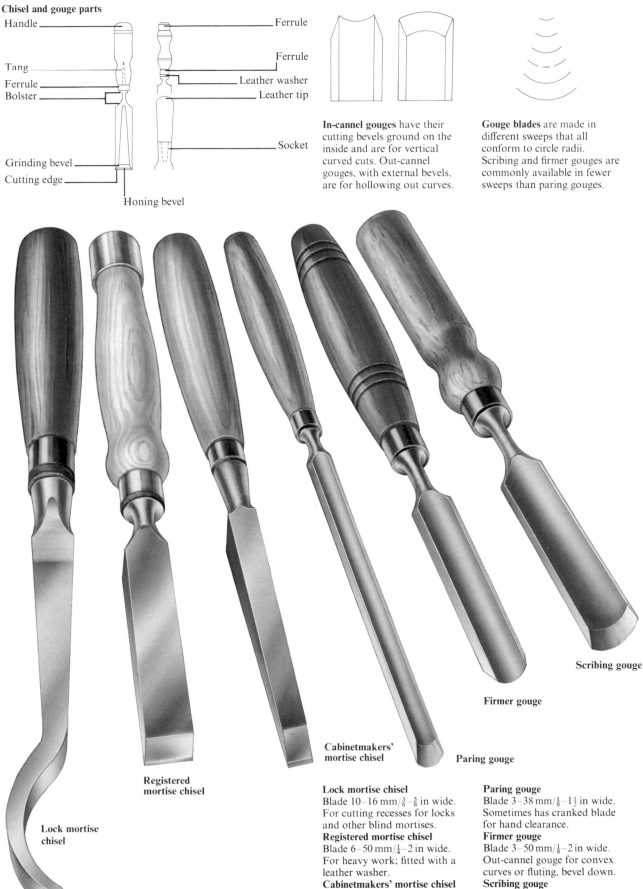

Chisel and gouge parts

Handle

Tang

Ferrule

Bolster

Grinding bevel

Cutting edge

Honing bevel

Ferrule

Ferrule

Leather washer

Leather tip

Socket

In-cannel gouges have their cutting bevels ground on the inside and are for vertical curved cuts. Out-cannel gouges, with external bevels, are for hollowing out curves.

Gouge blades are made in different sweeps that all conform to circle radii. Scribing and firmer gouges are commonly available in fewer sweeps than paring gouges.

Lock mortise chisel

Registered mortise chisel

Cabinetmakers' mortise chisel

Paring gouge

Firmer gouge

Scribing gouge

Lock mortise chisel
Blade 10–16 mm/$\frac{3}{8}$–$\frac{5}{8}$ in wide.
For cutting recesses for locks and other blind mortises.
Registered mortise chisel
Blade 6–50 mm/$\frac{1}{4}$–2 in wide.
For heavy work; fitted with a leather washer.
Cabinetmakers' mortise chisel
Blade 6–13 mm/$\frac{1}{4}$–$\frac{1}{2}$ in wide.
Sturdy socket chisel to withstand continual blows.

Paring gouge
Blade 3–38 mm/$\frac{1}{8}$–1$\frac{1}{2}$ in wide.
Sometimes has cranked blade for hand clearance.
Firmer gouge
Blade 3–50 mm/$\frac{1}{8}$–2 in wide.
Out-cannel gouge for convex curves or fluting, bevel down.
Scribing gouge
Blade 3–25 mm/$\frac{1}{8}$–1 in wide.
General purpose in-cannel gouge for concave curves.

Chisels and gouges 2

Vertical paring

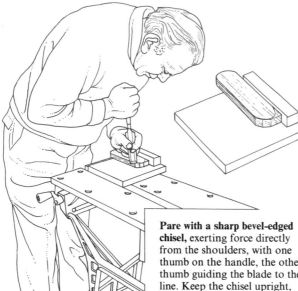

Pare with a sharp bevel-edged chisel, exerting force directly from the shoulders, with one thumb on the handle, the other thumb guiding the blade to the line. Keep the chisel upright, bevel out. Remove only small amounts of wood at a time.

Horizontal chiseling

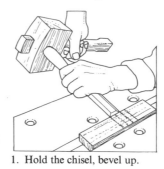

1. Hold the chisel, bevel up.

2. Remove the waste.

Horizontal chiseling
Saw on the waste side of the squared lines down to the gauged lines on each side. Then from one side chisel to the center of the housing. Steady one elbow on the bench with the hand holding the chisel near the blade, bevel up (1). Pointing the chisel slightly upwards, work down to half the housing depth, removing a little waste at a time (2). Reverse the wood and gradually chisel down to the gauged line. Then finish off the first side. Check for flatness. Level the base, using hand pressure only.

Cutting a rectangular recess

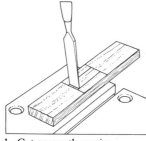

1. Cut across the grain.

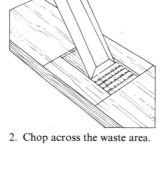

2. Chop across the waste area.

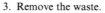

3. Remove the waste.

Cutting a rectangular recess
Chisel across the grain at one end, just inside the marked line, using a chisel held vertically, bevel inwards, and a mallet (1). Then chop on the line. Repeat at the other end. Cut along the grain near the lines, then directly on them. Chop across the wood grain at 6 mm/¼ in intervals (2). Lever out the waste working from the center, bevel down (3). Finish with a router.

Cutting a stopped rabbet

1. Chop into the waste.

2. Cut vertically.

3. Remove the waste.

Cutting a stopped rabbet
Start at the end of the rabbet, holding the chisel at an angle, bevel down. Chop into the waste, using a mallet (1). Move back about 10 mm/⅜ in at a time. Then chop vertically, close to the line (2). Clear out the waste with a slicing action (3). Trim the side and bottom of the rabbet, adjusting the work in the vise so as to keep the chisel cutting vertically all the time.

Drawer lock chisel

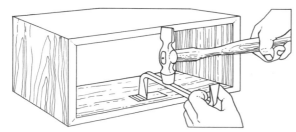

The drawer lock chisel can cut recesses and mortises in awkward places. Place the cutting edge in position. Supporting the chisel, strike its steel bar with a hammer as nearly as possible above the cutting edge. In very small spaces the side of the hammer may be used.

Trimming veneer

Veneer can be cut with a chisel as well as a knife. Hold the chisel with the bevel against the straightedge and the blade tilted away, so the cut is made with a pulling motion by the leading edge. Cut miters in veneer with the chisel held vertically to avoid splitting the wood grain.

Cutting a chamfer with mitered ends

Mark the chamfer in pencil by finger gauging or by using a simple waste-wood gauge (1). (Do not use a marking gauge, which will leave score marks in the wood.) Slice the mitered end in one stroke, holding the chisel, bevel up, and striking the handle with the palm of the hand (2). Then cut the chamfer from the other direction, if the direction of the grain allows, working towards the miter in small steps with the chisel held bevel down. Work with hand pressure, applying a pulling restraint with the other hand (3). Gradually work up to the miter and down to the limits of the chamfer. Take small chips first to establish the direction of the grain; then, if the grain permits, take longer strokes, aiming to leave a smooth surface to the chamfer so that it will need no more finishing.

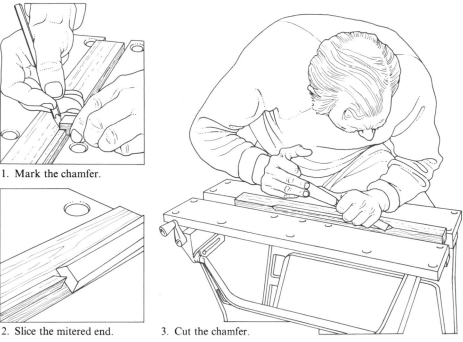

1. Mark the chamfer.

2. Slice the mitered end.

3. Cut the chamfer.

Cutting a chamfer with rounded ends

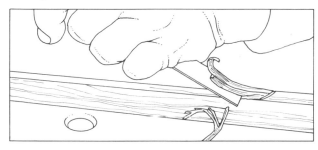

Cutting a chamfer on end grain

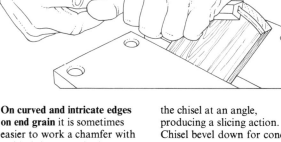

Mark the chamfer in pencil. Begin by cutting the rounded ends. Hold the chisel, bevel down, and cut each end in one stroke, with a scooping cut,

down as far as the limits of the chamfer. Then cut the body of the chamfer, working gradually in the same way as for the chamfer with mitered ends.

On curved and intricate edges on end grain it is sometimes easier to work a chamfer with a chisel than a spokeshave. Pencil in the chamfer. Work

the chisel at an angle, producing a slicing action. Chisel bevel down for concave curves and bevel up for flat surfaces and convex curves.

Out-cannel gouge

Use an out-cannel gouge to pare the edges of convex curves. The radius of the curve must match the sweep of the gouge. Cut away only a little at a time and guide the blade with a finger of one hand.

In-cannel gouge

1. Shape the ends.

2. Chop out the waste.

3. Lever out the waste.

In-cannel gouge

An in-cannel blade on, for example, a scribing gouge can be used to cut internal curves, such as the finger slots on sliding doors. The curve of the gouge must match the curve to be cut in the wood. Start by shaping the ends, holding the gouge vertically (1). Chop along the grain near the straight lines with a bevel-edged chisel, then cut directly on them. Chop out the waste with a series of 6 mm/$\frac{1}{4}$ in cuts, holding the gouge at an angle and moving it backwards after each cut (2). Lever out the waste with a chisel (3). Cut deeper, if necessary, repeating the sequence of cuts with a gouge and a chisel. Level the base of the recess with a small hand router.

Bench planes 1

A plane is a cutting tool for removing shavings to a precisely controlled depth. A bench plane has a flat sole and a wooden or metal body; it is used for reducing the width or thickness of a piece of wood, making it straight, and for smoothing a surface. The bench plane reached an advanced stage of development in Roman times and the modern metal version could be considered a definitive tool.

Although the wooden plane is still in use, the metal plane has a cutting edge that is more quickly and precisely adjusted. Once a plane has been correctly set, the flat sole controls the blade and the depth of cut so that the worker's full strength can be used for the driving force. It is important therefore to know how a plane is adjusted.

The cap iron stops the wood tearing by fracturing the shaving as soon as it is raised by the blade. The closer the cap iron to the cutting edge the sooner the shaving is broken, although there is greater resistance in the cut. The size of the mouth enables cuts of varying degrees of fineness to be made; a wide mouth allows a thicker shaving to be cut with a deeper set blade. Adjustment of the cap iron, mouth and blade should suit the grain and the required surface finish.

Adjusting a plane

 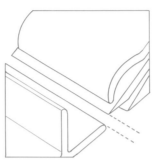  

Set the cap iron about 1.5 mm/$\frac{1}{16}$ in back from the cutting edge for a softwood. For a hardwood or a fine finish the gap should be reduced. This is done by realigning and securing the cap iron on to the back of the blade. Ensure a tight fit to prevent shavings wedging between the two.

The width of the mouth opening is governed by the position of the frog. For general use, line up the edge of the frog with the top edge of the mouth. For fine work the mouth should be narrower. Remove the cap iron and blade. Slacken the screws that fix the frog and move the frog with its adjustment screw.

The depth of cut is increased by pushing the blade out. This is done by turning the adjustment nut clockwise. Check the setting by holding the plane upside down with the sole level with the eye. The cutting edge should appear as a narrow black line. For fine work the blade should hardly show.

The lateral adjustment lever is used to ensure that the blade protrudes an equal amount across its width. Move the lever from side to side and check the alignment with the plane held upside down. Once the blade is set, take up any slack in the adjustment nut so the blade cannot move.

Metal plane parts

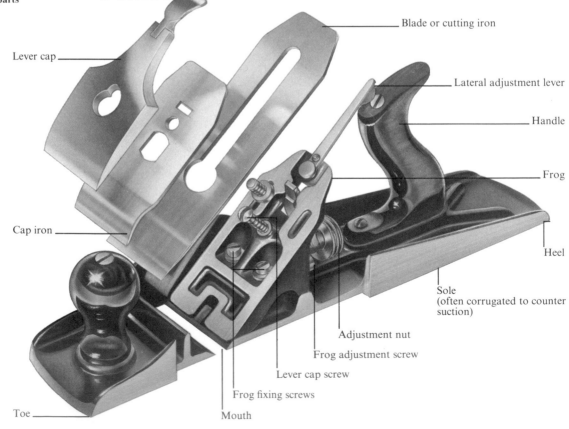

Lever

Lever cap

Cap iron

Toe

Mouth

Frog fixing screws

Lever cap screw

Frog adjustment screw

Adjustment nut

Sole (often corrugated to counter suction)

Heel

Frog

Handle

Lateral adjustment lever

Blade or cutting iron

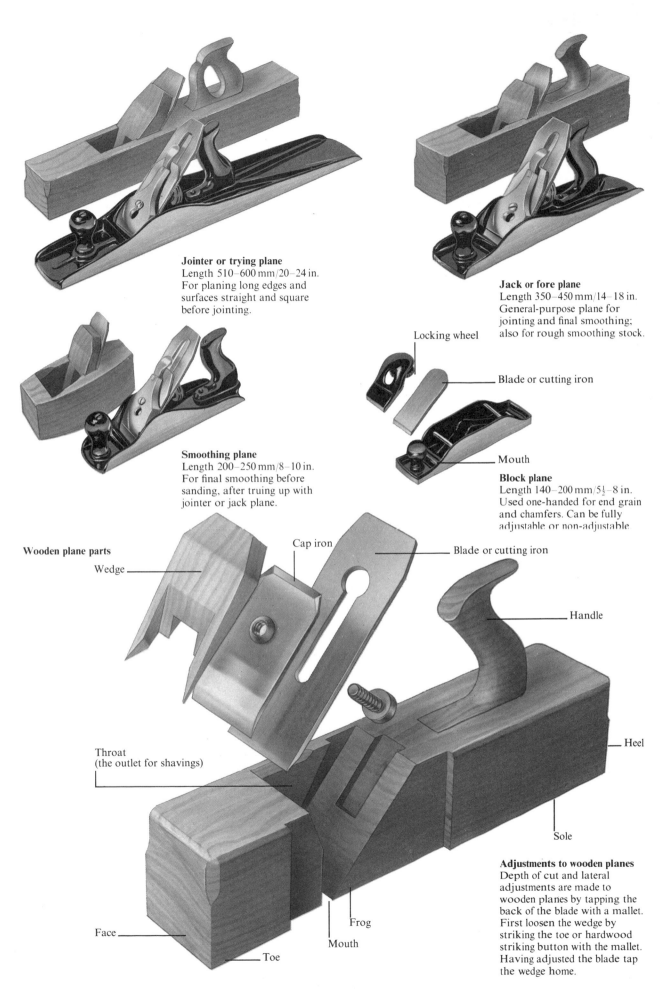

Jointer or trying plane
Length 510–600 mm/20–24 in.
For planing long edges and
surfaces straight and square
before jointing.

Jack or fore plane
Length 350–450 mm/14–18 in.
General-purpose plane for
jointing and final smoothing;
also for rough smoothing stock.

Locking wheel

Blade or cutting iron

Smoothing plane
Length 200–250 mm/8–10 in.
For final smoothing before
sanding, after truing up with
jointer or jack plane.

Mouth

Block plane
Length 140–200 mm/5½–8 in.
Used one-handed for end grain
and chamfers. Can be fully
adjustable or non-adjustable.

Wooden plane parts

Wedge

Cap iron

Blade or cutting iron

Handle

Heel

Throat
(the outlet for shavings)

Sole

Adjustments to wooden planes
Depth of cut and lateral
adjustments are made to
wooden planes by tapping the
back of the blade with a mallet.
First loosen the wedge by
striking the toe or hardwood
striking button with the mallet.
Having adjusted the blade tap
the wedge home.

Face

Frog

Mouth

Toe

79

Bench planes 2

Careful planing smooths the wood and makes surfaces square in preparation for accurate working out and cutting of joints. Always sharpen and adjust a plane before using it. Place the wood to be planed horizontally on the bench against a bench-stop; if gripped in the vise, planing may be uneven.

Keep a piece of candle on the bench or in the apron pocket and use it to lubricate the sole of the plane. Since resin from wood will make planing harder, wipe any residue off the plane sole using kerosene.

To square a piece of wood, begin by planing off the high spots; then take shavings from the whole surface. When planing wide surfaces it is easy to remove too much wood from diagonally opposite corners. Therefore check the shape of the board frequently.

To plane an edge straight, remove shavings from the middle first, then plane the whole edge. The success of all joints depends on accurate planing.

After using a plane, place it on its side on the bench. This will prevent the cutting edge coming into contact with other metal objects and being dulled. Store a plane on its side on a shelf or in its box. Wipe wooden planes with linseed oil occasionally to prevent drying.

The sole of a wooden plane will eventually wear down and the wedge-shaped mouth will widen. To restore the mouth to its original width, recess a piece of hardwood in front of the blade.

Correct stance

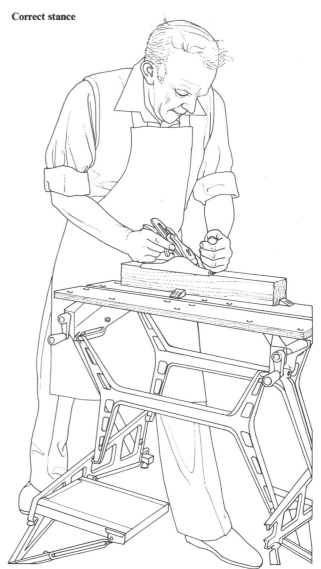

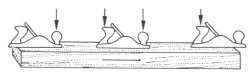

At the beginning of a stroke press down on the front of the plane; then press evenly. Finish by pressing at the rear.

Stand with the outer foot braced in front and place the shoulder, hip and plane in a straight line in order to exert full control.

Planing a wide board

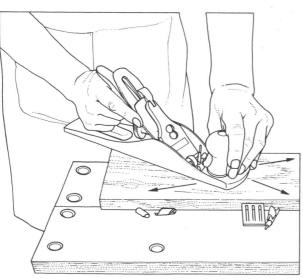

Checking a surface

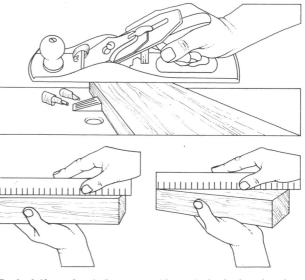

Move the plane diagonally across the wood in both directions, with and against the grain, overlapping strokes to cover the whole surface.

Finish by taking fine shavings in parallel strokes down the wood in the direction of the grain. Check the smoothness of the surface by hand.

To check if a surface is flat use the tilted edge of the plane as a straightedge. Place the straightedge diagonally to check against winding.

Alternatively check a planed surface with a steel rule, along the grain, across the grain at intervals and in both directions diagonally.

Planing a narrow edge

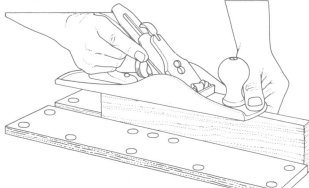

Planing square

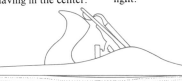

If an edge is out of square push the plane towards the high edge. If the blade is sharpened to a slight curve it will take a thicker shaving in the center.

To check if a face is square with one side push the stock of the try square tight against the side of the wood. Hold against the light.

To keep the plane in the center of a narrow surface place the thumb on the toe and run the fingers as a guide along the side of the wood.

A long plane rides on the high spots, leveling them out until one continuous shaving can be taken along the wood; a short plane follows the contours.

End grain and the edges of thin boards must be supported to prevent the wooden fibers on the edge breaking away. The stop on a shooting board will provide the necessary support. Hold the plane centrally at its point of balance and allow the wood to overhang very slightly. Alternatively, support end grain by gluing or clamping a piece of waste wood to the workpiece before planing, so any splintering will be from the waste wood and not from the workpiece.

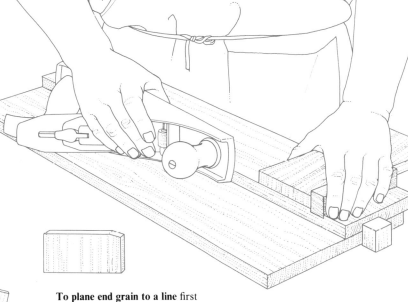

To plane end grain to a line first plane a chamfer at one end. Reverse the wood and plane towards the chamfer. Take care on the last strokes.

Planing chamfers

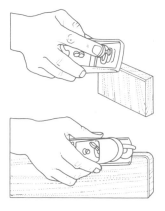

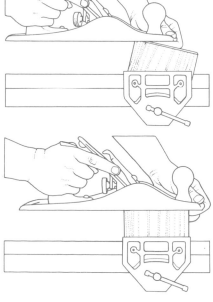

Planing halfway from each side is a good alternative, especially on wide boards. The plane must be sharp, finely set and firmly under control.

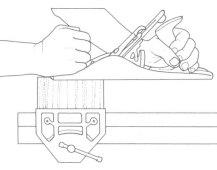

Instead of reversing the wood when planing from each side reverse the plane, holding it as shown. Secure the wood firmly against vibration.

For delicate work, such as putting a light chamfer on an edge, use a block plane held at an angle; a larger plane would obscure the work.

Planes for shaping 1

Shaping planes fall into two groups: those that cut grooves, rabbets and other shapes, and those that are used for trimming and smoothing shapes and awkward areas that bench planes cannot reach.

Among the first group are the plow, combination and multi-plane, which have interchangeable blades of different widths and profiles. The plow plane cuts with the grain; the fence enables the cut to be made a fixed distance from the edge of the work and the depth gauge limits the depth of cut. The combination plane is similar but has two knife-edge spurs for cross-grain cutting. The multi-plane is the most versatile of these planes and has the largest range of cutters.

Among the second group of shaping planes are the shoulder, bull-nose and compass planes and the hand router. The blades on the shoulder and bull-nose planes are set at a low angle bevel upwards, enabling them to be used on end grain.

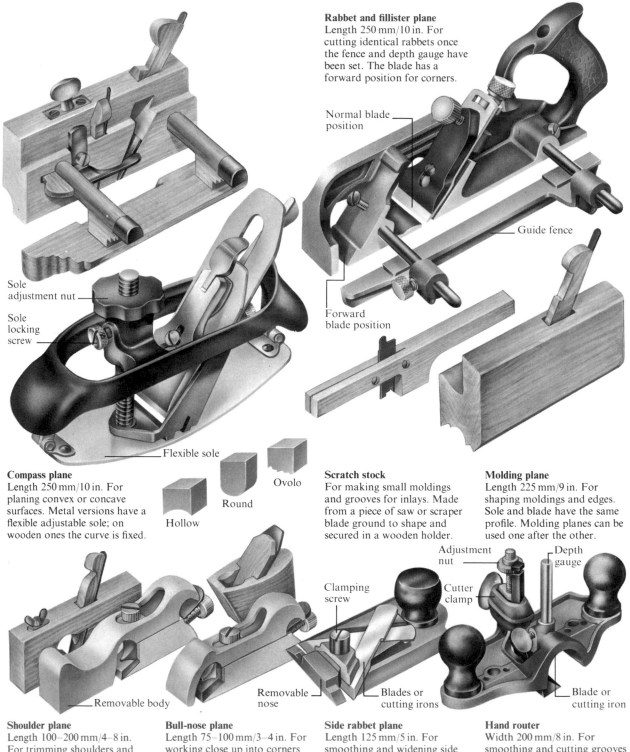

Rabbet and fillister plane
Length 250 mm/10 in. For cutting identical rabbets once the fence and depth gauge have been set. The blade has a forward position for corners.

Normal blade position

Guide fence

Forward blade position

Sole adjustment nut

Sole locking screw

Flexible sole

Ovolo

Round

Hollow

Compass plane
Length 250 mm/10 in. For planing convex or concave surfaces. Metal versions have a flexible adjustable sole; on wooden ones the curve is fixed.

Scratch stock
For making small moldings and grooves for inlays. Made from a piece of saw or scraper blade ground to shape and secured in a wooden holder.

Molding plane
Length 225 mm/9 in. For shaping moldings and edges. Sole and blade have the same profile. Molding planes can be used one after the other.

Clamping screw

Removable body

Removable nose

Adjustment nut

Cutter clamp

Depth gauge

Blades or cutting irons

Blade or cutting iron

Shoulder plane
Length 100–200 mm/4–8 in. For trimming shoulders and rabbets. Blade extends right across the sole. This plane can also be used on its side.

Bull-nose plane
Length 75–100 mm/3–4 in. For working close up into corners and ends of stopped rabbets. Blade is positioned at the toe of the sole.

Side rabbet plane
Length 125 mm/5 in. For smoothing and widening side walls of grooves and rabbets. Two blades enable plane to be used in either direction.

Hand router
Width 200 mm/8 in. For smoothing and cutting grooves and recesses. Cutters are set at a shallow angle to work with a paring action.

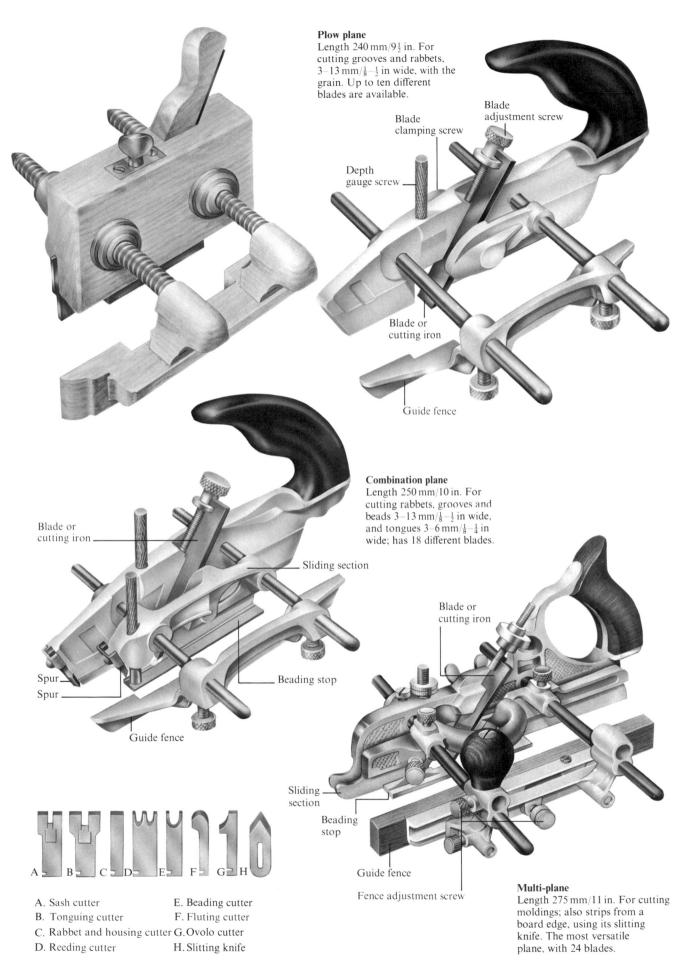

Plow plane
Length 240 mm/9½ in. For cutting grooves and rabbets, 3–13 mm/⅛–½ in wide, with the grain. Up to ten different blades are available.

Blade clamping screw

Blade adjustment screw

Depth gauge screw

Blade or cutting iron

Guide fence

Combination plane
Length 250 mm/10 in. For cutting rabbets, grooves and beads 3–13 mm/⅛–½ in wide, and tongues 3–6 mm/⅛–¼ in wide; has 18 different blades.

Blade or cutting iron

Sliding section

Spur

Spur

Beading stop

Guide fence

Blade or cutting iron

Sliding section

Beading stop

Guide fence

Fence adjustment screw

Multi-plane
Length 275 mm/11 in. For cutting moldings; also strips from a board edge, using its slitting knife. The most versatile plane, with 24 blades.

A. Sash cutter
B. Tonguing cutter
C. Rabbet and housing cutter
D. Reeding cutter
E. Beading cutter
F. Fluting cutter
G. Ovolo cutter
H. Slitting knife

Planes for shaping 2

To set a plow, combination or multi-plane, first fit the selected blade, adjusting it for depth of cut; then set the guide fence and the depth gauge. Blades are best set to cut a thicker shaving than a smoothing plane. The sliding section acts as a clamp to hold the blade against the body of the plane. A slot at the top of the blade locates on the shoulder of the adjustment screw. Very narrow blades must be set with the adjustment nut. The guide fence can be set either on the left- or right-hand side of the body; when on the right, however, it will cancel out the use of the depth gauge. The beading stop acts as a fence when cutting a bead on a tongued board, because the guide fence has nothing to bear against. For most operations the beading stop should be removed. Always hold the plane upright and keep the

guide fence tight against the wood to prevent the plane from wandering. To make any cuts across the grain, lower the spurs to score the grain ahead of the blade, thereby ensuring a clean cut. Keep the spurs sharp with an oilstone. Lightly oil all the adjustment screw threads occasionally. Rub candlewax on the fence and sole to help the plane run smoothly.

If possible always try to work a compass plane with the grain, holding it square to the wood and not at an angle. The blade is set and adjusted like a bench plane's. The blades on shoulder and bull-nose planes are set at a low angle, bevel upwards, enabling them to cut end grain. A scratch stock blade can be filed to any profile — a used hacksaw blade is ideal as the steel is tempered to just the right degree of hardness.

Combination and multi-planes

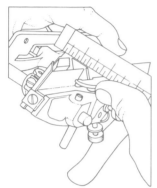

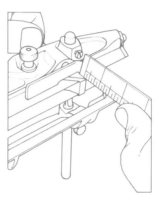

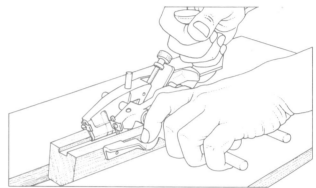

Set the guide fence the required distance from the plane body using a rule. Finger-tighten the two adjustment screws on the fence rods.

The depth gauge is a horizontal fence fitted to one side of the body of the plane. Use a rule to set the gauge to the required depth for the cut.

Start making all cuts at the far end of the work and take a few short cuts, moving a little farther back with each stroke. In this way, the plane will run

in a guide groove that it has cut. Then take continuous shavings the full length of the cut until the depth gauge rests on the wood.

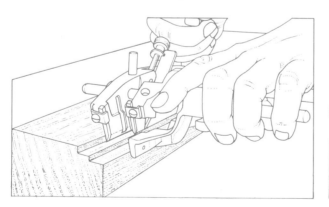

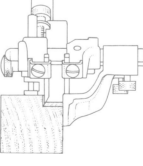

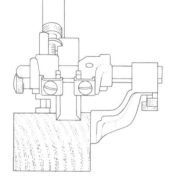

Cut a wide rabbet in two stages. Make the first cut on the outside edge to the required depth. Adjust the guide fence and cut the inside part of the

rabbet to complete the cut. If the inner part were cut first, the depth gauge would have no support when cutting the outside edge.

Cut a narrow rabbet using the method shown above. Choose a blade that is very slightly wider than the rabbet to cover the rabbet fully.

Cut a groove in the same way as a wide or narrow rabbet, but adjust the fence accordingly. The guide fence can be set on either side of the plane body.

When cutting a dado the fence cannot be used, because it will be working against end grain, which is not smooth enough for accurate work. Pin or clamp a batten in place of the fence, making allowance for the batten when setting the depth gauge. Lower the spurs. Use the body spur alone for a cross-grain rabbet.

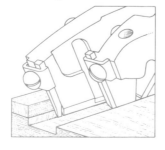

When cutting a tongue the depth gauge is not required as the depth of cut is limited by the blade itself, which has an adjustable stop in the center. Take extra care to set the guide fence so that the tongue will be cut exactly in the position required. Remove the beading stop from the sliding section on the plane.

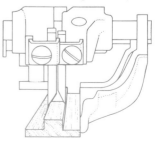

Shoulder plane

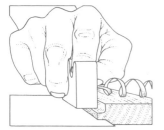

When trimming with this plane set the blade for a fine cut. Keep the side, which is square with the base, tight against the side or bottom of the rabbet.

Hand router
Insert the cutter from beneath. To set it, slacken the locking screw on the clamp; then raise or lower the cutter using the adjustment nut **(1)**. Tighten the locking screw. Remove most of the waste with a chisel. Lower the cutter until it rests on the bottom of the recess. Give the adjustment nut another half turn down; lock the cutter and plane at that level **(2)**. Continue, lowering the cutter a half turn at a time until the recess is the required depth.

Compass plane

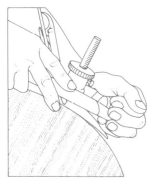

This plane is normally used with the grain, holding the blade square to the work, not at an angle. Set the blade when the sole is flat; then adjust the sole curvature.

Molding plane
Use a round plane for making hollows and a matching hollow plane for forming rounds and beads. First work a chamfer with a bench plane. Shape with the molding plane, starting at the front and working back. Then smooth with abrasive paper wrapped around a shaped block of wood. The plane has no guide fence so it requires skill to keep it at the correct angle. Hold it so the fingers of one hand act as a guide fence against the side of the work.

Bull-nose plane

Use a bull-nose plane for working rabbeted corners. It can be an integral tool or, as shown here, a shoulder plane with the nose removed.

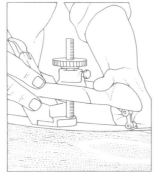

1. Adjust the cutter.

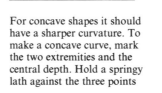

Hold the plane on the work and turn the adjustment nut until the sole is about the same shape as the work. For convex shapes, the sole should be slightly flatter than the work.

For concave shapes it should have a sharper curvature. To make a concave curve, mark the two extremities and the central depth. Hold a springy lath against the three points

Side rabbet plane

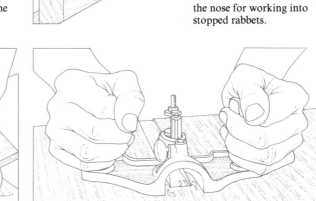

Adjust the depth gauge so the tip of the blade just touches the bottom of the work. Remove the nose for working into stopped rabbets.

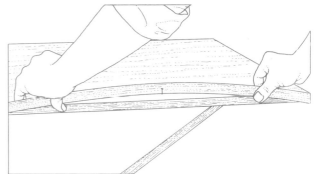

2. Plane a recess.

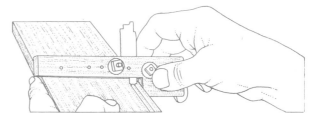

and get an assistant to draw the curve against the lath. Saw down to the line at intervals; remove most of the waste with a chisel, finish shaping to the line with the plane.

Scratch stock

This plane is homemade to any profile up to about 25 mm/1 in wide. Slot the cutter in a thick piece of wood or between two pieces screwed together. File

the cutting edge square and hone it like a scraper blade. Use only with a forward scraping action with the tool tilted away.

Power hand planes

The power plane takes the hard work out of smoothing large wood surfaces. The revolving cutter block holds two or three blades and is similar to the cutter block on the planer/thicknesser (*see* page 112). The cutter block extends across the full width of the plane sole and determines the maximum width of cut in one movement. The depth of cut can be adjusted in two ways. On some planes the soles are independently adjustable. The rear sole must be set level with the lowest point of the circle described by the blades, and the front sole is raised by the amount equal to the required depth of cut. Other models have a single sole pivoted at one end and this can be raised or lowered to set the depth of cut.

Always place the plane on the work before switching it on. The power plane must be used carefully on boards wider than the width of the sole, otherwise it will cut in a series of grooves the width of the sole, and ridges will be formed between cuts. Adjust the blades to take very shallow cuts on the final strokes, and plane slowly for a smooth finish.

The power router is capable of numerous cuts and it has largely superseded the hand router and molding planes. It cuts grooves, dadoes, rabbets, tenons and a wide variety of edge moldings. It can also cut a number of joints such as tongue and groove and dovetail, using a special dovetail template (*see* page 139). It consists of a motor fitted with a chuck that is raised or lowered above a flat sole plate. Bits held in the chuck protrude vertically down through the sole. The motor operates at very high speeds, from 18,000 rpm to 28,000 rpm, so the router leaves a very smooth finish on the work. The bits rotate in a clockwise direction, so the tool has a tendency to twist in that direction. For this reason, always feed the router into the work against the rotation of the bit.

Clamp the work firmly. Switch the tool on and allow it to reach full speed; then lower it or push it towards the work. Move it steadily through the work. Too fast a speed will overload the motor. Too slow a speed may burn the work and damage the cutters. The sound of the motor, practice will show, is the best guide to the correct rate of feed. Bits with a very wide range of profiles can be fitted into the power router. Some types, with pilot tips, are used for shaping and molding edges. The pilot tip does not cut, but rides against the edge of the work to act as a guide for the bit.

Power plane

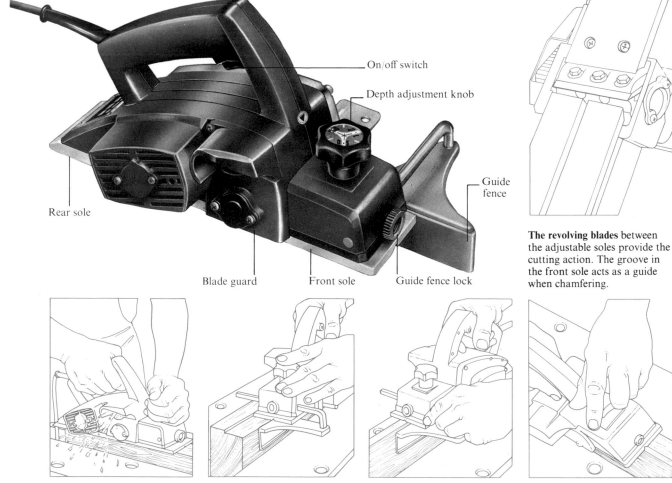

On/off switch

Depth adjustment knob

Guide fence

Rear sole

Blade guard

Front sole

Guide fence lock

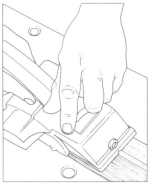

The revolving blades between the adjustable soles provide the cutting action. The groove in the front sole acts as a guide when chamfering.

To plane a surface place the tool on the work, switch it on and apply a steady, even pressure. Do not push too hard through the work.

When planing a rabbet adjust the fence so it limits the sole to the rabbet width. Make a series of shallow cuts to the required depth of the rabbet.

For a narrow surface adjust the fence so the cutting area is in the middle of the sole, making it easier to balance the plane on the work.

When planing a chamfer place the groove on the front sole onto the edge of the work. Plane in several passes to the desired depth.

Power router

Depth-of-cut lock

Depth-of-cut adjuster

Sole

Handle

On/off switch

Depth-of-cut scale

Chuck

Hole for cutter

Router bits

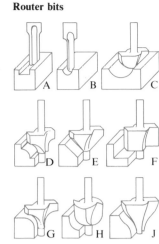

Router bits can have carbide tips for longer life. Bits A-C cut into the thickness of the wood. Bits D-J have pilot tips for shaping edges. The straight cutter (**A**) is for grooves and routing. There are also bits for veining (**B**), wide, shallow grooves (**C**), beading edges (**D**), chamfering (**E**), rabbeting (**F**), rounding edges (**G**), cutting coves (**H**) and ogees (**J**).

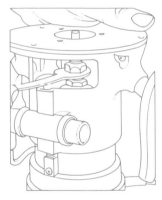

To fit a bit disconnect the router, loosen the chuck with the wrench, insert the bit shank and tighten the chuck. Adjust the sole for depth of cut.

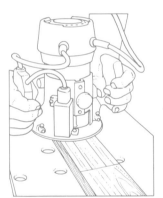

Freehand routing requires careful control as varying wood grain directions as well as the motion of the bits tend to misdirect the cut.

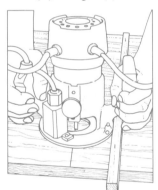

When cutting near an edge fit the guide fence. Hold the fence firmly against the edge of the work, which must be straight and smooth.

When cutting away from the edge pin a batten to the left of the router to act as a fence, because of the clockwise rotation of the bit.

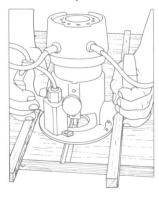

For a wide dado use two guide battens. Position each so it allows the router to cut the outer sides of the dado; then rout out the center.

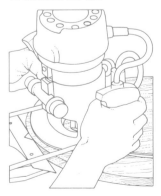

When cutting near a curved edge remove the straightedge plate on the front of the fence. Rest the V-cut recess in the fence edge against the work.

When molding the edge of a circular board or cutting a circle, use the fence as a trammel point by pinning it to the center of the work.

To convert the router to a spindle molder clamp it upside down beneath a special conversion table. Push the work past the projecting bit.

87

Shaping and smoothing tools 1

Shaping tools are used for the intermediate processes between preliminary forming and finishing. They are employed extensively to shape and smooth curved work that is inaccessible to planes. Rasps and files, with their bludgeoning abrasive action against wood, are rarely used by cabinetmakers, who feel that wood should be cut and not just worn away. The modern Surform, however, has a blade surface that contains numerous cutting edges, which have a long life even when used on chipboard, plastic laminates and gritty surfaces. It has the added advantage that there is clearance for the dust and shavings to fall away.

A half-round file has a mild action when shaping wood. The round side will smooth the edges of a hole and sandpaper can be wrapped around it for final finishing.

A drawknife is a specialist tool that will remove surplus wood much more quickly than a plane, but it needs experience and a close understanding of grain to produce good-quality work.

A spokeshave is used with a pushing action to smooth narrow sections of curved wood. It produces the same results as a smoothing plane. A metal spokeshave is adjusted much like a plane and it has largely supplanted the older spokeshave version, which is set by tapping the tanged wooden blade through the framework.

A cabinet scraper is a piece of saw steel with two edges turned to form sharp hooks. It removes minute shavings, leaving a cleaner finish than abrasives. A scraper plane does a similar job.

File parts

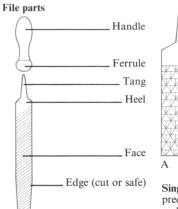

Handle — Ferrule — Tang — Heel — Face — Edge (cut or safe) — Point

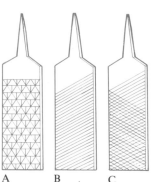

Single cut file teeth (B) are for precision work. Double cut **(C)** and rasp teeth **(A)** are for fast removal of wood.

To remove a file handle strike the ferrule of the handle sharply with the "safe" (uncut) edge of another file.

To fix a handle on a file place the tang in the handle. Hold both vertically and strike the handle on the bench.

Surform block plane
Length 140 mm/5½ in.
Use with one hand for numerous roughing-off jobs on wood, man-made and laminated boards.

Surform plane
Length 250 mm/10 in.
Use as a rough plane for fast preliminary shaping of the work.

Surform round file
Length 225 mm/9 in.
For fast removal of wood. Can be used with or without its front handle.

Surform flat file
Length 250 mm/10 in.
Similar to a Surform plane but more flexible. May break fibers if worked across the grain.

Drawknife
For quick
rough-shaping
before using a
plane or a
spokeshave.

Spokeshave
Has round sole
for concave
work, flat for
straight and
convex work.

Scraper plane
A cabinet scraper
held in a frame and
bowed in the middle.
Use with the blade
sloping forwards.

Cabinet scraper
For a fine finish
on flat hardwood
surfaces. Is made
from good-quality
saw steel.

Smooth flat file
Length 100–350 mm/4–14 in.
For all types of work except
curves. Clean frequently with
a file card.

Cabinetmakers' half-round file
Length 100–350 mm/4–14 in.
Combines features of both
round and flat files. Tempered
for working only on wood.

Round file
Length 100–350 mm/4–14 in.
Obtainable in various degrees
of coarseness. For wood
carving and for enlarging holes.

File card
For cleaning up dust
from file teeth. Consists of
fine steel wires bedded upright
in a wooden base.

Shaping and smoothing tools 2

To work successfully with shaping and smoothing tools requires skill and careful attention to the direction of the wood grain. When using these tools, which are all worked with a pushing or pulling action, it is important that the work is held very securely in the vise or with clamps, or is butted against a bench stop.

The cabinet scraper is a versatile shaping tool. The wire edge, or hook, with which it cuts removes very fine shavings when it is correctly sharpened and used, enabling the cabinet scraper to give a very smooth finish to wood. Unlike an abrasive, it should produce shavings, not dust, which can clog the grain. It can also deal with awkward areas of wood such as knots, burls and wild grain, both on solid wood and veneers, but be careful not to leave a small hollowed area as it will be evident when work is polished. Scrape over a wide area so that the hollow cannot be felt by the flat of the hand. Check by holding the work to the light.

The cabinet scraper is not an easy tool to operate, because its cutting edge is so fine and because the control must come entirely from the hands and fingers. It must be held at the correct angle; otherwise it will not cut effectively. The area of the cut is controlled by the pressure of the thumbs, which bow out the blade, varying the curve and localizing the cut. Work in the direction of the wood grain or diagonally across it. This tool can also remove old surface finishes without the aid of solvents. When scraping away a finish, angle the cabinet scraper so that the hook bites between the surface finish and the wood, and work the tool at an angle along the grain. The cabinet scraper will remove gummed paper tape from veneers and excess glue that has been allowed to dry on the surface of veneer or solid wood.

A scraper plane is much easier to operate than the cabinet scraper alone, although it is not possible to achieve such a fine degree of control. It is, however, less hard on the fingers and thumbs.

In contrast to cabinet scrapers, files and rasps are used for much rougher and coarser work. Never smooth with a file or rasp without first fitting a handle, because the tang is very dangerous and it can easily pierce the hand. Always work from short grain to long grain when smoothing surfaces. Surform tools can be worked with or against the grain, as the individually formed teeth slice away the wood.

Spokeshaves have rather limited uses and are best operated on narrow, curved shapes. They are rather similar to a smoothing plane, but require more skill to be worked successfully. On a wooden spokeshave there is no cap iron, so the grain will tear if the spokeshave is not always used from short grain to long grain. When smoothing curves with both metal and wooden spokeshaves, always stroke the grain down the curve.

The blade of a drawknife is beveled like that of a chisel. It can be used for a range of tasks, including rapidly removing wood from an over-wide board, debarking a log and rough shaping of turning stock.

Cabinet scraper

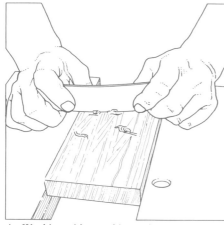

A. Working with a pushing action.

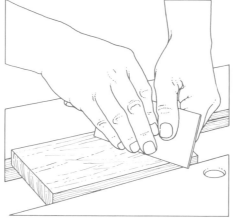

B. Working with a pulling action.

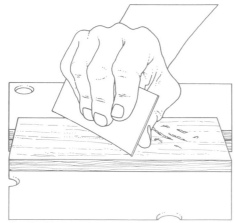

C. Working one-handed.

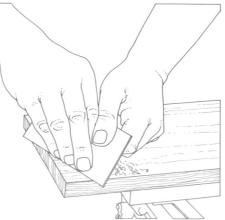

D. Scraping at the end of a board.

Cabinet scraper
The cabinet scraper can be operated in a variety of ways. The most common method is with a pushing action. Hold the scraper in both hands, and bow the center slightly outwards by pressing with the thumbs (A). Start with the scraper vertical; then tilt it away until the hook on the edge just bites into the wood — this will be at an angle of about 70°. Push the scraper away taking fine shavings from the surface. If dust, and not shavings, is removed, the scraper needs sharpening. Should the corners of the blade dig into the wood, press more firmly with the thumbs to bow out the blade more.

The scraper can also be operated with a pulling action. Hold the tool in both hands with the thumb of one hand and the fingers of the other bowing the blade (B). Alternatively, hold it in one hand, bowing the blade with pressure from the three middle fingers pushing against the little finger and thumb (C). For the best and most even cut, hold the cabinet scraper at a slight angle across the grain, and push or pull it parallel to the grain. When scraping wood at the end of a board, work the scraper diagonally across the wood (D).

File, rasp and Surform tool

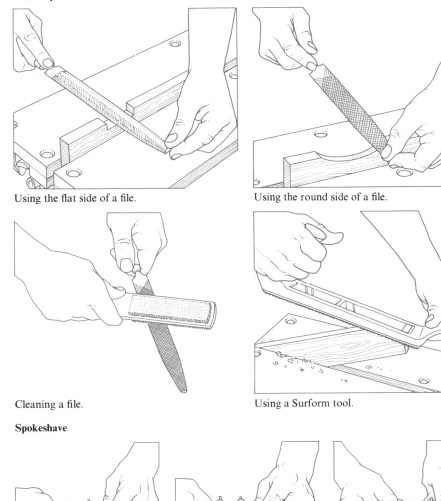

Using the flat side of a file.

Using the round side of a file.

Cleaning a file.

Using a Surform tool.

Spokeshave

Shaving concave work.

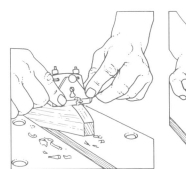

Shaving convex work.

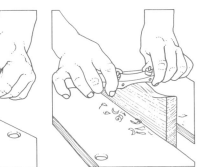

Using a wooden spokeshave.

Drawknife

Shaping convex work.

Shaping concave work.

File, rasp and Surform tool
All of these tools cut on the forward stroke only; lift the tool on the return stroke. The cabinetmakers' half-round file is the most common wood file. The flat side is handy for leveling out recesses after chiseling. Hold it diagonally across the recess and take care that the file's thin edges do not damage the sides of the recess. The flat side can be worked on convex curves. Level out concave curves with the round side of the file. Keep the file teeth clean and free of clogging dust by cleaning the file with a file card. Stroke the wires of the card down the file parallel to the teeth. Rasps and rasp planes, such as Surform tools, rapidly remove large amounts of wood and are ideal tools for preliminary shaping before filing or planing.

Spokeshave
Use a spokeshave to smooth narrow curved pieces. All spokeshaves, whether with a rounded sole for concave work, a flat sole for convex work or with a wooden frame, are operated in the same way. Use with a pushing action and keep the wrists very flexible. The angle will frequently need to be changed. Control the angle of the tool with the thumbs pressing on the handles. Work from short to long grain, stroking the grain down.

Drawknife
Operate the drawknife with a pulling action. The tool requires skill and experience to work accurately as the depth of cut depends on the angle at which the beveled blade is presented to the work and all control comes from the wrists. Greater control can be exerted when working concave curves if the bevel is held downwards. For straight and convex cuts, wood can be removed more easily with the bevel upwards. As much heavier chips can be cut than with a plane, it is essential to watch the direction of the grain. Keep the drawknife very sharp.

Grinding and honing 1

Dull tools are unsafe and hard to use. Sight down the blade: a dull edge or one with tiny defects in it reflects the light and shows up as a white line.

New chisels, gouges, planes and shaping tools have a bevel of 25° ground on the cutting edge by the manufacturer. Before the tool can be used, a bevel of 30° must be honed on the oilstone. A smaller angle will cut wood more easily but its fragile edge will break readily; a larger angle will hold its edge longer but will create more resistance. The honing bevel of 30° can be maintained over a number of honing sessions. The grinding bevel of 25° must be restored on the tool only after repeated honing, or if the edge has become chipped. After grinding back to 25°, the edge must be honed again to 30°.

All sharpening stones work in the same way: the particles of the stone remove the softer particles of metal. Coarse particles cut fast, but leave a coarse edge on the tool. Fine particles cut slowly and leave a finer, sharper edge.

Oilstones are used for honing. Most oilstones are composite man-made stones that are of even density throughout their thickness.

Artificial stones are made from abrasive particles, crushed, graded and then bonded together. They are available in coarse, medium and fine grades. Combination stones are available with different grades for each half thickness.

Oilstones must never be used dry — the pores soon choke with metal particles. A lubricant carries away the particles and prevents the tool becoming too hot through friction, which can draw the temper of the cutting edge, shown by the edge turning blue. If this happens, the metal must be ground back to remove all the softened metal. With oilstones, a good, thin machine oil is suitable. (Vegetable oils can clog the pores.) Kerosine can be used on artificial stones. Always soak a new stone in oil until it is saturated. Keep it in a box. If a stone becomes clogged with metal particles, scrub it with a stiff brush dipped in kerosine

Care of oilstones

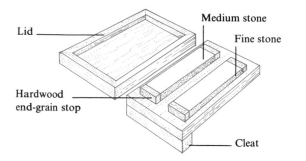

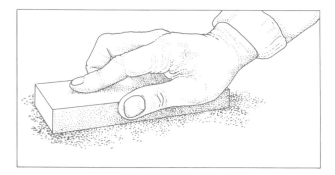

Make a box to hold oilstones to keep out dust. With a Forstner bit, drill out two recesses in a solid block. Fit hardwood end-grain stops at each end, level with the top of the stones. Screw a cleat to the bottom to hold the box in the vise. Then make a well-fitting lid from five-ply and 19mm/¾in battens.

When an oilstone wears hollow, grind it flat again on a paving stone or marble slab, using silver sand and water for natural stones and carborundum powder and water for artificial stones. Work the stone in a circular movement. A hollow stone cannot produce a straight edge.

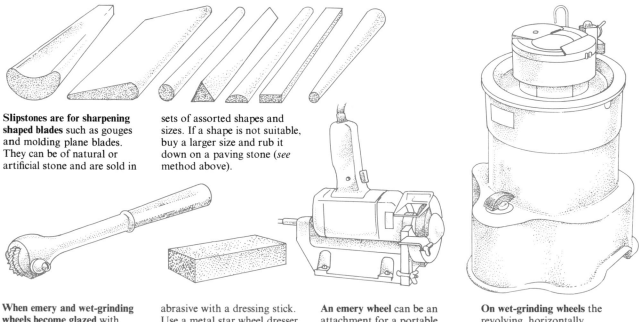

Slipstones are for sharpening shaped blades such as gouges and molding plane blades. They can be of natural or artificial stone and are sold in sets of assorted shapes and sizes. If a shape is not suitable, buy a larger size and rub it down on a paving stone (*see* method above).

When emery and wet-grinding wheels become glazed with accumulated dust and particles, scrape off this layer and expose a new layer of abrasive with a dressing stick. Use a metal star wheel dresser for emery wheels and a carborundum or silicon carbide stick for wet-grinding stones.

An emery wheel can be an attachment for a portable power drill held in a horizontal stand, or it can be an integral power tool.

On wet-grinding wheels the revolving, horizontally mounted abrasive cylinder is constantly washed with a flow of water during operation.

or gasoline; soak a badly clogged stone overnight, then scrub it. Soak the stone in oil again.

Many woodworkers use an electrically powered emery wheel for grinding. This has many disadvantages. It is used dry and revolves at around 3,000 rpm so there is danger of drawing the temper from the blade unless the tool is constantly dipped in water to cool it. It is difficult to produce an even edge on a narrower edge than the plane blade or chisel surface, and it is dangerous as it can chip or shatter. If it is used, always wear goggles and work with great care. Wet-grinding machines with horizontal wheels are much more suitable for both grinding and honing cabinetmakers' tools. The artificial stone is either a coarse 50 grit, a medium 180 grit or a fine 280 grit abrasive. The stone revolves at only 150 rpm and is constantly saturated with water or oil, so it is not dangerous and the tool is kept cool. Honing is completed on a fine stone.

To hone a tool at the correct angle, hold the tool in both hands, place the grinding bevel flat, then raise the hands slightly. Hone on an oilstone with a forward and backward motion. Push forwards, keeping pressure constant; return using no pressure. Keep the wrists rigid — rocking will cause the bevel to round over. After a few strokes, check that the bevel is forming evenly. Adjust the pressure if it is not. Keep the stone lubricated and keep honing until a wire edge, or burr, appears. Wipe the oil from the tip and test with the thumb to feel if the wire edge has formed along the whole width of the tool. To remove the wire edge, lay the blade absolutely flat and gently rub once on the stone. Wipe off the oil, and check. Finish the edge by stropping the blade across the palm or on leather.

Sharpening follows the same basic procedure for all edge tools. When sharpening narrow blades, it is vital that grooves are not worn in the oilstone or blades will not be sharpened flat. Move blades about in a figure of eight to utilize the whole stone. Be careful not to dig the corners of blades into the stone. Sharpen very narrow blades on the edge of the oilstone.

Using a honing guide

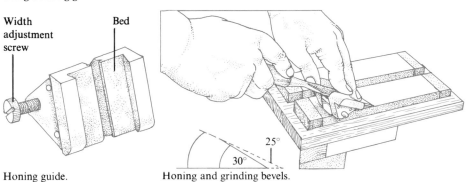

Width adjustment screw

Bed

Honing guide.

30° 25°

Honing and grinding bevels.

Sharpening plane blades

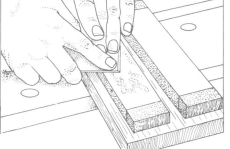

1. Hold the blade at an angle.

2. Remove the wire edge.

Replacing plane blades

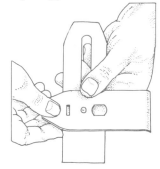

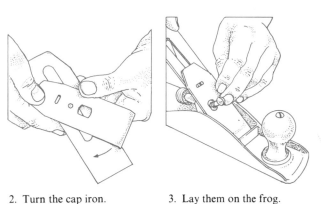

1. Position the cap iron.

2. Turn the cap iron.

3. Lay them on the frog.

Using a honing guide
Honing guides hold chisel and plane blades at the correct angle. The guide consists of a frame mounted on wheels or rollers. The frame is adjustable to take blades of varying widths and has a bevel scale so it can be set at the angle required. To hone, push the guide backwards and forwards along the oilstone. Woodworkers eventually dispense with these guides but they are useful as an aid for beginners.

Sharpening plane blades
Hold a wide blade at an angle so its whole edge is in contact with the stone (1). Remove the wire edge when it appears on the reverse side (2). Hone jack plane and smoothing plane blades with a very slight curvature for fast cutting. Make a few figure eights to produce the slight curve. Hone jointer plane blades dead square; round the corners slightly to prevent the edges digging in. Hone rabbet, block and combination plane blades dead square.

Replacing plane blades
Hold the cap iron at right angles to the blade (1). Insert the lever cap screw. Slide the cap iron back along the slot in the blade and turn it in line with the blade (2). Move it forwards until it is the required distance from the cutting edge. Hold the two together and tighten the screw. Lay them on the frog (3). On a wooden plane, replace the wedge; on a metal plane, replace the lever cap. Then adjust the blade projection as explained on page 78.

Grinding and honing 2

Sharpening chisels

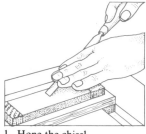

1. Hone the chisel.

2. Remove the wire edge.

3. Strop the chisel.

Sharpening chisels

To hone a chisel, hold it bevel down at the correct 30° angle on the oilstone. Hold the end of the handle, with the fingers of the other hand applying pressure towards the cutting edge (1). Chisel blades must be absolutely square across. An off-square blade cannot make a true cut and is dangerous as it tends to slip sideways. Turn the chisel over to remove the wire edge on the fine oilstone (2). Keep the blade absolutely flat; otherwise, a secondary bevel will be formed on the flat back of the blade and when paring vertically the cut, instead of going straight down, will follow the line of the bevel. To finish the edge, strop the chisel on the palm of the hand or on a piece of leather (3).

Sharpening in-cannel gouges

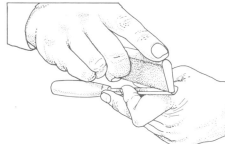

1. Hone an in-cannel gouge with a slipstone.

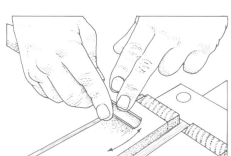

2. Remove the wire edge on the oilstone.

Sharpening in-cannel gouges

To sharpen an in-cannel gouge, first hone the bevel on the inside of the blade with a slipstone (1). Use a slipstone that matches the curve of the blade or has a slightly sharper curve. When the wire edge appears, remove it on the fine oilstone, working from side to side with a rocking motion (2). Strop the gouge on the palm or on a piece of leather glued to a wooden board.

Sharpening out-cannel gouges

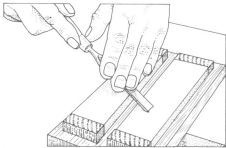

1. Hone an out-cannel gouge on the oilstone.

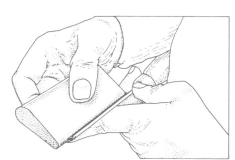

2. Remove the wire edge with a slipstone.

Sharpening out-cannel gouges

To sharpen an out-cannel gouge, hone the bevel first on the oilstone until the wire edge forms (1). Move the gouge in a figure-eight motion and roll it throughout its width during the movement. This should produce a smooth curve on the bevel. Remove the wire edge by rubbing a slipstone of the appropriate section, back and forth within the curve (2).

Sharpening drawknives

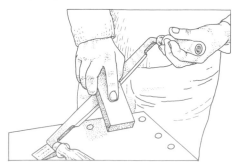

1. Hold the tool, bevel up.

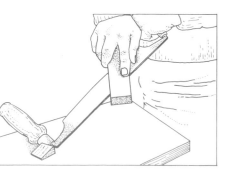

2. Remove the wire edge with the oilstone.

Sharpening drawknives

The drawknife has such a large blade that it is more convenient to bring the oilstone to the blade. Brace one handle against a bench stop, holding the tool bevel up (1). Take a firm grip on the oilstone and rub it the full length of the blade, holding the stone flat against the bevel. Remove the wire edge on the reverse side by turning the drawknife over, putting the oilstone against the flat of the blade and rubbing back and forth (2).

Sharpening spokeshaves

1. Hold the blade by the tangs.

2. Rub the blade lengthways.

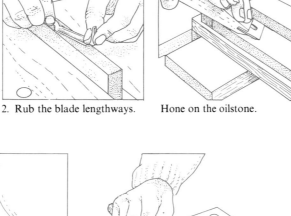

Hone on the oilstone.

Sharpening cabinet scrapers

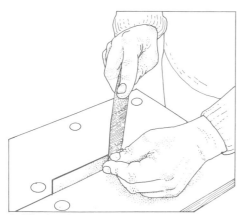

1. File the edges square.

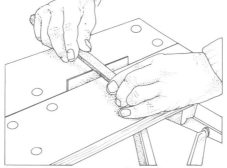

2. Drawfile to remove the file marks.

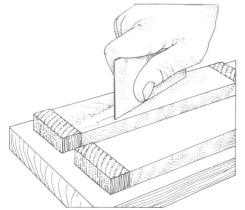

3. Hold the scraper upright across the oilstone.

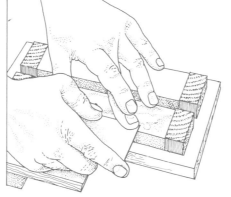

4. Remove the wire edge.

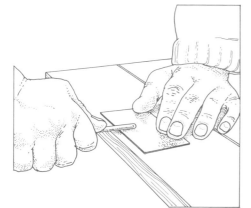

5. Consolidate the metal.

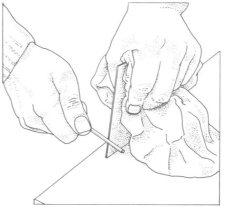

6. Form the hook.

Sharpening spokeshaves

Remove the blade from a wooden spokeshave by tapping the two tangs with a hammer. Grip the oilstone on its edge in the bench vise. Put the blade bevel down, and hold it by the tangs **(1)**. Hone it like a chisel or plane blade. Remove the wire edge by rubbing the blade lengthways **(2)**. Metal spokeshave blades are similar to plane blades, but being shorter are difficult to hold. Make a wooden holder with a kerf at one end to take the blade. Sharpen it in the same way as a plane blade.

Sharpening cabinet scrapers

File the edges square with a smooth metal file **(1)**. Hold the file at an angle so its teeth are at right angles to the scraper edge. Move the file diagonally both across and along the edge. Drawfile to remove the first file marks **(2)**. Hold the file lightly and at right angles to the blade. Then hone the edge square on the oilstone, holding the cabinet scraper upright at an angle across the oilstone to avoid forming a groove **(3)**. Rub back and forth using strong pressure. To counteract any tendency to lean more to one side than the other, turn the blade around and repeat. Now lay the blade flat on the oilstone and remove the wire edge by rubbing first one side of the blade, then the other, applying pressure with the fingers along its length **(4)**. Form the hook with a burnisher. This is simply a piece of silver steel rod harder than the cabinet scraper. It can be a special tool with a handle, or the back of a gouge can be used. First consolidate the metal **(5)**. Lay the scraper flat on the bench and, with very strong pressure, rub the burnisher along the edge. This presses the molecules together, so, when the razor edge hook is formed, it will be maintained longer. Then form the hook. Put a drop of oil on the blade; then hold it vertically on the bench, gripping the top in a soft cloth. Form the hook in two strokes. On the first stroke, hold the burnisher 5° away from the right angle, and on the second stroke, a further 5° away **(6)**. Use firm pressure with each stroke. Treat the other three edges in the same way. The hook can be burnished down and re-formed a number of times before it is necessary to repeat the complete process using the file and oilstone.

Abrasives

After shaping, wood will require further cutting with an abrasive before it is ready for the final finish.

Abrasives, which are backed with paper, cloth or metal, are graded according to the coarseness of the grit used: 800 grit is the finest and 12 is the coarsest. Gradings used to be from Flour, 180 grit, through to Strong, about 50 grit. Grits spaced close together are called close-coat abrasives, while those spread more thinly are known as open-coat abrasives. The latter cut much faster because they clog up with dust less quickly.

Work an abrasive along the grain. Progress from a coarse abrasive, through medium, to a fine one.

Power sanders save considerable work when smoothing a piece of wood. Always switch them on before lowering them onto the wood. A belt sander is a heavy machine that will smooth rough stock fast. An orbital sander is much lighter and can produce a very fine finish. A drum sander can be used for smoothing hollows and contoured wood.

Type of abrasive	Sandpaper	Garnet paper	Aluminum oxide	Silicon carbide
Type of grit	powdered quartz	crushed garnet	bauxite	coal and quartz
Abrasive quality	fairly soft	medium	hard	extremely hard
Suitable materials	hard- and soft-woods	hard- and soft-woods	hard-woods	hardwoods, particle boards
Uses	paintwork; bare wood finishing	by hand and with machine sander	general smoothing; bare wood finishing	wet on paintwork; dry on bare wood
General comment	cuts slowly	obtainable in very fine grit size	its hard grains take fast machine smoothing	also known as carborundum paper and wet-and-dry paper

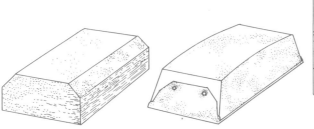

When smoothing with an abrasive use a block to produce an even surface. A cork block is firm and resilient.

Hand sanding block

A metal-backed tungsten carbide abrasive lasts much longer than a conventional abrasive but is harsher.

For economy and a good fit around the block, tear a standard-sized sheet in six against the workbench.

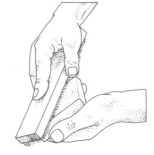

Clear the dust on open-coat papers with a file card; rinse wet-and-dry papers under running water.

Smooth with the shoulder and block in line, with the chamfers on top. The paper should be tight around the block.

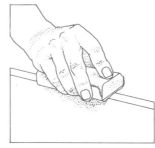

To maintain the sharpness on an edge hold the paper tightly around the block. On a wider surface hold it diagonally.

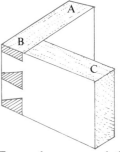

To smooth a corner work the abrasive between points A and B. Then work between C and B, avoiding surface AB.

To work an abrasive against a concave curve use the chamfered rather than the sharp edges on the block.

To smooth a molding use a specially shaped sanding block with an abrasive paper held tightly around it.

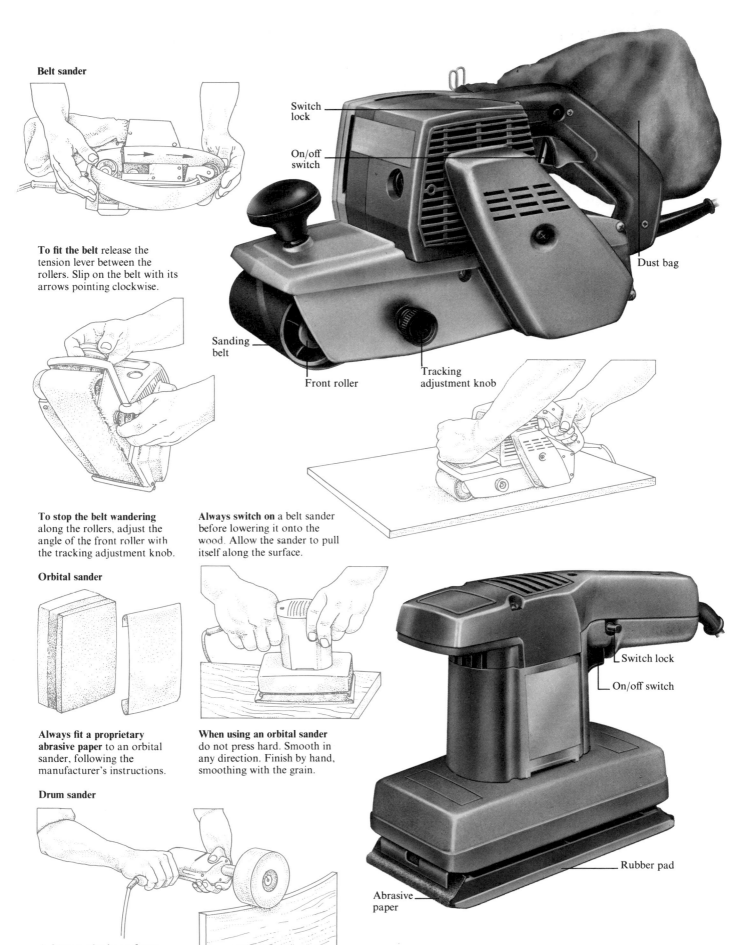

Belt sander

To fit the belt release the
tension lever between the
rollers. Slip on the belt with its
arrows pointing clockwise.

Switch
lock

On/off
switch

Dust bag

Sanding
belt

Front roller

Tracking
adjustment knob

To stop the belt wandering
along the rollers, adjust the
angle of the front roller with
the tracking adjustment knob.

Always switch on a belt sander
before lowering it onto the
wood. Allow the sander to pull
itself along the surface.

Orbital sander

**Always fit a proprietary
abrasive paper** to an orbital
sander, following the
manufacturer's instructions.

When using an orbital sander
do not press hard. Smooth in
any direction. Finish by hand,
smoothing with the grain.

Switch lock

On/off switch

Drum sander

Rubber pad

Abrasive
paper

A drum sander has a foam
rubber drum that follows the
contours of the work being
smoothed.

Drills and braces 1

An auger bit is the basic wood-boring tool and is used in a brace. At its tip is the threaded feed-screw that centers the bit and pulls it into the wood. Two spurs scribe the diameter of the cut and give precise centering so the bit remains accurate in grain of any direction. After the spurs, the cutters bite into the wood, lifting the wood chips and feeding them up the throat to the twist, which conveys them up and out of the hole.

A woodworkers' twist drill bit has a throat specially designed to clear away the wood chips on both hard- and softwood when boring at high speed. An engineers' twist drill bit is, however, the bit in common use even though its twist is specifically designed to clear the swarf produced when drilling metal. It is unable to clear wood chips efficiently at high speeds; nor is it suitable for unseasoned softwoods.

Because they are the same diameter all the way up the shank, auger bits and twist drills are more accurate than bits that are relieved behind the cutters.

A Forstner bit, like many others, has a central point to lead the bit into the wood. The actual cutting is done by its circular rim and two cutters; the bit is therefore unaffected by knots and grain direction.

One expansive bit can stand in for a range of center bits. To check its size, after adjusting the cutter, measure from the tip of the feed-screw to the outside edge of the spur. At the top of the feed-screw is a cutter that severs the wood ahead of the extended cutter.

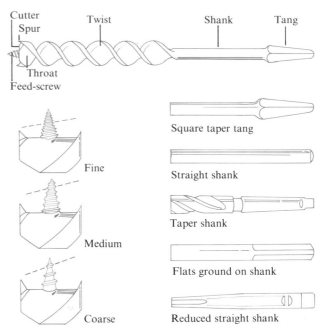

Cutter · Spur · Twist · Shank · Tang · Throat · Feed-screw

Fine

Medium

Coarse

Square taper tang

Straight shank

Taper shank

Flats ground on shank

Reduced straight shank

The pitch of the feed-screw thread determines the cut made by an auger bit. Use coarse and medium threads for general work; fine for a smooth finish.

A straight, taper or flattened shank fits a three-jaw self-centering chuck on a hand or power drill. A square-taper tang fits only a brace jaw.

Bradawl
Tip width 2–3 mm/$\frac{3}{32}$–$\frac{1}{8}$ in. For boring small holes and starting bits, drills, nails and screws.
Automatic push drill
Length 240 mm/9$\frac{1}{2}$ in. For rapid boring of small holes.
Archimedes drill
Length 290 mm/11$\frac{1}{2}$ in. For tiny holes. Bits are stored in the hollow handle.

Bradawl

Automatic push drill

Archimedes drill

Power drill
Chuck capacity 6–16 mm/ $\frac{1}{4}$–$\frac{5}{8}$ in. For general-purpose boring. Graded by chuck capacity, which indicates the maximum size hole the drill can make in steel; this figure can be more than doubled when drilling wood. Various speed models are available.

Power drill

Brace

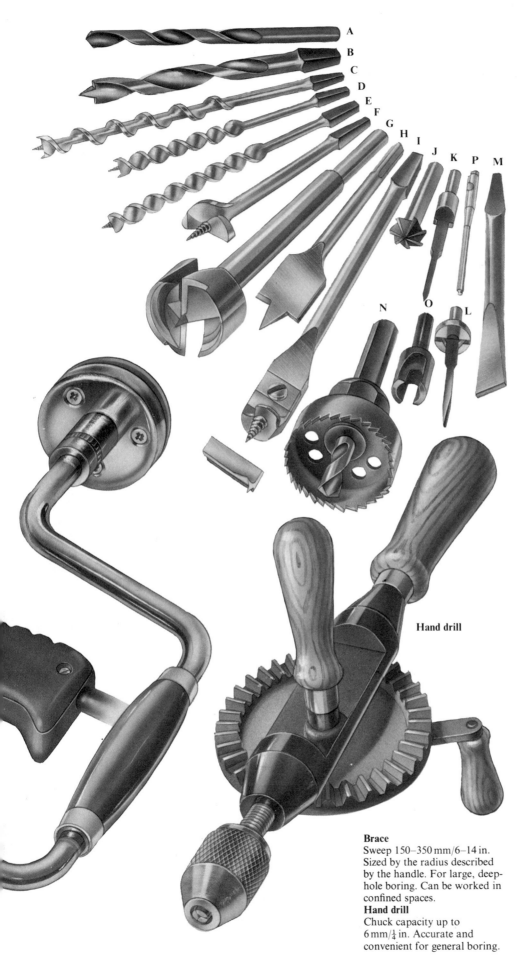

Twist drill (A)
Diameter 0.38–38 mm
$\frac{1}{64}$–1$\frac{1}{2}$ in for general drilling of woods, metals and plastics in hand and power drills.

Woodworkers' twist drill (B)
Diameter 6–25 mm/$\frac{1}{4}$–1 in. For general drilling of hardwoods and softwoods in a brace.

Solid-center auger bit (C)
Diameter 5–38 mm/$\frac{3}{16}$–1$\frac{1}{2}$ in. A strong, fast-cutting general-purpose bit.

Jennings-pattern auger bit (D)
Diameter 5–38 mm/$\frac{3}{16}$–1$\frac{1}{2}$ in. More accurate than the solid-center auger bit and better for fine work because of the smooth finish it leaves.

Scotch auger bit (E)
Diameter 5–38 mm/$\frac{3}{16}$–1$\frac{1}{2}$ in. For boring hardwoods; also end grain.

Center bit (F)
Diameter 5–56 mm/$\frac{3}{16}$–2$\frac{1}{4}$ in. Fast-cutting bit for shallow, large-diameter holes.

Forstner bit (G)
Diameter 6–50 mm/$\frac{1}{4}$–2 in. For accurate flat-bottomed holes and, where the auger cannot be used, for end grain, thin wood and three-ply.

Flat bit (H)
Diameter 19–38 mm/$\frac{3}{4}$–1$\frac{1}{2}$ in. For general large-hole boring. The lead point enables angled holes to be started easily.

Expansive bit (I)
Diameter 13–150 mm/$\frac{1}{2}$–6 in. For fast cutting shallow, large-diameter holes only in softwood. Has adaptable interchangeable cutters.

Countersink bit (J)
Diameter 6–25 mm/$\frac{1}{4}$–1 in. For widening the mouth of a countersunk screw-hole, allowing the screw-head to be set flush or below the surface.

Combination drill bit (K)
Sized to suit common sizes of screw. For drilling, countersinking and counterboring to the required depth all in one.

Drill and countersink bit (L)
Sized to suit common sizes of screw. Bores a screw-hole and countersinks.

Screwdriver bit (M)
Blade width 6–16 mm/$\frac{1}{4}$–$\frac{5}{8}$ in. Use in a brace for inserting or extracting screws.

Hole saw (N)
Diameter 16–100 mm/$\frac{5}{8}$–4 in. For large-diameter holes. A central pilot bit is surrounded by interchangeable saw blades.

Plug cutter (O)
Diameter 10–16 mm/$\frac{3}{8}$–$\frac{5}{8}$ in. For cutting wood on dowel plugs and pellets to conceal screw-holes and nail-heads.

Fluted bit (P)
Diameter 2–16 mm/$\frac{3}{32}$–$\frac{5}{8}$ in. For boring tiny holes. Fits only the automatic push drill.

Hand drill

Brace
Sweep 150–350 mm/6–14 in. Sized by the radius described by the handle. For large, deep-hole boring. Can be worked in confined spaces.

Hand drill
Chuck capacity up to 6 mm/$\frac{1}{4}$ in. Accurate and convenient for general boring.

Drills and braces 2

Brace

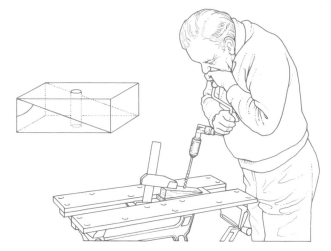

Use a try square to check if a drill is leaning forwards or backwards. The drill leaning to one side is more noticeable and easier to correct.

To bore a series of holes at an angle, make a simple jig. Place a block of wood against the rod (*see* page 37); mark in the angle of the slope. Bore a vertical hole right through the wood. Then saw down the slope and plane it smooth. Hold the jig securely on the work with a clamp.

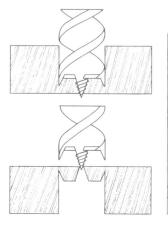

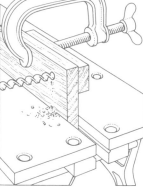

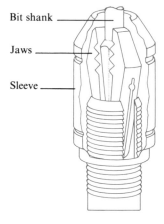

Bit shank

Jaws

Sleeve

A through hole bored with an auger bit will splinter. As soon as the feed-screw tip emerges, turn the work around and bore from the other side.

Another way to prevent splintering is to clamp a piece of waste wood to the back of the work to support the fibers as the bit emerges.

To fit a bit, lock the ratchet. Holding the sleeve, swing the frame clockwise until the shank enters the jaws. Then tighten counter-clockwise.

The corners of the bit shank must fit into the V-grooves in the jaws. Brace jaws can be replaced if they slacken after considerable use.

Power drill

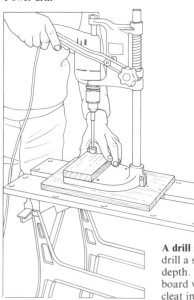

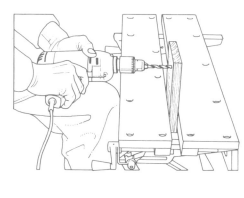

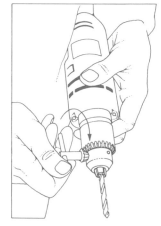

A drill stand makes it easy to drill a series of holes to a fixed depth. Screw the base to a board with a cleat and grip the cleat in the vise.

To use the drill position the bit on the wood and start it while applying light pressure. Remove from the wood before switching the drill off.

To fit a drill bit, open the jaws by turning the chuck key counter-clockwise. Turn it clockwise to secure the bit, tightening at all three holes.

Hand drill

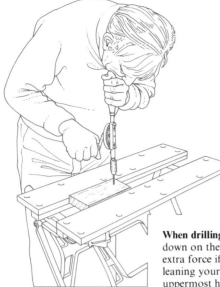

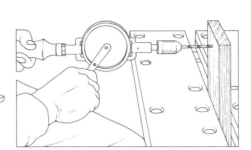

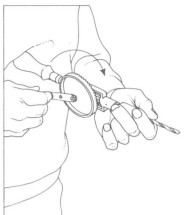

When drilling vertically, push down on the handle; apply extra force if necessary by leaning your head against your uppermost hand.

When drilling horizontally, extra pressure can be brought to bear by pushing the body weight against the hand that is holding the drill.

To fit a drill bit, hold the sleeve and turn the handle counter-clockwise. Insert the bit and turn the handle clockwise to tighten the chuck.

Special bits

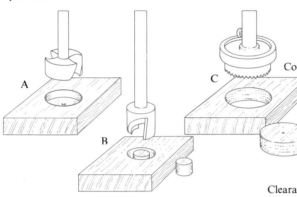

Countersink

Counterbore

D   E

Clearance hole   Pilot hole

Depth stops

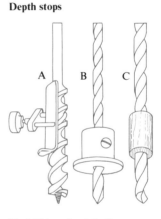

A   B   C

Use a Forstner bit (A) to cut clean flat-bottomed holes. This bit is best used in a drill stand with a depth stop, and it is ideal for removing the bulk of waste from a recess. A plug cutter (**B**) bores to a fixed depth. After boring, break the pellet at the root, using a small screwdriver. Glue and insert

the pellet into the hole in line with the grain; then plane it flush. Plug cutters are sized to match standard bit sizes. To bore with a hole saw (**C**), first secure the required size of blade to the pilot bit. Unlike other large diameter drills, the hole saw removes the waste as a complete disk.

Screws require a pilot hole to take the threaded portion, a clearance hole for the threadless portion, and a countersink if the head is to be flush with the surface. The drill and countersink bit (**D**) does all this in one. A combination drill bit (**E**) is shaped to form a counterbore as well.

To drill to a fixed depth without a drill stand, a depth stop is fitted to the bit itself. This can be a metal clamp (**A**), which locates in the flutes of an auger bit, or a screw-on sleeve (**B**) for a twist drill. Simplest of all is a short length of dowel, with a hole bored through the center (**C**).

Sharpening bits

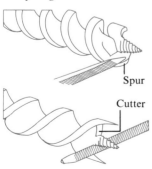

Spur

Cutter

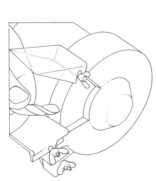

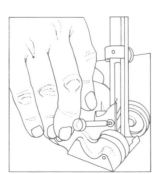

Sharpen spurs on an auger bit from the inside, not lower than the cutters, using a smooth file. Sharpen cutters on the edge facing the shank.

Place the cutting edge of a twist drill against the wheel; rotate the cutting edge, lifting at the same time. Repeat for the other edge.

A jig will hold a twist drill at the correct angle for sharpening. Push the jig over a sheet of 180-grit abrasive paper to sharpen the drill.

Twist drill sharpeners powered by the drill itself are made to fit some drills. On some the chuck is removed (as above); others fit into the chuck.

Hammers, mallets and screwdrivers 1

The first hammer was a stone held in the hand; the first mallet, a cudgel. The Romans used a hammer with a claw peen for pulling nails, while medieval pictures show a hammer with a square head that is similar to current European hammers.

Hammers are sized by the weight of their head. A hammer should be well-balanced with a forged head; a cast head tends to shatter. The claw should be carefully ground, tapering to a fine V so it can pull nails of all sizes. A hammer handle is traditionally made from ash or hickory—both tough, long-grained woods. The end of the wooden shaft is impregnated with oil to prevent moisture loss and splitting. Hammers with metal or fibreglass shafts are a modern development. The head is securely wedged onto a wooden handle or is forged and locked onto a metal handle. A wooden-handled claw hammer usually has an extended eye, like the eye of an adze, to prevent excessive strain on the shaft when pulling large nails.

Every workshop needs at least four screwdrivers to fit all sizes of screw likely to be used. A traditional screwdriver has a beech or boxwood handle strengthened with a metal ferrule. A modern plastic handle is molded directly onto the blade. The bulbous end to most handles enables more torque to be applied, although some workers prefer fluted plastic handles.

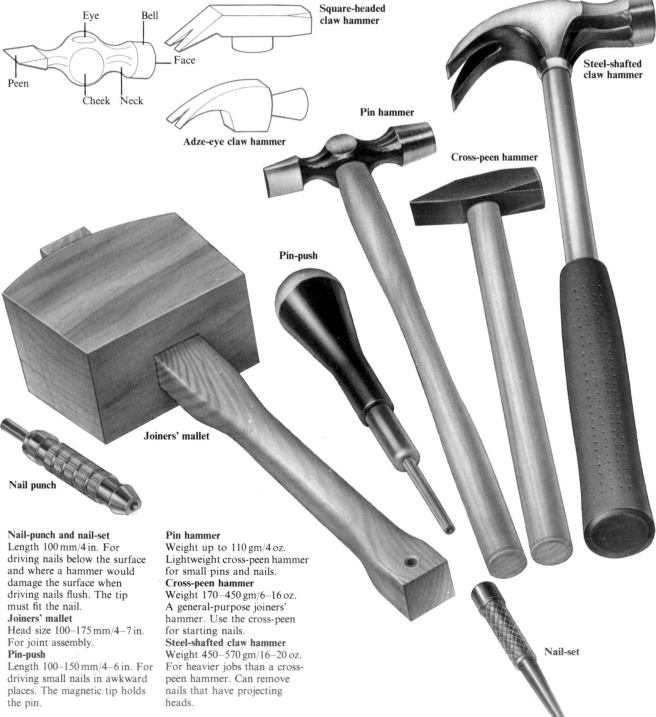

Hammer parts

Eye Bell Peen Face Cheek Neck

Square-headed claw hammer

Adze-eye claw hammer

Pin hammer

Cross-peen hammer

Steel-shafted claw hammer

Pin-push

Joiners' mallet

Nail punch

Nail-set

Nail-punch and nail-set
Length 100 mm/4 in. For driving nails below the surface and where a hammer would damage the surface when driving nails flush. The tip must fit the nail.

Joiners' mallet
Head size 100–175 mm/4–7 in. For joint assembly.

Pin-push
Length 100–150 mm/4–6 in. For driving small nails in awkward places. The magnetic tip holds the pin.

Pin hammer
Weight up to 110 gm/4 oz. Lightweight cross-peen hammer for small pins and nails.

Cross-peen hammer
Weight 170–450 gm/6–16 oz. A general-purpose joiners' hammer. Use the cross-peen for starting nails.

Steel-shafted claw hammer
Weight 450–570 gm/16–20 oz. For heavier jobs than a cross-peen hammer. Can remove nails that have projecting heads.

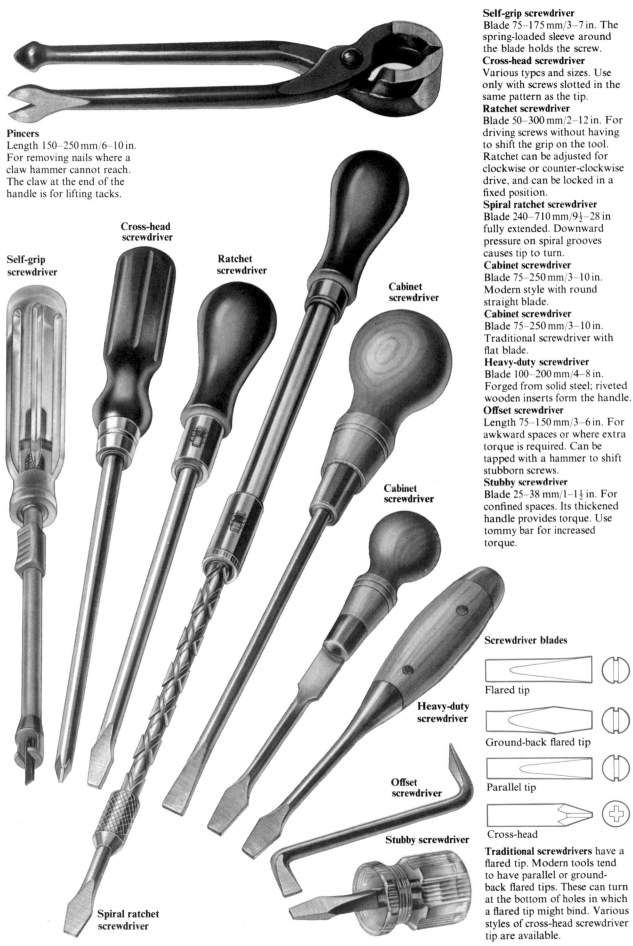

Pincers
Length 150–250 mm/6–10 in.
For removing nails where a
claw hammer cannot reach.
The claw at the end of the
handle is for lifting tacks.

**Self-grip
screwdriver**

**Cross-head
screwdriver**

**Ratchet
screwdriver**

**Cabinet
screwdriver**

**Cabinet
screwdriver**

**Heavy-duty
screwdriver**

**Offset
screwdriver**

Stubby screwdriver

**Spiral ratchet
screwdriver**

Self-grip screwdriver
Blade 75–175 mm/3–7 in. The
spring-loaded sleeve around
the blade holds the screw.
Cross-head screwdriver
Various types and sizes. Use
only with screws slotted in the
same pattern as the tip.
Ratchet screwdriver
Blade 50–300 mm/2–12 in. For
driving screws without having
to shift the grip on the tool.
Ratchet can be adjusted for
clockwise or counter-clockwise
drive, and can be locked in a
fixed position.
Spiral ratchet screwdriver
Blade 240–710 mm/9½–28 in
fully extended. Downward
pressure on spiral grooves
causes tip to turn.
Cabinet screwdriver
Blade 75–250 mm/3–10 in.
Modern style with round
straight blade.
Cabinet screwdriver
Blade 75–250 mm/3–10 in.
Traditional screwdriver with
flat blade.
Heavy-duty screwdriver
Blade 100–200 mm/4–8 in.
Forged from solid steel; riveted
wooden inserts form the handle.
Offset screwdriver
Length 75–150 mm/3–6 in. For
awkward spaces or where extra
torque is required. Can be
tapped with a hammer to shift
stubborn screws.
Stubby screwdriver
Blade 25–38 mm/1–1½ in. For
confined spaces. Its thickened
handle provides torque. Use
tommy bar for increased
torque.

Screwdriver blades

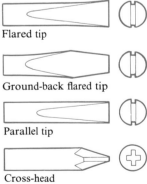

Flared tip

Ground-back flared tip

Parallel tip

Cross-head

Traditional screwdrivers have a
flared tip. Modern tools tend
to have parallel or ground-
back flared tips. These can turn
at the bottom of holes in which
a flared tip might bind. Various
styles of cross-head screwdriver
tip are available.

103

Hammers, mallets and screwdrivers 2

The process of hammering nails and pins and driving screws is not that simple. Skill is required to hammer in a nail perfectly every time, and careful preparation and choice of screwdriver are necessary for flawless work. Keep the hammerhead free from grease and rub off rust spots with a fine abrasive. When nailing near end grain, the danger of splitting is reduced if the nail tip is blunted by cutting off the tip or banging it on metal. Use clinch nailing only where strength is more important than appearance.

A screwdriver blade becomes rounded through use, so restore the flat sides and tip occasionally by honing on an oilstone. Grind any sharp corners off the screwdriver blade tip to avoid damaging the wood on the last few turns, when driving countersunk screws.

Hammers and nailing

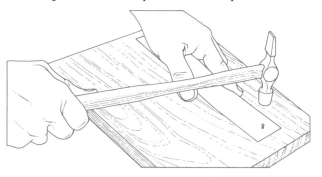

Hold a hammer by the end of the handle, not in the middle. Swing your arm, pivoting from the elbow and keeping your wrist straight but flexible. Keep an eye on the nail, hitting with a firm, clean stroke so that the hammerhead meets the nail-head at right angles at the moment of impact.

To start a nail in the wood tap it gently with the cross-peen. When it is able to stand unsupported, drive it in with the hammer face.

Short nails and pins that cannot be held in the fingers can be gripped with a small pair of pliers while they are tapped into position in the wood.

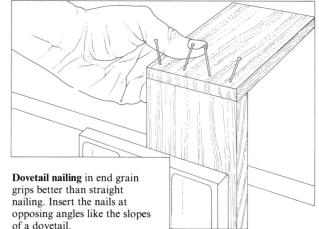

Short nails and pins can also be held by being pushed through a strip of cardboard. Then hammer the nail through the card into the wood; when the nail grips, tear away the card. This method is useful if the nail is to be inserted in an awkward place, such as a corner or into the ceiling.

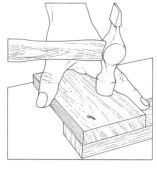

If only one hand can be used start a heavy nail by gripping its head against the side of the hammer and drive both in together until the nail holds.

For clinch nailing tap a long nail through both parts of the wood. Hammer the tip over and then flush, with the work resting on a firm surface.

Dovetail nailing in end grain grips better than straight nailing. Insert the nails at opposing angles like the slopes of a dovetail.

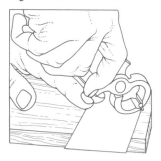

Remove a nail if it starts to bend. Pincers and claw hammers work by leverage so pressure on the handle is magnified at the nail end.

The force on the pivoting section can damage the work, so place a cabinet scraper or waste wood beneath the tool to protect the surface.

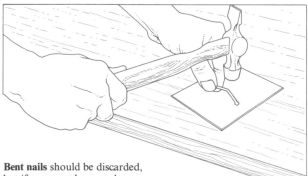

Bent nails should be discarded, but if one must be reused place it on a piece of scrap metal and straighten it with a few taps from a hammer.

Fitting a new hammer handle

1. Plane the handle.

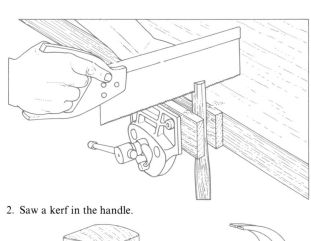

2. Saw a kerf in the handle.

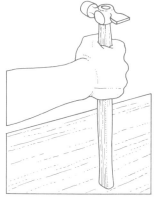

3. Tap the handle firmly.

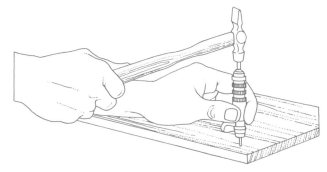

4. Finish with a mallet.

5. Hammer the wedge flush.

Fitting a new hammer handle
If a wooden hammer handle
breaks or splinters, it can be
replaced with a new one,
secured by a metal wedge. Saw
off the old handle close to the
head. Drive out the remainder
with a chisel or punch,
working from the underside
towards the top, because the
eye is tapered in that direction.
Plane the top of the new
handle to fit the hammer eye
(**1**). Tap it gently partway into
the eye. Remove and plane any
shiny parts, which indicate
where the wood is binding. In
the top of the handle, saw a
kerf in line with the hammerhead to the depth of a metal
wedge (**2**). Place the head on
the handle and tap the butt end
firmly on the bench until the
head is driven onto the shaft
by its own weight (**3**). Turn the
hammer over and use a mallet
for the final blows (**4**). If the
kerf closes, open it with a
chisel or screwdriver. Insert the
wedge and hammer it flush (**5**).

Punches

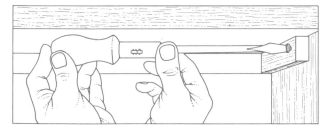

To use a nail-punch hold it
vertically just above the nail-
head. Hit the top so the spring
tip punches the nail below the
surface. Fill the punched hole

with stopping and smooth the
surface flush for an invisible
finish. The smaller the hole
made by the tapered punch tip,
the neater the finish will be.

Hold a nail-set just above the
nail against the fleshy part of
the middle finger. After impact
the nail-set will automatically
spring up again.

To use a pin-push drop the nail
into the sleeve — the magnetic
tip will hold it, even upside
down. Press down on the
handle to operate.

Screwdrivers

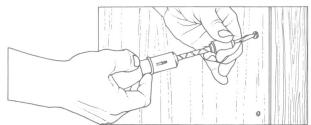

**When using a ratchet
screwdriver** in awkward places,
it can be difficult to get
sufficient pressure on the blade.
For right-handed workers, in

such cases, it may be easier to
push on the screwdriver handle
with the left hand and support
the blade with the right; do this
vice versa if left-handed.

**When using a spiral ratchet
screwdriver** apply downward
pressure on the handle. This
pressure is transferred to the
spiral grooves to turn the tip.

Small models have a mild
spring that is less dangerous
than the strong spring in large
models, which can damage the
woodworker and the work.

Table saw

The basic function of a table saw is ripping and cross-cutting wood, both fast and accurately. The saw is sized by the diameter of its blades, which are similar to but larger than those used on the portable circular saw. A 200–250 mm/8–10 in saw is suitable for the home workshop. The blade is fitted on a revolving arbor and can be adjusted to project as required above the flat, iron table. The work is fed into the revolving blade, with the aid of fences and jigs, by the operator. An adjustable metal rip fence is positioned parallel to the blade, acting as a guide when ripping. A miter fence slides in one of two grooves in the table, on either side of the blade, to act as a guide when cross- and miter-cutting. The fences aid accurate and safe working and should always be used. For beveled cuts, a tilting arbor saw is much safer than one where the table tilts.

A table saw should always have a blade guard and a spreader and anti-kickback pawls. The spreader is a metal tongue behind the blade, which keeps the kerf open and prevents the wood binding on the blade; the pawls prevent it from being kicked back towards the operator. The spreader must be used for all ripping. If the blade guard must be removed, always use other safety devices such as a push stick, a holding jig or a guide. Turn off the power isolator switch before making any major adjustment to the table saw. When operating the saw, allow it to reach full speed before beginning a cut.

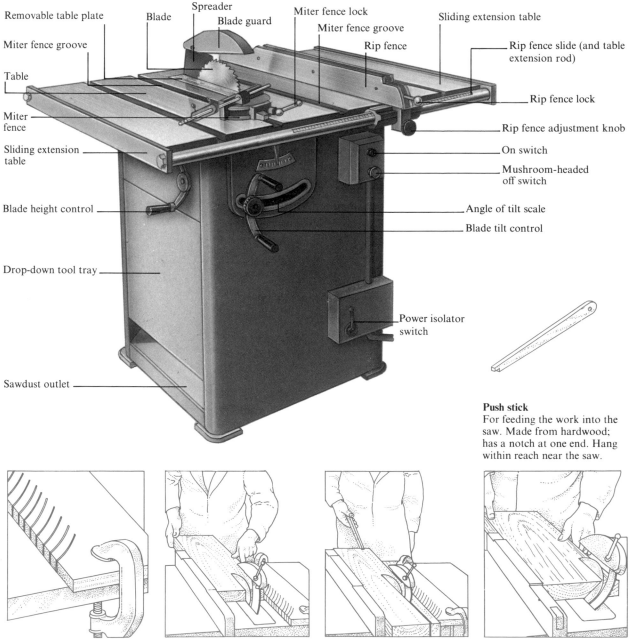

Removable table plate · Blade · Spreader · Blade guard · Miter fence lock · Miter fence groove · Sliding extension table

Miter fence groove · Rip fence · Rip fence slide (and table extension rod)

Table · Rip fence lock

Miter fence · Rip fence adjustment knob

Sliding extension table · On switch · Mushroom-headed off switch

Blade height control · Angle of tilt scale · Blade tilt control

Drop-down tool tray

Power isolator switch

Sawdust outlet

Push stick
For feeding the work into the saw. Made from hardwood; has a notch at one end. Hang within reach near the saw.

A springboard, kerfed on the band saw, is clamped against the work, providing pressure and flexibility. Always hold the work against the fence.

When ripping hold the work against the fence with one hand; feed it in with the other, which should stay near the table edge.

Ripping is much safer if a push stick and a springboard are used. Stand slightly to one side in case the board should suddenly kick back.

When bevel ripping set the saw with the blade tilt control. The tilt scale gives the angle of cut; for accuracy, check the blade angle with a sliding bevel.

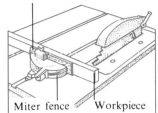

Wooden auxiliary fence

Miter fence   Workpiece

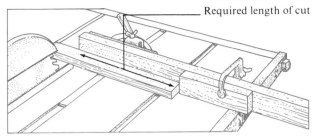

Required length of cut

When crosscutting and miter-cutting, use the miter fence. Bolt or screw on a higher wooden auxiliary fence so it is easier to hold the work.

When crosscutting use both hands to hold the work tightly against the fence. If only one hand is used, the work could slide during the cut.

To crosscut several boards to the same length, clamp a stop-block to the miter fence. A C-clamp is ideal for this. Glue abrasive cloth to the face of the

wooden auxiliary fence to prevent the wood slipping along the fence during the cut. This makes accurate cross-cutting much easier.

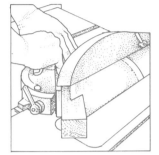

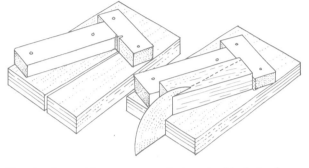

When making miter cuts at any angle feed the work slowly into the saw to minimize movement, which would result in inaccurate miters.

To make beveled miter cuts use a miter fence and tilted blade. Before starting the saw, check that the fence will not touch the blade.

To cut wedges and tapers, use a jig. Make a kerf in the baseboard parallel to one long side. Draw in the shape of the wedge from this datum cut. Pin

the two stop-blocks in place. Put the work in the angle between the blocks. Rest the baseboard against the rip fence and feed the jig into the blade.

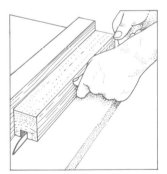

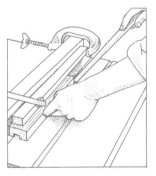

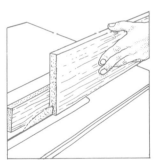

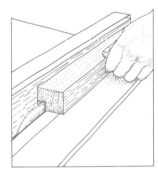

To cut a groove make a kerf. Then reset the fence, a fraction away from the blade, and make another kerf to widen the original one.

If the blade guard must be removed, a length of wood clamped to the fence holds the work down and also prevents the hands going near the blade.

To cut a groove in an edge, hold the board against the fence. Use a sliding jig to hold work that is more than twice the fence height.

Cut rabbets in two passes, tongues in four. For the second cut, hold the work so that the waste is against the fence, covering the blade.

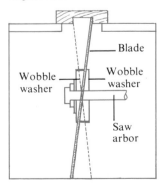

Blade

Wobble washer   Wobble washer

Saw arbor

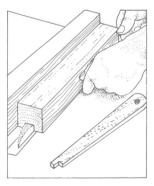

Tapered wobble washers are used in pairs and are fitted on either side of the blade to angle it so that a wider kerf can be made in one pass.

When cutting rabbets, grooves and tongues use wobble washers. Run the saw slowly to ensure the blade does not bow inwards, forming curved sides.

Safety pointers for fixed power tools

○ Disconnect the tool when making adjustments to blades.
○ Always use the guards and fences provided. If they must be removed, improvise with push sticks and jigs.
○ Never sweep waste off the table with the hand. Switch off the tool and use a stick or brush.
○ Never reach over blades; always walk around the tool.
○ Check when buying that the tool has an emergency mushroom-headed stop button or foot-press stop button.
○ When working long boards, enlist the aid of an assistant or use a roller on a stand to support the work as it comes off the tool.
○ Where possible, check the movement fully by hand before switching on.
○ Never hurry. Always be on guard against any potential danger.
○ Never leave the tool running unattended.
○ Avoid wearing clothing that may become caught in the tool.

Band saw

The band saw blade is a flexible steel strip, running continuously around two, or sometimes three, wheels. The blade is toothed on one edge like a handsaw blade. The band saw's primary function is to cut curves, although it also makes straight ripping cuts, crosscuts and miter cuts. It is also the best tool for resawing and for cutting shapes with compound curves. The saw is usually sized by the wheel diameter; one of less than 350 mm/14 in is suitable for a home workshop. Sometimes, however, it is sized by throat depth, that is, the distance between the blade and the frame of the machine. This affects the size of wood that can be swung across the table on curved cuts. Before use, the blade must be tensioned correctly so that it is just possible to flex it slightly. This is done by raising or lowering the top wheel. The blade must also be adjusted so that it tracks correctly between the blade guides. This is done by angling the top wheel. The guides keep the blade from deflecting to either side when cutting curves. The back-thrust wheel rotates with the blade and supports it against the feed pressure of the work. If the blade guides are not correctly set, the blade will be unable to make a true cut. The guide assembly is adjustable for height to allow for the thickness of the work. When in use, always follow the basic safety pointers listed on page 107.

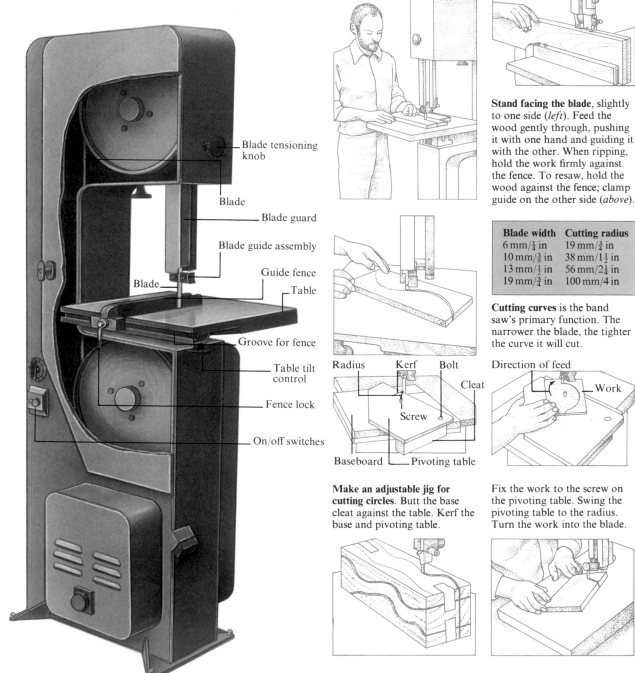

Blade tensioning knob

Blade

Blade guard

Blade guide assembly

Guide fence

Blade

Table

Groove for fence

Table tilt control

Fence lock

On/off switches

Stand facing the blade, slightly to one side (*left*). Feed the wood gently through, pushing it with one hand and guiding it with the other. When ripping, hold the work firmly against the fence. To resaw, hold the wood against the fence; clamp guide on the other side (*above*).

Blade width	Cutting radius
6 mm/$\frac{1}{4}$ in	19 mm/$\frac{3}{4}$ in
10 mm/$\frac{3}{8}$ in	38 mm/1$\frac{1}{2}$ in
13 mm/$\frac{1}{2}$ in	56 mm/2$\frac{1}{4}$ in
19 mm/$\frac{3}{4}$ in	100 mm/4 in

Cutting curves is the band saw's primary function. The narrower the blade, the tighter the curve it will cut.

Radius   Kerf   Bolt

Cleat

Screw

Baseboard   Pivoting table

Make an adjustable jig for cutting circles. Butt the base cleat against the table. Kerf the base and pivoting table.

Direction of feed

Work

Fix the work to the screw on the pivoting table. Swing the pivoting table to the radius. Turn the work into the blade.

To cut compound curves draw the profile of the shape on all four sides. Tape the waste in place after each cut.

The table on most band saws tilts up to 45° on one side for making beveled cuts. It tilts only slightly on the other side.

Radial arm saw

The radial arm saw is extremely versatile. It has an arm that can be swung through 360° to the right or to the left of an upright column. To adjust the depth of cut, the arm is raised or lowered on the column by turning the elevating handle. Each complete circle moves the arm up or down 3 mm/⅛ in. The yoke is suspended from the arm and holds the motor and blade. The yoke can be moved along the arm when crosscutting and miter-cutting, or can be locked at any point on it. The yoke can also be swiveled around to turn the blade parallel to the fence. The motor and blade can be swung within the yoke to make beveled or even horizontal cuts. The saw is sized by the diameter of its blades. These are similar to the blades fitted to a circular table saw. A diameter of around 250 mm/10 in is suitable for a home workshop. Although the radial arm saw can perform many tasks, it is safest and most efficient when crosscutting. This is because the blade rotates down and away from the operator, pushing the work in against the fence. When ripping, the blade rotates up and towards the operator, and so the anti-kickback pawls must always be used to prevent the wood from being shot out of the saw back towards the operator. The saw can also be fitted with accessories, including shaping cutters and sanding disks. Always observe the safety pointers listed on page 107.

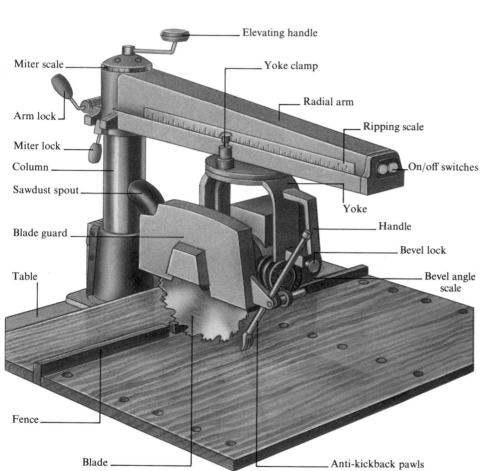

Elevating handle

Miter scale

Yoke clamp

Radial arm

Arm lock

Ripping scale

Miter lock

Column

On/off switches

Sawdust spout

Yoke

Blade guard

Handle

Bevel lock

Table

Bevel angle scale

Fence

Blade

Anti-kickback pawls

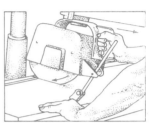

When crosscutting hold the work against the fence and pull the saw to the end of the arm. Return the saw to the fence.

When ripping, set the yoke the required distance from the fence, using the ripping scale. Feed the work through.

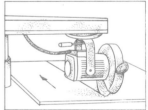

When ripping wide materials, cut with the yoke turned around and the blade locked at the outer end of the arm.

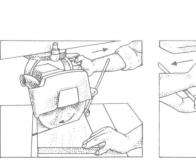

When miter-cutting swing the arm to right or left of the column and lock it. Make the cut like a crosscut.

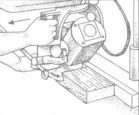

For beveled cuts tilt the motor and blade to the required angle. Rip, miter- and cross-cuts can be made.

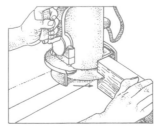

The motor can be swung so that the blade is horizontal within the yoke. Use this position for rabbeting and grooving.

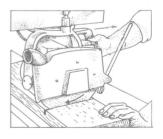

To make grooves by kerfing adjust the depth of cut with the elevating handle and make a series of crosscuts.

Spindle molder

The spindle molder cuts and shapes the edges of straight and curved pieces of wood. It can make moldings, rabbets, tenons and grooves. It consists of a flat table with a hole in the center near the back edge. A rotating spindle projects through the hole, and a cutter is fitted on or through the spindle. The spindle can be raised and lowered to adjust the height of the cutter. The table has a groove, to take a sliding miter fence, and threaded holes, to take two wooden fences on either side of the spindle. The fences move independently backwards and forwards, to adjust the amount by which the cutter projects, and from side to side, to provide clearance on each side of the spindle. The work rests against the fences and is fed into the cutter against the direction of rotation. The spindle revolves at up to 10,000 rpm. Cutters are available in a wide range of profiles. Some cutters are sold in matching positive and negative pairs. Blank cutters are available for filing to any profile. Split collar cutters consist of a pair of cutters that are fixed into parallel slots in grooved collars. French cutters consist of a single cutter passing through the spindle shaft. Disk cutters make cuts to a greater depth in the wood. They can be used singly or in pairs separated by adjustable spacing collars (*see* page 128). Before making any cut, turn the spindle by hand to check that it clears all guides and fences. Always make a test cut on waste wood first. To shape the edge of curved work, fit the ring fence. The uncut portion on the edge of the work rides against the ring fence. When the complete edge of curved work is to be shaped, pin a wooden template to the work; the ring fence will ride against the template. When the spindle molder is in use, always follow the basic safety pointers listed on page 107.

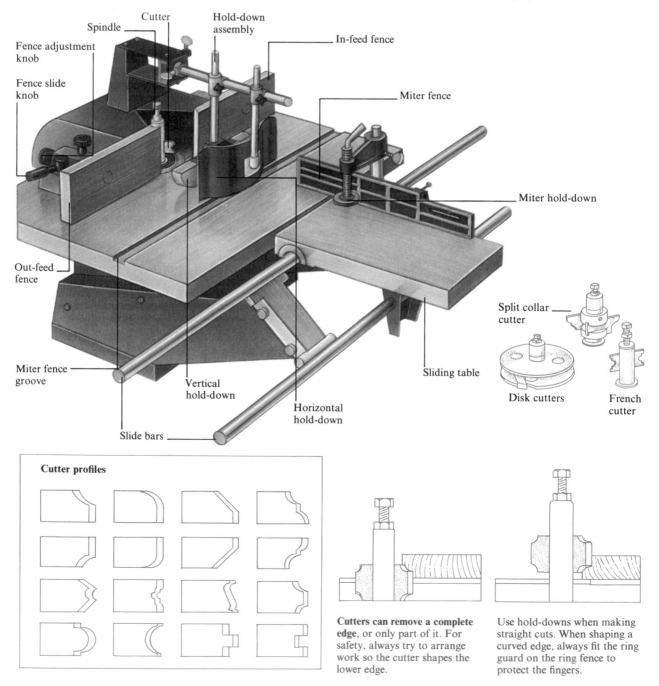

Cutter profiles

Cutters can remove a complete edge, or only part of it. For safety, always try to arrange work so the cutter shapes the lower edge.

Use hold-downs when making straight cuts. When shaping a curved edge, always fit the ring guard on the ring fence to protect the fingers.

Sight along the fences and adjust them so the cutter protrudes to produce roughly the profile required. Make test cuts on waste wood before making the final adjustments. The fences should just clear the cutter on either side.

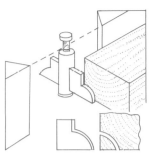

The work must be supported by both the in-feed and the out-feed fences along its whole length during the course of the cut. If only part of the edge is being cut away, then both fences should be set in line with each other.

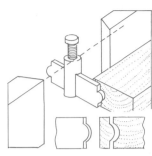

When the whole of an edge is to be cut away, set the out-feed fence forwards so the newly shaped edge is supported as it clears the cutter. If the cut is deep, first make a thin cut; readjust the fences and complete the cut.

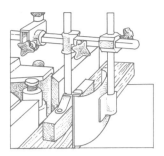

Use the hold-downs wherever possible. An overhead arm clamped behind the cutter supports a wooden shoe to hold the work down, while still allowing it to move, and a spring steel plate holds the work against the fence.

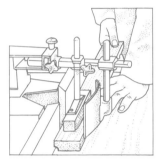

To shape the edge of a straight board, let the cutter reach full speed; then press the wood against the in-feed fence and push it forwards. Allow the

wood to bridge the cutter gap. Then transfer one hand to press it against the out-feed fence. Keep movement smooth and continuous.

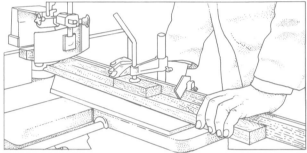

To shape end grain, the wood must have the extra support supplied by the sliding table and the miter fence. Clamp the work to the miter fence. Place

some waste wood beneath the work to support the cutter, and place another piece of waste wood behind the work to prevent the grain from splintering.

To shape end grain at an angle, hold the work at the required angle against the miter fence and push it past the cutter using the sliding table.

When shaping a wide board, use the vertical hold-down. Raise the horizontal hold-down in front of the cutter and press the work against the fence.

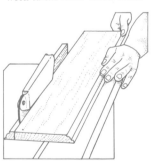

To make molding strips, shape both edges of a board. Then rip off the strips on a table saw. Repeat, until sufficient molding is made.

Combine cutters to make a deep molding. Position the second cutter for continuity with the first and readjust the fences as necessary.

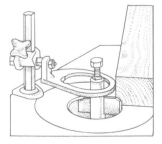

To shape curved edges, remove the wooden fences and hold-downs. Press the work against the ring fence and allow the cutter to protrude beneath it.

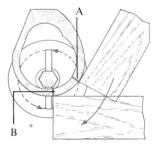

Mark the minimum (A) and maximum (B) cutter projection on the ring fence. Introduce the work at A and pivot it around to B.

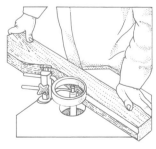

Make the cut along the length of the work at point B. Always fit the ring guard. (For clarity it has been omitted in the illustrations *left* and *far left*.)

To smooth shaped work, fit a drum sander. Fit a sleeve of abrasive paper over the rubber cylinder. Screw down the cap so the rubber expands.

Combined planer & thicknesser

The planer and thicknesser perform related tasks and are often combined. The planer smooths and trues up faces and edges on a piece of wood. When one wood face is flat and true it can register against the flat bed of the thicknesser, the machine smooths the other face, resulting in a piece of wood of a uniform thickness. Cutting in both planer and thicknesser is done by a rotating cylindrical metal cutter block.

On the planer, the work is fed over the cutter block between two metal tables. These are raised and lowered independently. The out-feed table must be the same height as the cutters at their highest point so the newly planed surface rides smoothly onto it. The in-feed table

is set lower — by the amount of wood to be removed. The maximum depth of cut should be no more than $3 \, mm/\frac{1}{8}$ in. Make deeper cuts by a series of passes. To plane a cupped board, place the concave side downwards and take several thin cuts. If a board is "in winding", remove wood from the high corners.

Wood is fed into the thicknesser on a flat bed which is raised or lowered to adjust the depth of cut. Corrugated in-feed rollers grip the work and draw it beneath the rotating cutter block. The feed rollers hold the wood flat while it is being smoothed.

When using these tools, always observe the safety pointers listed on page 107.

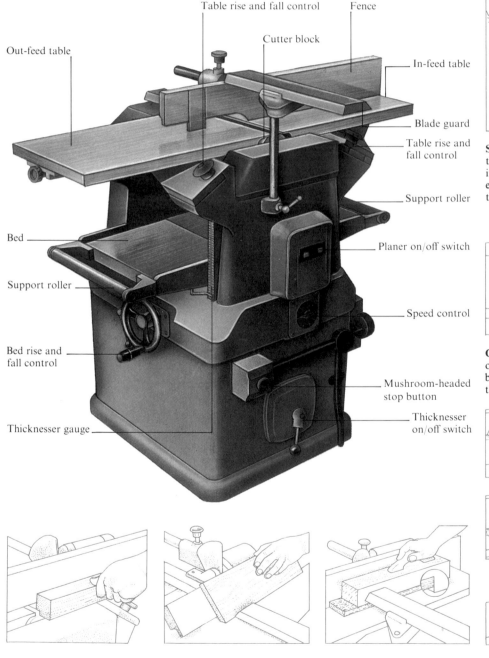

Table rise and fall control · Fence · Cutter block · Out-feed table · In-feed table · Blade guard · Table rise and fall control · Support roller · Bed · Planer on/off switch · Support roller · Speed control · Bed rise and fall control · Mushroom-headed stop button · Thicknesser gauge · Thicknesser on/off switch

Stand to one side when feeding the work into the thicknesser in case of kickback. Rollers at each end of the bed support the work.

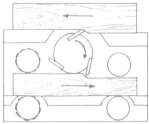

On simple combination tools, one cutter block serves for both parts. Feed rollers in the thicknesser pull the wood in.

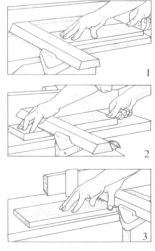

1

2

3

To plane a rabbet, remove the blade guard and adjust the fence to the rabbet width. Cut the full depth of the rabbet in steps in a number of passes.

To plane bevels and chamfers tilt the fence in- or outwards (*above*), using the miter scale behind the fence. Check the setting with a sliding bevel.

Use a push block for short or thin wood. Wood less than 300 mm/12 in long is too short to be safely planed with power tools and must be hand planed.

Start planing with both hands holding the wood on the in-feed table (**1**). Gradually move the hands over the blade guard to the out-feed table (**2, 3**).

Drill press

The drill press can bore and drill a large number of holes quickly and accurately to a predetermined depth. It is available in both bench and floor-standing models, the only difference being in the length of column. The size of the drill press is determined by measuring the distance from the chuck center to the column and then doubling that measurement: this is the maximum diameter that can be accommodated on the table. The table can be moved up and down the column and can be locked to hold work of varying thicknesses. It may also be tilted to bore angled holes or to support wood for boring into end grain. Set the table by aligning the bit with a sliding bevel.

Place the work beneath the bit and lower the bit directly, and with even pressure, into the work, using the feed lever. Drill press speeds range from around 300 rpm to 6,000 rpm. Follow the manufacturer's instructions for the correct speed for different materials; the fastest speeds are for using accessories such as router cutters and a drum sander.

Open and close the three-jaw chuck with a key. The chuck capacity is usually up to 13 mm/$\frac{1}{2}$ in. The drill press takes the same range of bits and drills as the portable power drill (*see* pages 98–9).

When using the drill press, it is extremely important to observe the safety pointers listed on page 107.

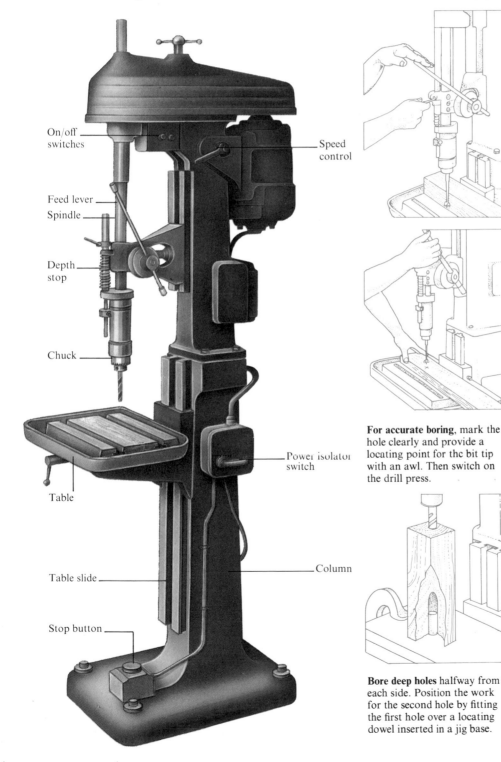

On/off switches

Speed control

Feed lever

Spindle

Depth stop

Chuck

Table

Power isolator switch

Table slide

Column

Stop button

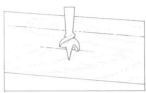

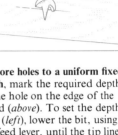

To bore holes to a uniform fixed depth, mark the required depth of the hole on the edge of the wood (*above*). To set the depth stop (*left*), lower the bit, using the feed lever, until the tip lines up with the mark on the wood. Secure the locking nuts on the depth stop.

For accurate boring, mark the hole clearly and provide a locating point for the bit tip with an awl. Then switch on the drill press.

Lower the feed lever until the bit is above the work. Move the work so the bit will be centered accurately. Hold the work down firmly.

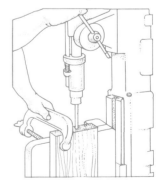

Bore deep holes halfway from each side. Position the work for the second hole by fitting the first hole over a locating dowel inserted in a jig base.

To bore holes in end grain and in board edges tilt the table until it is at right angles to the bit. Clamp the wood to the table; then bore.

Methods

A comprehensive demonstration of all the
essential techniques in woodworking

Joints

A joint must be strong enough to resist the stresses that will be imposed upon it. (For uses of joints *see* pages 150–65 and pages 8–29; for terms peculiar to each joint *see* under the appropriate joint.) The design of most joints has evolved over the centuries to solve the particular problems posed by solid wood construction. Few of these joints work successfully with manufactured boards except for multi-ply (*see* page 149). All joints, except those that are dismantled or pivoted, must be glued. The greater the area on the side and edge available to be glued, the greater the strength of the joint. End grain does not hold glue well.

Consider the grain direction carefully before starting to make a joint. There should be no knots in the joint area. Decide how the pieces are to be arranged (*see* page 151). Choose the best surfaces of the wood to be the face side and face edge. Cut and plane the wood to size and square it up, marking the faces (*see* pages 56–7). Then cut the joint. The surfaces of the joint left by the saw will be quite rough; this is an advantage as it helps the glue to grip. Remove ragged edges; smooth faces that will be inaccessible when the joint is assembled, but leave penciled reference marks on until the whole project has been glued and assembled. Before gluing, a joint can be tested for fit — known as a dry assembly. A well-made joint will be friction tight. Glue all the inside surfaces of the joint on both pieces. Then knock the joint pieces home with a hammer. The hammer has the same weight as a mallet but the impact area is smaller and is consequently easier to control. Use waste wood as a buffer. Wipe off excess glue before it dries, using a damp cloth. When dry, level all surfaces with a smoothing plane and smooth with sandpaper.

Halved joints

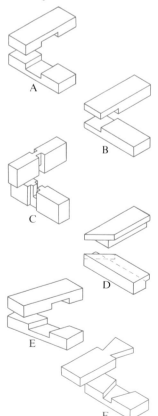

Halved joints
Halved joints provide a simple way of joining two pieces that cross or meet in an L or T formation. They are used for small simple frames and for intermediate frames and divisions within a carcass. They do not have the strength and rigidity of joints such as mortise and tenons, dovetails or housed joints. The halved cross lap joint (**A**) can be cut in manufactured boards crossing in their width, making it suitable for internal dividers in boxes and small carcasses. The corner half-lap joint (**B**) needs to be reinforced by screwing underneath, or by dowels or bolts. Variations on the halved joint include the housed and halved joint (**C**) and the mitered and rabbeted corner half-lap (**D**) — a useful joint for picture and mirror frames. Secure the miter by doweling or screwing from the back. The straight shoulders on the oblique halved joint (**E**) give it strength and resistance to twisting. In the half-lap dovetail (**F**), the strength of a dovetail joint is added to that of a middle lap joint. Use it for jointing carcasses and frames.

Mortise and tenon joints

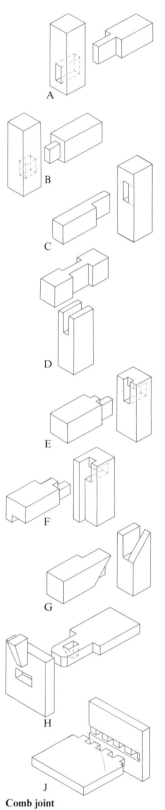

Comb joint

Mortise and tenon joints
These are the strongest end to side edge joints and they are used in cabinetmaking for frame jointing. The through mortise and tenon joint (**A**) is the basic form of this joint. The mortise is cut right through the wood revealing the end grain of the tenon. Shoulders on the tenon provide stops for the tenon and a resistance to twisting. Wedges can be inserted into either tenon or mortise to give additional strength. In the blind mortise and tenon joint (**B**), the tenon is shortened to avoid showing end grain on the outer surface. The bare-faced mortise and tenon joint (**C**) is used when a thin slat must be tenoned into a thicker upright to finish flush on one side. The tenon has only one shoulder. This allows the mortise to be displaced safely away from the front edge. The bridle joint (**D**) is often used for leg and rail joints where a thick leg must be connected to a thinner rail. On the haunched mortise and tenon joint (**E**) the width of the tenon is reduced to prevent the joint being withdrawn upwards. Instead of a complete strip of the tenon being removed, a small portion near the shoulder is left — called a haunch. The haunch helps resist any tendency to twist. The haunch can be tapered and the mortise piece cut to a matching slope. The long and short shoulder mortise and tenon joint (**F**) is necessary when joining rabbeted members. The mortise and tenon joint with a mitered front (**G**) gives a neat appearance and is useful when working moldings on the inside edge. The projected and wedged mortise and tenon joint (**H**) forms the leg and rail connection on a table that might need to be dismantled. The tenon passes right through the leg, or truss, and is wedged on the outside. The pin joint (**J**) is extremely strong. There is a very large gluing area and the joint can resist downward and twisting pressure.

Comb joint
This joint is used for boxes and drawers, particularly when many drawers are needed. The tongues make a tight fit in the kerfs, and the joint has a large gluing area. A comb joint is formed quickly and accurately with a table saw.

Dovetail joints

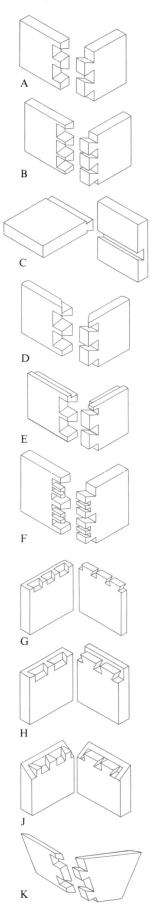

A

B

C

D

E

F

G

H

J

K

Dovetail joints

Dovetail joints are the strongest way of connecting pieces of wood together in their width. They are used to join carcasses rigidly and to make boxes and drawers. They have a comparatively large gluing area and are therefore particularly strong. Generally, the longer the tails, the better the grip. Properly made tails only fit once, so a joint should only be half entered on a dry assembly. The carpenters' through multiple dovetail joint (A) is the basic form of the joint. End grain shows on both pieces. When handmade, the tails are wider than the pins. The fineness of the pin is judged as a sign of quality workmanship, and in the cabinetmakers' dovetail joint (B) the pin often comes to a point. In the housed dovetail joint (C), the tail is cut across the width of the wood. The joint is usually tapered on assembly to avoid excessive friction across such a wide area. This is a particularly strong joint to use for securing shelves. The through dovetail joint with miter (D) has the top edge mitered and usually rounded over for a neat appearance on trays and boxes. The through dovetail joint with rabbet (E) is used with a rabbeted frame. The decorative through dovetail joint (F) makes a feature of the fact that end grain shows on both pieces. The small tails add to the strength of the joint by increasing the gluing area on the very large tails. The lap dovetail joint (G) is used when the end grain is required to be hidden on one of the pieces. The through tails are modified by shortening the tails and letting the socket piece lap over the end. The joint is often used for drawers where the end grain is not desired to show on the front. The stopped lap dovetail joint (H) is used where tails are not desired to show on either piece. One piece presents a flush surface while the other shows only a thin line of end grain. No end grain shows at all on the blind miter dovetail joint (J), which from the outside looks like a simple butted miter joint. The beveled dovetail joint (K) is used to join the sides of boxes where all four sides slope outwards; it is one of the most difficult of all joints to make. Through and lap dovetails can be cut by machine, using a power router and jig. This is useful if many joints are to be cut.

Housed joints

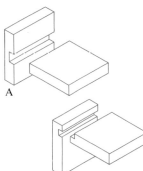

A

B

Dowel joints

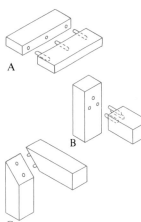

A

B

C

Edge joints

A

B

C

D

Pivoting joints

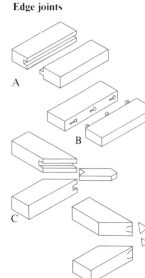

A

B

Housed joints

The through dado joint (A) is used for fixing shelves in carcasses and for holding intermediate rails in carcass frames. The dado can be stopped, so there is an unbroken vertical line on the carcass. In the rabbet and dado joint (B), used for jointing a top to a side, the horizontal piece is reduced in thickness by having a tongue cut in it and only this is housed. This makes the joint better able to resist twisting.

Dowel joints

Dowels can be used to join wood edge to edge (A) or at right angles, edge to end (B). They can also be used for miter joints (C) and to secure solid-wood or manufactured-board shelves. In traditional work they were rarely used as the dowels had to be cut by hand. Dowel joints are now often substituted for mortise and tenon joints, as they are easier to make and almost as strong. They are especially suitable when awkwardly shaped or angled pieces of wood meet as in canted or round table legs joining a rail.

Edge joints

To build up a wide surface in solid wood, two or more boards can be joined edge to edge. The most basic way is to use the rubbed glue joint. It is a perfectly strong joint if the mating edges are planed and glued together correctly. The tongue and groove joint (A) has a large gluing area. The tongue can be cut in one operation using a tonguing cutter or be formed by cutting a rabbet in from each face. Both pieces can be grooved and a loose tongue, or spline, glued in. The slot-screwed joint (B) may be used as an alternative. A miter joint can be strengthened with a spline glued in grooves (C) or by keys inserted after assembly (D).

Pivoting joints

The knuckle joint (A) is cut when a pivoting joint is required. It is traditionally formed on the brackets supporting the flaps on Pembroke tables. The rule joint (B) is traditionally used for fall flaps such as on a gateleg table. It looks neat when the flap is both up and down and adds strength as pressure on the flap is transmitted to the main top when the flap is up. It must be used with a special rule joint hinge (*see* page 258).

Halved

A halved cross lap joint

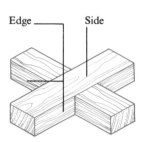

Edge ____ Side

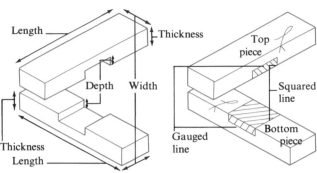

Length ____ Thickness
Depth Width
Thickness
Length

Top piece
Squared line
Gauged line
Bottom piece

The finished joint. The joint pulled apart. The wood marked up.

Making a halved cross lap joint

1. Mark the width of the top.

2. Square down both edges.

3. Gauge half the thickness.

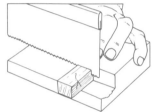

4. Saw to the gauged line.

5. Check the marked lines.

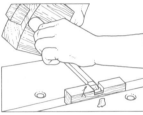

6. Chisel out the waste.

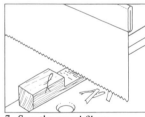

7. Saw the wood fibers.

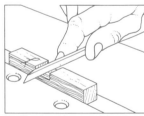

8. Check with a straightedge.

9. Check the joint for fit.

Variations on the halved cross lap joint

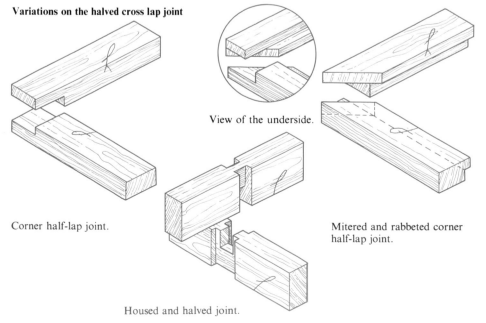

Corner half-lap joint.

View of the underside.

Mitered and rabbeted corner half-lap joint.

Housed and halved joint.

Making a halved cross lap joint
The pieces must be of equal thickness. Place the top piece on the bottom piece, face sides uppermost, and mark the width of the top piece on the bottom piece (1). Square the marks across the bottom piece and halfway down both edges (2). Place the pieces in their original position and mark the width of the bottom piece onto the underside of the top piece. Set the marking gauge to half the thickness of the wood and gauge between the marks on the edges of both pieces from the face sides (3). Hold one piece in the bench hook and saw across the grain just inside the squared line, down to the gauged line (4). Put the two pieces together and check that they will make a tight fit when the second cut is made to the marked line (5). Move the saw in relation to the squared line if necessary. Make one or two kerfs in the center of the waste to enable it to be removed more easily. Secure the piece in the vise. With a bevel-edged chisel and mallet, chisel out the waste from the dado in stages (6). Work halfway in from each edge. With a saw, remove the fibers that do not come away cleanly (7). Check the flatness of the dado with a straightedge (8). Level out any remaining unevenness, chiseling with hand pressure only. Cut the second dado. Check the joint by putting the end of one piece into the dado on the other and vice versa (9). The joint should be tight enough to enable one piece to be lifted up by the other when pressed together.

Variations on the halved cross lap joint
The corner half-lap joint is one of the easiest joints to make for frames. It needs reinforcing either with a through dowel or by gluing and screwing from the underside. The housed and halved joint is another variation on the halved cross lap joint and is similar to the bridle joint. Although more wood is removed than in the halved cross lap joint, the variation is stronger because of the double dadoes, which eliminate lateral movement. The mitered and rabbeted corner half-lap joint is a picture or mirror frame joint that is stronger than the plain miter. The lap allows a screw to be inserted from the underside. Drill the screw-hole near the miter shoulder for strength.

An oblique halved joint

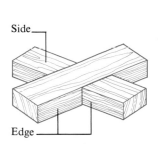

Side

Edge

The finished joint.

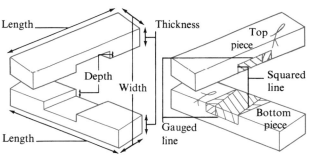

Length

Thickness

Depth

Width

Top piece

Squared line

Bottom piece

Length

Gauged line

The joint pulled apart.

The wood marked up.

Making an oblique halved joint

1. Place the sliding bevel on the rod.

2. Mark the length of the dado.

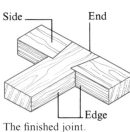

3. Mark in the dado.

4. Make the sawcuts.

5. Chisel away the waste.

A half-lap dovetail joint

Side   End

The finished joint.

Edge

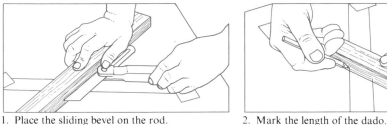

Length

Tail piece

Width

Thickness

Depth

Housing piece

Thickness

Length

Squared line

Gauged line

Gauged line

The joint pulled apart.

The wood marked up.

Making a half-lap dovetail joint

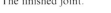

1. Gauge the wood.

2. Saw across the grain.

3. Mark in the tail slope.

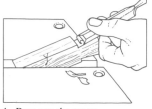

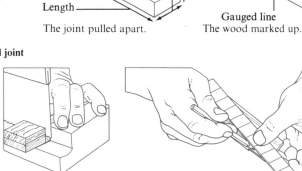

4. Remove the waste.

5. Mark the dado.

6. Cut the dado.

Making an oblique halved joint
The pieces must be of equal thickness. Take the oblique angle from the rod of the whole project. Hold the top piece in position on the rod, face side up. Place a sliding bevel against the edge, aligning the stock with the marks for the bottom piece (1). Set the bevel blade to this angle. Mark the width and position of the oblique dado on the edges of the top piece (2). Check the marks against the sliding bevel and use it to draw the oblique dado on the underside of the top piece (3). Square the marks halfway down the edges. Then mark the dado on the face side of the bottom piece in the same way. Gauge the depth of the dado on both pieces from the face sides. Remove the waste at the center of each dado, where it is squared across, using a saw and bevel-edged chisel. To make the oblique sawcuts, use a bench hook and clamp the wood where necessary so that the cut can be made at the easiest angle. Make two straight sawcuts (4). Then chisel away the waste in the oblique corner pieces (5).

Making a half-lap dovetail joint
Mark the housing piece width on the tail piece. Square the marks across the underside and down both edges. Set the marking gauge to half the thickness of the wood and gauge along both edges and around the end (1). Remove half the thickness of the dovetail piece from the underside; saw a 3 mm/⅛ in kerf with the wood held in the vise. Then clamp the wood at an angle and saw down the line gauged along one edge. Turn the wood around and repeat. Then put the wood upright in the vise again and complete the cut to the squared line. Put the wood on the bench hook and saw across the grain (2). Square the line of the cut edges across the face of the tail piece. With a pencil, mark the angle of the tail slope by eye, or with a rule measuring 1 in 6 or 1 in 8 units (3). Repeat the slope on the other side. Saw the tail shoulders across the grain to the pencil marks. Remove the waste with a chisel, working along the grain towards the sawcut (4). Mark the dado from the tail (5). Saw and chisel out the waste (*see* method above) (6).

Mortise & tenon 1

A through mortise and tenon joint

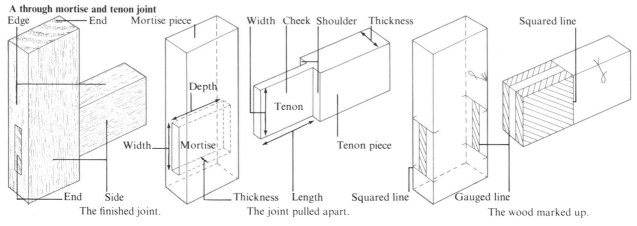

The finished joint.   The joint pulled apart.   The wood marked up.

Marking up a through mortise and tenon

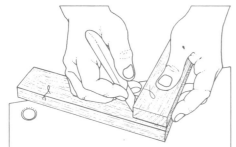

1. Mark the length of the tenon.

2. Square the line all around for the tenon.

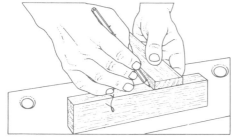

3. Mark the mortise width from the tenon piece.

4. Square the line around for the mortise.

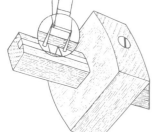

5. Set the spurs of the gauge.   6. Set the gauge stock.   7. Check from the other side.

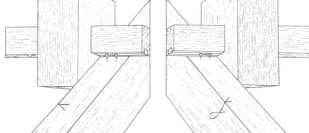

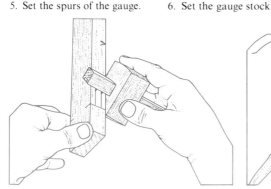

8. Mark with gauge from the face side.

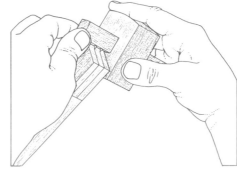

9. Mark around the tenon end.

Marking up a through mortise and tenon

A tenon should generally be one third the thickness of the tenon piece. The grain of the mortise piece should run along the width of the mortise so that the mortise can be chopped out against the grain. This will ensure a clean cut that is the exact width of the chisel being used. Mark the length of the tenon from the mortise piece, using a marking knife (**1**). Square the line all around the tenon piece (**2**). Mark the width of the mortise from the width of the tenon piece (**3**). Square the line around to both edges (**4**). To mark the mortise thickness set the spurs of the mortise gauge to the width of the chisel to be used for chopping out the mortise (**5**). Set the gauge by eye so the marks are in the center of the mortise piece (**6**). Check by placing the stock against the other side to see if the spur marks coincide (**7**). If they do not coincide, readjust the stock. When the mortise gauge is correctly set, mark the mortise on both edges from the face side (**8**). To set out the tenon leave the spurs at the same setting but adjust the stock if joining woods of different thicknesses. Grip the tenon piece in the vise and mark with the gauge around one edge and the end from the face side (**9**). Then turn the wood around and mark the other edge from the face side.

Chopping out a mortise

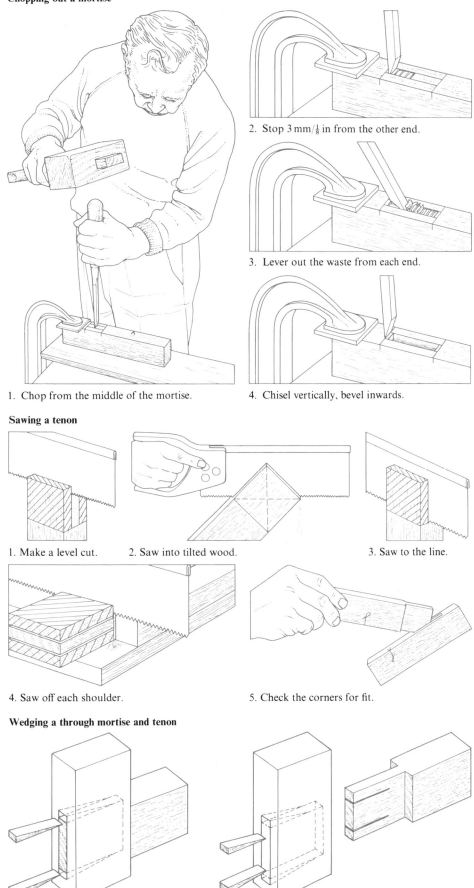

2. Stop 3 mm/⅛ in from the other end.

3. Lever out the waste from each end.

1. Chop from the middle of the mortise.

4. Chisel vertically, bevel inwards.

Sawing a tenon

1. Make a level cut.   2. Saw into tilted wood.   3. Saw to the line.

4. Saw off each shoulder.   5. Check the corners for fit.

Wedging a through mortise and tenon

A. Outside wedges.   B. Inside wedges.

Chopping out a mortise
Place the mortise piece on a piece of protective waste wood and clamp to the workbench. To prevent the wood splitting, the mortise must be chopped halfway through from each side. Begin chopping out by making an upright cut in the middle, then move the chisel backwards about 3 mm/⅛ in at a time with the bevel towards the middle (1). Stop 3 mm/⅛ in from the squared line (2). This allows for compression when levering out the waste. Reverse the chisel and continue from the middle to within 3 mm/⅛ in of the other end. Turn the chisel round and lever out the waste from each end to the center (3). Finish by chiseling vertically on the end lines, bevel facing inwards (4). Unclamp the mortise piece and shake out any loose wood chips. Sweep the chips off the bench. Placing the uncut edge uppermost, reclamp the mortise piece to the bench. Repeat the chiseling and waste removal process as necessary to meet the cut on the opposite edge. Square the ends.

Sawing a tenon
Clamp the tenon piece upright in the vise. At each stage of sawing work on both gauged lines on each side of the tenon. Position the saw against the waste side of the gauged line. Tilting it so it bites into the edge of the wood, saw down about 3 mm/⅛ in. Level out the saw and cut about 3 mm/⅛ in down (1). Regrip the tenon piece at a 45° angle. Saw to the squared line (2). Turn the tenon piece around and saw the other edge, again at 45°. Clamp the tenon piece upright in the vise and saw straight down to the gauged line (3). Then, with the tenon piece flat, saw off the waste wood at each shoulder (4). To check the fit, try each corner of the tenon in the mortise (5).

Wedging a through mortise and tenon
Wedges give the joint extra strength. They can either be inserted on the outside of the tenon (A) or into kerfs in the tenon (B). Extend the ends of the mortise on the outside edge by chiseling at an angle 3 mm/⅛ in from the squared line to about half the mortise depth. If the tenon is a very tight fit, grip the mortise piece in the vise. Insert the tenon and tap in the wedges alternately. Saw off the excess and plane flush.

Mortise & tenon 2

A blind mortise and tenon joint

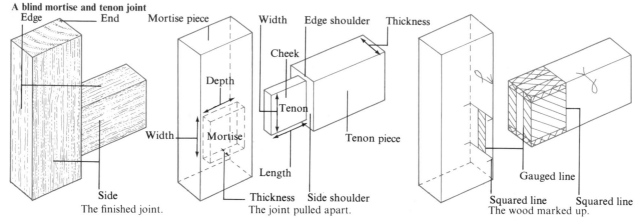

The finished joint.

The joint pulled apart.

The wood marked up.

Making a blind mortise and tenon

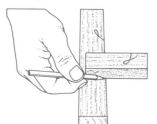

1. Establish the tenon length.

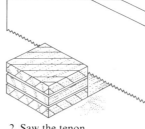

2. Saw the tenon.

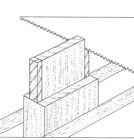

3. Saw the edge shoulders.

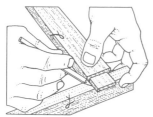

4. Mark the mortise width.

5. Set the mortise depth.

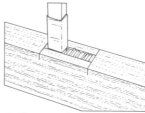

6. Chop out the mortise.

Fox-tailed mortise and tenon

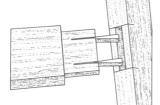

1. Insert the wedged tenon.

2. Clamp the joint.

Bare-faced and offset mortise and tenon joints

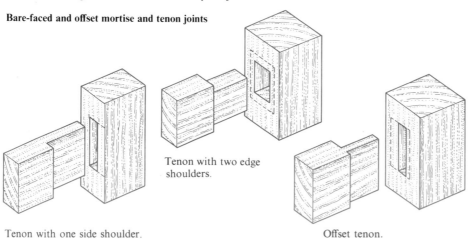

Tenon with one side shoulder.

Tenon with two edge shoulders.

Offset tenon.

Making a blind mortise and tenon

Establish the required length of the tenon by holding it against the mortise piece (1). Mark up and saw the tenon (*see* pages 120 and 121) (2). With a marking gauge, mark the edge shoulders, which should be minimal so as not to weaken the tenon. (The edge shoulders will ensure that a gap does not appear when the wood shrinks.) Saw down to the squared lines and cut off the waste (3). Mark the mortise width from the completed tenon (4). Holding the chisel against the tenon, wrap masking tape around the blade to mark the depth of the mortise to be cut (5). Chop out the mortise to a little more than the required depth (6).

Fox-tailed mortise and tenon

Cut two wedges the thickness of the tenon. In the tenon make kerfs slightly longer than the wedges. Undercut the mortise at its base by 3 mm/⅛ in at each end. Glue and insert the wedges into the kerfs. Glue and insert the tenon into the glued mortise (1). Tighten with a sash clamp. This is safer than hammering the joint together (2).

Bare-faced and offset mortise and tenon joints

When joining a thin rail to a thick leg a "standard" tenon, with its equal side shoulders, will not give a strong joint at the front of the leg if the rail and leg are to finish flush. Therefore a bare-faced mortise and tenon with only one side shoulder on the tenon is used. Make the mortise and the tenon (*see* pages 120-1). If the rail is to finish in the middle of the leg the tenon can have two edge shoulders. If the rail is to be inset from the front of the leg a slightly offset bare-faced tenon can be used.

A bridle joint

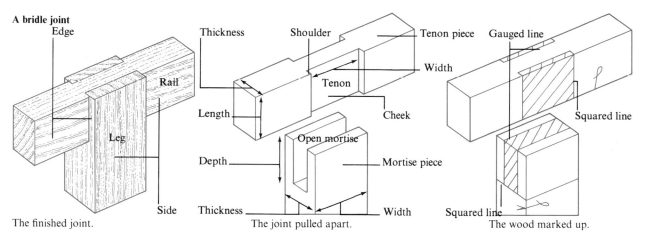

Edge
Rail
Leg
Side

The finished joint.

Thickness
Shoulder
Tenon piece
Width
Tenon
Length
Cheek
Open mortise
Depth
Mortise piece
Thickness
Width

The joint pulled apart.

Gauged line
Squared line
Squared line

The wood marked up.

Making a bridle joint

1. Mark the width of the mortise.

2. Mark the length of the tenon.

3. Saw down to gauged line.

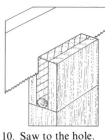

4. Chisel the waste from the tenon.

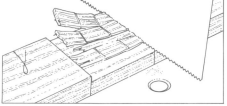

5. Sever the fibers cleanly.

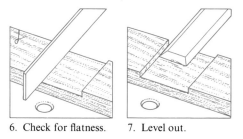

6. Check for flatness.　7. Level out.

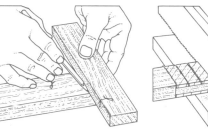

8. Set the mortise gauge for the mortise.

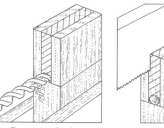

9. Bore out the base.　10. Saw to the hole.

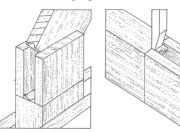

11. Remove waste.　12. Square up.

Making a bridle joint

A bridle joint is a modified open mortise and tenon and is used where the leg is thicker than the rail. Mark in pencil the width of the mortise onto the tenon (1). Then, to ensure a tight fit, square all around with a marking knife slightly inside the pencil lines. Mark the length of the tenon onto the top end of the mortise (2). Square all around the mortise piece. To form the tenon, mark a 5 mm/$\frac{3}{16}$ in recess either side of the tenon piece. Hold the tenon piece on the bench hook and saw to the gauged line on the waste side of the line (3). Make one or two sawcuts in the center of the waste wood, which can then be removed more easily. Secure the tenon piece in the vise. With a bevel-edged firmer chisel and a mallet begin removing the waste wood. Rest your elbow on the bench and hold the chisel bevel upwards with your hand near the cutting edge. Hold the mallet just below its head. Chisel away a bit at a time halfway across; chiseling directly at the gauged lines may cause the wood to break away along the grain to a greater depth than required (4). If the fibers do not come away cleanly, saw slightly deeper into the waste wood to sever them (5). Check the tenon for flatness (6). Level out any remaining unevenness, chiseling with hand pressure only — not using a mallet (7). Set the mortise gauge to the thickness of the tenon (8). Adjust the stock on the top of the mortise piece and gauge around for the open mortise. Start cutting the mortise by boring a hole at the base, drilling in to the middle from both sides (9). Saw along each gauged line down to the hole (10). Remove the waste wood (11). Square up the mortise with a mortise chisel (12).

Mortise & tenon 3

A haunched mortise and tenon joint

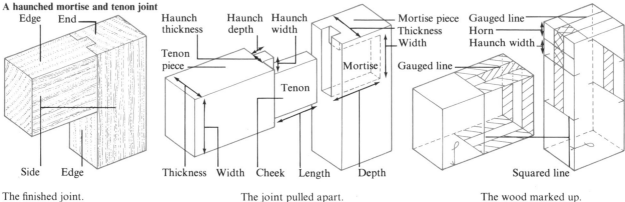

The finished joint.

The joint pulled apart.

The wood marked up.

Making a haunched mortise and tenon

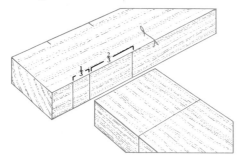

1. Mark out the mortise.

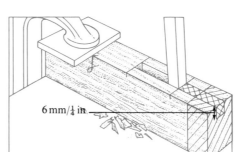

2. Chop out the through mortise.

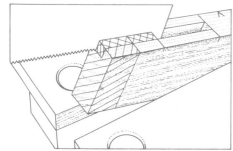

3. Saw on both sides of the crosshatched area.

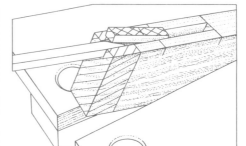

4. Chisel out the waste.

5. Mark the haunch depth.

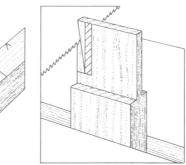

6. Mark the tenon width.

7. Cut wedges from the waste.

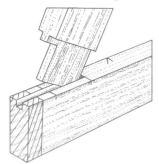

8. Check the joint for fit.

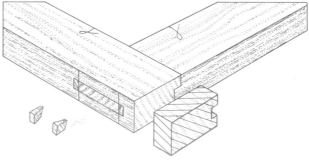

9. Saw off the horn.

Making a haunched mortise and tenon

Chiseling a mortise near the end of a piece of wood is likely to split the wood. To prevent this, mark up the mortise 19 mm/$\frac{3}{4}$ in from the end. This waste, or horn, is cut off when the joint is complete. Mark the tenon-piece width on the mortise piece, allowing for the horn and square across one edge. From the top mark come in the haunch width, which should be no greater than one third the width of the tenon piece, and again square across the edge. Transfer this haunch mark and the lowest mortise mark to the outer edge, to give the limits of the through mortise (1). Set the mortise gauge and mark the thickness of the mortise, continuing 6 mm/$\frac{1}{4}$ in around the end of the horn to allow for the depth of the haunch. Crosshatch the waste to distinguish it from the through mortise. Chop out the through mortise (*see* page 121) (2). Extend ends of the mortise on the outside edge to take the wedges (*see* page 121). To make the entry for the haunch saw down to the 6 mm/$\frac{1}{4}$ in gauge mark (3). Then chisel out the waste on both sides (4). Mark up and saw the tenon (*see* page 120). Mark the depth of the haunch (5). Then mark the width of the tenon against the through mortise on the mortise piece (6). Cut two wedges from the waste on the tenon (7). Then cut off the remaining waste to form the haunch. Check the corners of the tenon for fit (8). If there is a gap at the shoulders check that the haunch is not too deep. If it is, deepen the haunch recess on the mortise piece rather than shortening the haunch. Glue and tap the mortise and tenon together. Glue and tap in the wedges alternately from the outside edge. Then saw them off flush. Saw off the horn (9).

A long and short shoulder mortise and tenon joint

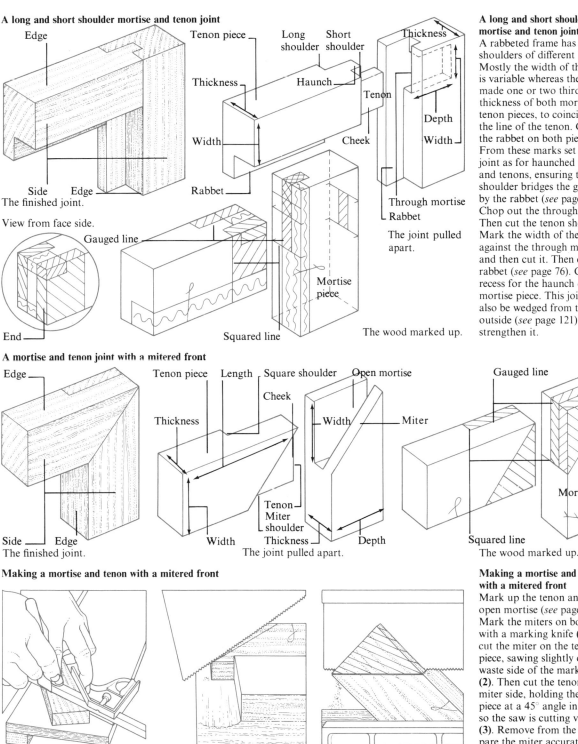

Edge

Side   Edge
The finished joint.

View from face side.

Gauged line

End

Tenon piece

Thickness

Width

Rabbet

Long
shoulder

Short
shoulder

Haunch

Thickness

Tenon

Cheek

Depth

Width

Through mortise

Rabbet

The joint pulled
apart.

Squared line

Mortise
piece

The wood marked up.

A long and short shoulder mortise and tenon joint

A rabbeted frame has tenon shoulders of different lengths. Mostly the width of the rabbet is variable whereas the depth is made one or two thirds the thickness of both mortise and tenon pieces, to coincide with the line of the tenon. Gauge the rabbet on both pieces. From these marks set out the joint as for haunched mortise and tenons, ensuring that one shoulder bridges the gap made by the rabbet (*see* page 124). Chop out the through mortise. Then cut the tenon shoulders. Mark the width of the haunch against the through mortise and then cut it. Then cut the rabbet (*see* page 76). Cut the recess for the haunch on the mortise piece. This joint can also be wedged from the outside (*see* page 121) to strengthen it.

A mortise and tenon joint with a mitered front

Edge

Side   Edge
The finished joint.

Tenon piece

Thickness

Length

Cheek

Square shoulder

Width

Open mortise

Miter

Tenon
Miter
shoulder

Width   Thickness

Depth

The joint pulled apart.

Gauged line

Mortise piece

Squared line

The wood marked up.

Making a mortise and tenon with a mitered front

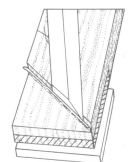

1. Mark both miters.

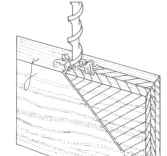

2. Saw close to the line.

3. Saw the tenon.

4. Chisel the tenon miter.

5. Bore out the mortise base.

6. Trim the mortise base.

Making a mortise and tenon with a mitered front

Mark up the tenon and the open mortise (*see* page 123). Mark the miters on both pieces with a marking knife (**1**). Then cut the miter on the tenon piece, sawing slightly on the waste side of the marked line (**2**). Then cut the tenon on the miter side, holding the tenon piece at a 45° angle in the vise so the saw is cutting vertically (**3**). Remove from the vise and pare the miter accurately on the marked line, using a chisel (**4**). Saw on the waste side of the tenon down to the square shoulder; then saw off the waste (*see* page 121). On the mortise piece, start by cutting the miter just over a third of the way through the wood, finishing the cut accurately with the chisel. To form the mortise, first bore out the base, drilling halfway in from each edge (**5**). Then saw down to the hole on the gauged lines. Remove the waste and trim the base with the chisel (**6**).

Mortise & tenon 4

A projected and wedged mortise and tenon joint

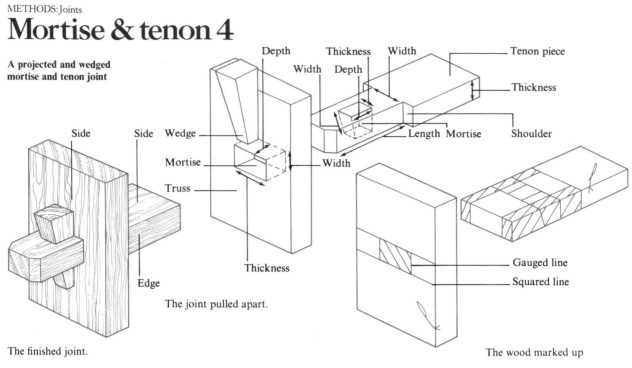

The joint pulled apart.

The finished joint.

The wood marked up

Making a projected and wedged mortise and tenon

1. Place the tenon edge on the truss edge.

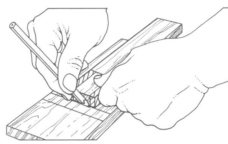

2. Mark the tenon position.

3. Gauge the mortise on the other side.

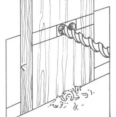

4. Bore out the waste.

5. Finish with a chisel.

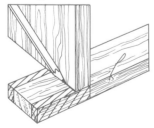

6. Mark the truss thickness.

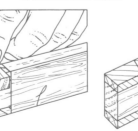

7. Mark the wedge angle.

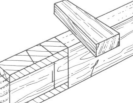

8. The mortise shape.

9. Gauge the mortise thickness.

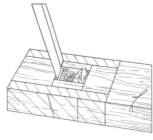

10. Cut the slanting side.

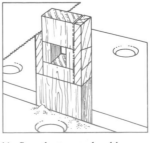

11. Saw the tenon shoulders.

Making a projected and wedged mortise and tenon

Mark the position of the mortise on the truss. The mortise width must be equal to the thickness of the tenon piece. Square these lines on both sides. Mark the tenon shoulders. The width of the tenon should be one third the width of the truss. Position the edge of the tenon against the edge of the truss (1). Measure the remaining width and divide by two. This is the distance of the tenon from each edge (2). Mark the mortise on the other side of the truss using the gauge (3). Bore out the waste from the mortise with a brace and bit (4). Complete the mortise using a chisel (5). Mark the thickness of the truss on the tenon (6). Then mark the tenon mortise slightly in from this mark. Mark the depth of the tenon mortise, so, at its deepest, it is about half the protruding tenon length. Cut a wedge for the tenon mortise. The wedge length should be about four times the thickness of the tenon. Mark the profile of the mortise by holding the wedge with its straight side tight against the first mortise mark. Then mark along the tapered side (7). The mortise and wedge shapes (8). Gauge lines for the thickness of the mortise on both sides of the tenon (9). Remove the waste with a brace and bit and finish with a chisel. Cut the straight side first, then the slanting side (10). Saw the tenon shoulders (11). Lightly chamfer the edges of the tenon and the wedge. Use this joint for collapsible structures, so do not apply glue.

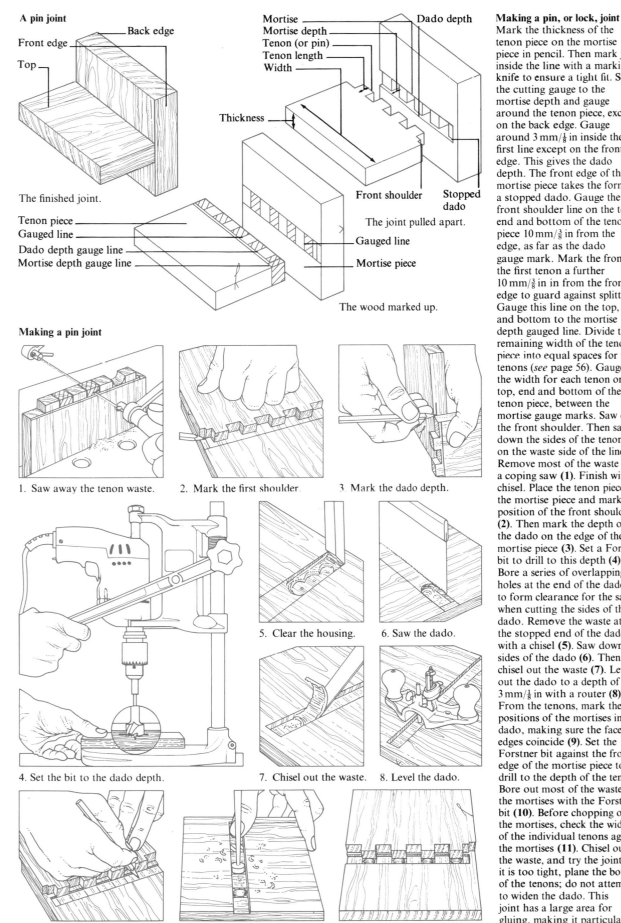

A pin joint

Front edge
Back edge
Top

The finished joint.

Tenon piece
Gauged line
Dado depth gauge line
Mortise depth gauge line

Mortise
Mortise depth
Tenon (or pin)
Tenon length
Width
Dado depth

Thickness

Front shoulder
Stopped dado

The joint pulled apart.

Gauged line
Mortise piece

The wood marked up.

Making a pin joint

1. Saw away the tenon waste.

2. Mark the first shoulder.

3. Mark the dado depth.

4. Set the bit to the dado depth.

5. Clear the housing.

6. Saw the dado.

7. Chisel out the waste.

8. Level the dado.

9. Mark the mortises.

10. Bore out the mortise waste.

11. Check the tenon width.

Making a pin, or lock, joint
Mark the thickness of the tenon piece on the mortise piece in pencil. Then mark just inside the line with a marking knife to ensure a tight fit. Set the cutting gauge to the mortise depth and gauge around the tenon piece, except on the back edge. Gauge around 3 mm/⅛ in inside the first line except on the front edge. This gives the dado depth. The front edge of the mortise piece takes the form of a stopped dado. Gauge the front shoulder line on the top, end and bottom of the tenon piece 10 mm/⅜ in from the edge, as far as the dado gauge mark. Mark the front of the first tenon a further 10 mm/⅜ in in from the front edge to guard against splitting. Gauge this line on the top, end and bottom to the mortise depth gauged line. Divide the remaining width of the tenon piece into equal spaces for the tenons (see page 56). Gauge the width for each tenon on the top, end and bottom of the tenon piece, between the mortise gauge marks. Saw off the front shoulder. Then saw down the sides of the tenons on the waste side of the lines. Remove most of the waste with a coping saw (1). Finish with a chisel. Place the tenon piece on the mortise piece and mark the position of the front shoulder (2). Then mark the depth of the dado on the edge of the mortise piece (3). Set a Forstner bit to drill to this depth (4). Bore a series of overlapping holes at the end of the dado to form clearance for the saw when cutting the sides of the dado. Remove the waste at the stopped end of the dado with a chisel (5). Saw down the sides of the dado (6). Then chisel out the waste (7). Level out the dado to a depth of 3 mm/⅛ in with a router (8). From the tenons, mark the positions of the mortises in the dado, making sure the face edges coincide (9). Set the Forstner bit against the front edge of the mortise piece to drill to the depth of the tenons. Bore out most of the waste for the mortises with the Forstner bit (10). Before chopping out the mortises, check the width of the individual tenons against the mortises (11). Chisel out the waste, and try the joint. If it is too tight, plane the bottom of the tenons; do not attempt to widen the dado. This joint has a large area for gluing, making it particularly strong and firm.

127

Mortise & tenon 5

Power tools are particularly useful if a number of joints of the same size are to be cut. To speed up the marking-out process, clamp the pieces together and square the marks across. Once the tools have been set up, the joints can be cut quickly and accurately. The hollow square mortise chisel and bit, shown here in a power drill fitted in a special mortising stand, can be used in many different power tools, from factory production mortisers to drill presses and portable drills. Not all drill manufacturers make mortising stands.

Mortising stand for power drill

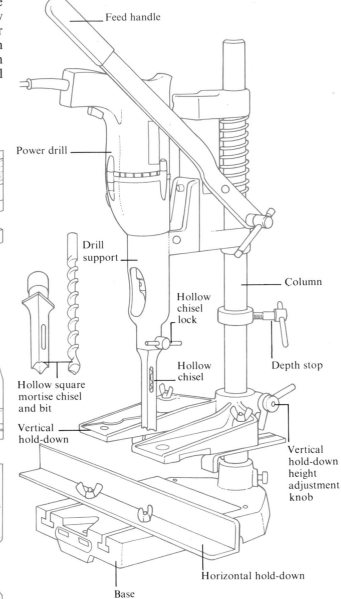

Feed handle

Power drill

Drill support

Column

Hollow chisel lock

Hollow chisel

Depth stop

Hollow square mortise chisel and bit

Vertical hold-down

Vertical hold-down height adjustment knob

Horizontal hold-down

Base

Using a hollow square mortise chisel and bit

The bit, which is similar to an auger, fits inside the hollow chisel. It bores just ahead of the chisel, which squares up the hole as it is lowered into the work; it does not revolve. The chisel and bit cut mortises ranging from $6\,mm/\frac{1}{4}$ in to $13\,mm/\frac{1}{2}$ in. The mortise is cut as a series of overlapping square holes from one end to the other. First adjust the mortiser for depth of cut. A through mortise should be cut halfway through from each side. Position the work with the chisel lined up to make the first cut at one end of the mortise (**1**). Set the hold-downs so it is possible to slide the work under them. Start the drill. Lower the chisel and bit. Cut the mortise (**2**).

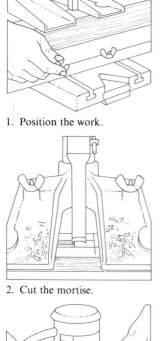

1. Position the work.

2. Cut the mortise.

Using a power router

Mortises can be cut quickly using a power router and square cutter. The wood thickness must be no more than twice the maximum depth of the cutter for a through mortise. The depth of a blind mortise is restricted by the cutter depth. Set up guides to restrict the side to side travel of the router. Then switch on and cut the mortise.

Using a power router.

Cutting tenons

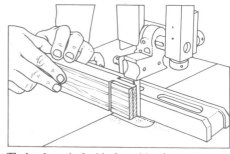

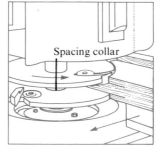

Spacing collar

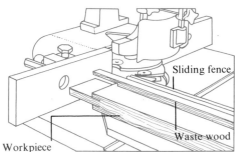

Sliding fence

Waste wood

Workpiece

The band saw is the ideal machine for cutting tenons. To cut a series of tenons quickly and accurately, set the guide fence and make all the cross grain shoulder cuts first. Readjust the fence and then make all the cuts along the grain for the tenon cheeks at the same time.

A disk cutter on a spindle molder makes deep cuts. Form a series of tenons using two disk cutters separated by spacing collars adjusted to the tenon thickness.

Set the disk cutters and make a sample tenon; check for fit in the mortise. When the cutters are correctly set, back each tenon piece with waste wood to prevent the wood from splintering. Cut each tenon, supporting the wood on the end grain extension table.

Comb

A comb joint

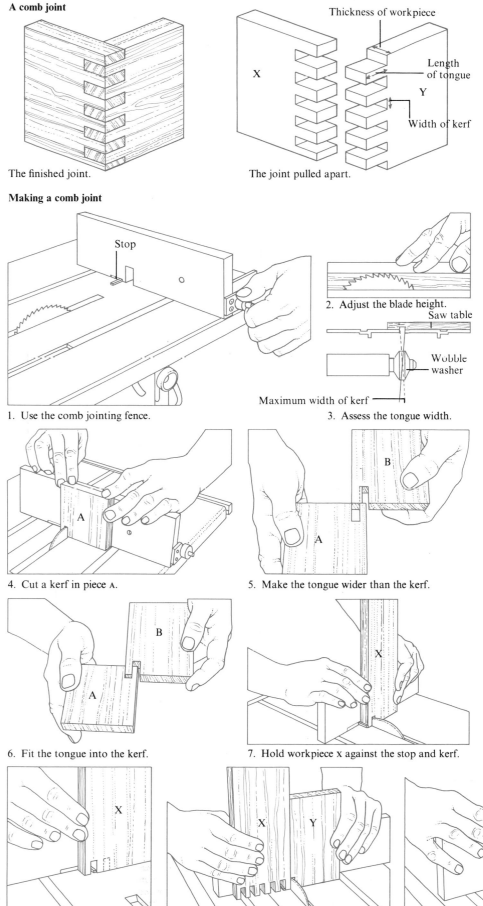

The finished joint.

Thickness of workpiece

X

Length of tongue

Y

Width of kerf

The joint pulled apart.

Making a comb joint

Stop

1. Use the comb jointing fence.

2. Adjust the blade height.

Saw table

Wobble washer

Maximum width of kerf

3. Assess the tongue width.

B

A

4. Cut a kerf in piece A.

5. Make the tongue wider than the kerf.

B

A

6. Fit the tongue into the kerf.

X

7. Hold workpiece X against the stop and kerf.

Making a comb joint
Because this joint is cut on the table saw, the pieces are all exactly the same shape. The large gluing area gives the joint its strength. The comb jointing fence must be used. This has a kerf to clear the blade and a metal stop that is finely adjusted by a knob at one side (1). It is attached to the miter fence, which is set at 90°. The tongues should be square so their length is equal to the thickness of the workpiece. Set the blade height to this thickness (2). Fix wobble washers to the blade to enable it to make a wide kerf. The tongues are the same width as the spaces between them. This distance must be no more than the maximum width of the kerf (3). Use two pieces of waste wood (A and B) that are the same thickness as the work-pieces (X and Y) to determine the width of the tongues. Adjust the stop on the comb fence, and make a kerf on piece A so that the tongue left on the edge is narrower than the kerf (4). Adjust the stop, and make a kerf on piece B so that the tongue is wider than the kerf (5). Then gradually diminish the tongue on piece B until it makes a sliding fit into the kerf on piece A (6). Do this by bringing the stop in closer to the blade and taking fine cuts from the tongue. Lock the comb fence in position when the fit is right. Then take work-piece X. Hold it face side towards the blade and, with its edge tight against the stop, make a kerf (7). Lift the piece over the stop and, with the edge of the kerf tight against the stop, make a second kerf (8). Repeat this process across the edge of piece X. Reverse piece X so its face side rests against the fence. Place the last kerf against the stop. Set work-piece Y face side towards the fence, and hold it tight against piece X. Make the first kerf in piece Y (9). Remove piece X. Continue making kerfs across the edge of piece Y to complete the joint (10).

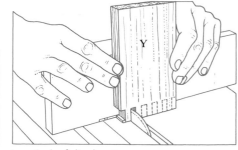

X

X Y

Y

8. Make a second kerf.

9. Make a kerf in piece Y.

10. Continue to make kerfs in piece Y.

Dovetail 1

A carpenters' through multiple dovetail joint

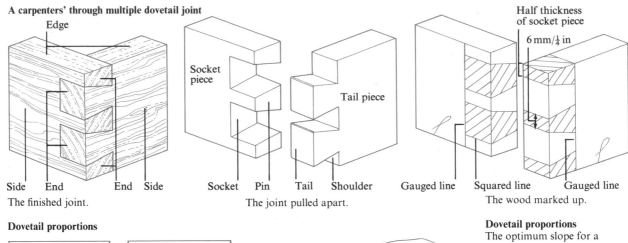

Edge

Side   End   End   Side

The finished joint.

Socket piece

Tail piece

Socket   Pin   Tail   Shoulder

The joint pulled apart.

Half thickness of socket piece

6 mm/¼ in

Gauged line   Squared line   Gauged line

The wood marked up.

Dovetail proportions

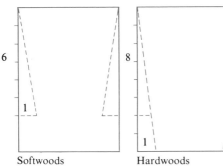

6

1

Softwoods

8

1

Hardwoods

Using a dovetail template.

Dovetail proportions

The optimum slope for a dovetail is 1 in 6 for softwoods and 1 in 8 for hardwoods. Softwood cells compress more easily and so require a steeper slope. Too steep a slope produces a sharp dovetail, leaving weak short grain at the corners. Too shallow a slope would be ineffective as the joint could be pulled apart. For repeated marking, make a dovetail template from hardwood or aluminum.

Making a carpenters' through multiple dovetail

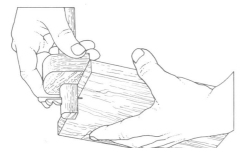

1. Set the gauge to the wood thickness.

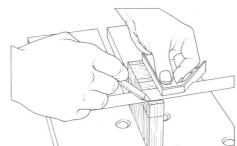

2. Set out the tails.

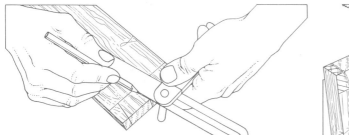

3. Mark the tail slopes.

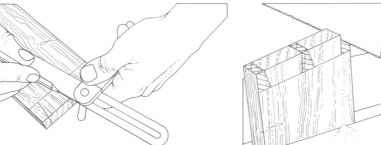

4. Saw the tails.

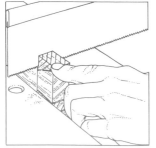

5. Cut the shoulders carefully.

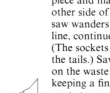

6. Remove most of the waste.

Making a carpenters' through multiple dovetail

Set the cutting gauge to the wood thickness (**1**). Gauge the thickness of the socket piece all around the tail piece, and the thickness of the tail piece on both sides of the socket piece. On each end of the tail piece mark in from the edge a distance equal to half the thickness of the socket piece. Measure the distance between the lines marked on the end as there should be a 6 mm/¼ in gap to accommodate a pin between each tail. Subtract 6 mm/¼ in for each pin. Divide the remainder by the required number of tails to give the width of the tails (**2**). Mark the tail slopes, using a dovetail template or sliding bevel (**3**). Secure the wood in the vise at a slant so the proposed saw-cuts are vertical. Saw on the waste side of the lines. Then change the angle of the tail piece and make the cuts on the other side of the tails (**4**). If the saw wanders from the marked line, continue in a straight line. (The sockets will be cut to fit the tails.) Saw off the shoulders on the waste side of the line, keeping a finger against the waste to prevent it from suddenly falling away and causing the saw to slip (**5**). Remove most of the waste between the tails quickly and neatly with a coping saw (**6**). Chisel out the remainder of the

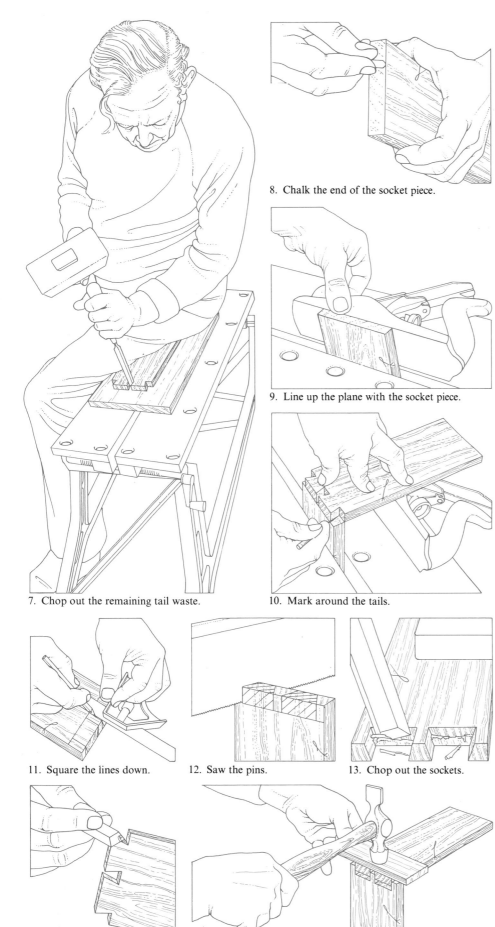

7. Chop out the remaining tail waste.

8. Chalk the end of the socket piece.

9. Line up the plane with the socket piece.

10. Mark around the tails.

11. Square the lines down.

12. Saw the pins.

13. Chop out the sockets.

14. Remove the sharp corners.

15. Tap the joint home.

waste. Using as wide a chisel as possible, chop halfway through the tail piece, first from one side and then from the other. Chop near the line first, chisel upright, bevel outwards. Then chop precisely on the line. When only a few tails are being made, a quick and simple method is to hold the wood by sitting on it (7). Mark the sockets directly from the tails. Use the following method whenever possible as it is simple, quick and accurate. Rub chalk all over the end of the socket piece so that any scribed marks will show up clearly (8). Lay a plane on its side close to the vise and secure the socket piece in the vise with the end level with the top of the plane (9). Tighten the vise and push the plane away. Rest the tail piece on the plane and the end of the socket piece, keeping the edges flush and the shoulders in line with the inside surface of the socket piece. Mark round the tails with a scriber (10). Square the marked lines down to the cutting gauge mark (11). Place the socket piece back in the vise, allowing it to project only slightly to minimize vibration. Make vertical sawcuts just inside the scribed lines (12). Remove most of the waste with a coping saw. Finish by chopping out the sockets with a chisel held at an angle, following the slope of the tail (13). Chisel halfway from each side. Remove the inside sharp corners of the tails with the chisel to give the sockets a lead-in (14). Dovetail joints fit correctly only once, so try the joint by tapping only partly into position. Remove, smooth the inside surfaces and glue both pieces. Then, using a piece of waste wood to spread the pressure, hammer the joint home (15).

Dovetail 2

A cabinetmakers' dovetail joint

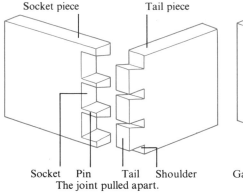

Edge

Side   End   End   Side
The finished joint.

Socket piece   Tail piece

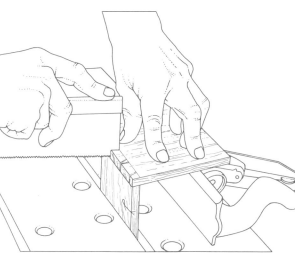

Socket   Pin   Tail   Shoulder
The joint pulled apart.

Squared line

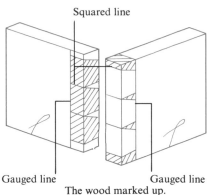

Gauged line   Gauged line
The wood marked up.

Making a cabinetmakers' dovetail

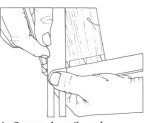

1. Square the tail marks.

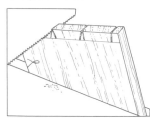

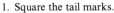

2. Make the vertical sawcuts.

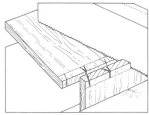

4. Saw the socket piece.

3. Draw the saw through the kerfs.

5. Clean out the waste with the saw tip.

Experienced cabinetmakers' method

1. Hold the saw at an angle.

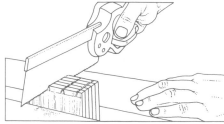

2. Saw off the shoulders.

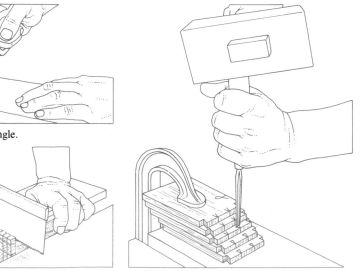

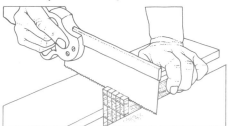

3. Chop out the tails.

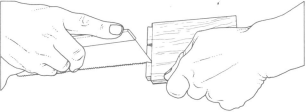

Making a cabinetmakers' dovetail

Set the cutting gauge to the thickness of the tail piece and gauge around the socket piece. Then set it to the thickness of the socket piece and gauge around the tail piece. Rub chalk all over the tail-piece end grain; then mark and square across 6 mm/¼ in from each end. Divide the distance between by the number of tails required. Square across and mark in the tail slopes (**1**). Make vertical sawcuts on the tail piece (**2**). Chalk the end of the socket piece and set it upright in the vise. Rest the tail piece on the plane and the edge of the socket piece. Draw a dovetail saw through each of the kerfs (**3**). Square the marks down the face side. Align the tail piece with the socket piece as a guide for an accurate cut; then saw the sockets (**4**). Remove the waste with a coping saw and finish cutting the sockets with a chisel. Use a 3 mm/⅛ in bevel-edged chisel to chop out the tails. Clean out the waste using the chisel and the tip of the dovetail saw (**5**).

Experienced cabinetmakers' method

Mark up the tails on all the relevant pieces. Put all the tail pieces together upright in the vise. Make all the sawcuts for one side of the tails first, holding the saw at the angle of the tail slope (**1**). Then make all the cuts for the other side of the tails. Chalk and mark the socket pieces, pairing them with individual tail pieces. Clamp the tail pieces back in the vise and saw off the shoulders (**2**). Stagger the tail pieces one on top of the other, and clamp. Chop the tails halfway through on each piece. Turn the wood over. Stagger and clamp the pieces and complete the tails (**3**). Stagger and clamp the socket pieces similarly and chop out the pins and sockets.

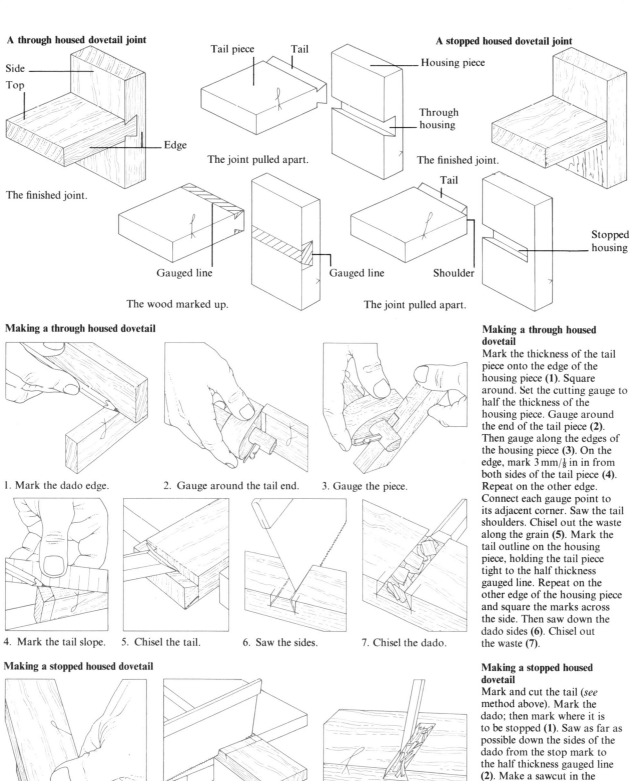

A through housed dovetail joint

Side

Top

Edge

The finished joint.

Tail piece — Tail

The joint pulled apart.

Gauged line

The wood marked up.

Gauged line

A stopped housed dovetail joint

Housing piece

Through housing

The finished joint.

Tail

Shoulder

Stopped housing

The joint pulled apart.

Making a through housed dovetail

1. Mark the dado edge.

2. Gauge around the tail end.

3. Gauge the piece.

4. Mark the tail slope.

5. Chisel the tail.

6. Saw the sides.

7. Chisel the dado.

Making a stopped housed dovetail

1. Mark the stopped dado.

2. Saw down the dado sides.

3. Chisel out the dado.

4. Mark the tail length.

5. Saw off the waste.

Making a through housed dovetail

Mark the thickness of the tail piece onto the edge of the housing piece (1). Square around. Set the cutting gauge to half the thickness of the housing piece. Gauge around the end of the tail piece (2). Then gauge along the edges of the housing piece (3). On the edge, mark 3 mm/⅛ in in from both sides of the tail piece (4). Repeat on the other edge. Connect each gauge point to its adjacent corner. Saw the tail shoulders. Chisel out the waste along the grain (5). Mark the tail outline on the housing piece, holding the tail piece tight to the half thickness gauged line. Repeat on the other edge of the housing piece and square the marks across the side. Then saw down the dado sides (6). Chisel out the waste (7).

Making a stopped housed dovetail

Mark and cut the tail (*see* method above). Mark the dado; then mark where it is to be stopped (1). Saw as far as possible down the sides of the dado from the stop mark to the half thickness gauged line (2). Make a sawcut in the center of the waste. Chisel away the waste horizontally, going no deeper than the saw cuts. (Alternatively bore a series of overlapping holes along the housing, using a Forstner bit set to the half-thickness gauged line.) To clear the waste from the dado, start chiseling vertically in the middle; work towards the sides (3). Mark the dado length onto the tail (4). Saw off the waste at the tail shoulder (5). Saw across the grain first.

Dovetail 3

A through dovetail joint with miter

Edge

Miter shoulder

Tail piece

Socket piece

6 mm/¼ in

Side

Side

Tail
Shoulder
Socket
Pin

End

Gauged line
6 mm/¼ in
Squared line

Gauged line

The finished joint.

The joint pulled apart.

The wood marked up.

Making a through dovetail with miter

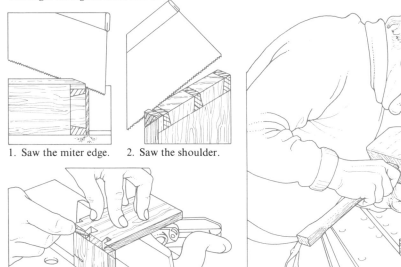

1. Saw the miter edge. 2. Saw the shoulder.

3. Mark sockets from the tail piece.

4. Finish the miter with a chisel.

Making a through dovetail with miter

Both pieces must be the same thickness. Set the gauge to this thickness and mark the sides and bottom edge of the tail piece. Gauge both sides of the socket piece. Mark the miter on the face edge of both pieces with a marking knife. Mark 6 mm/¼ in in from the face edge of the tail piece for the miter shoulder and then another 6 mm/¼ in in. Mark 6 mm/¼ in in from the bottom edge. Set out the tails between these last two marks. Saw the miter edge on the tail piece near the miter (**1**). Finish with a chisel. Then saw down to the miter shoulder (**2**). Finish with a chisel. Saw and remove the waste between the tails. Mark the sockets from the tail piece (**3**). Saw the miter. Finish with a chisel (**4**). Round the miter edge with a plane and abrasive.

A through dovetail joint with rabbet

Edge

Rabbet

Tail piece

Stop

Tail

Socket piece

Shoulder

Socket

Pin

End

Rabbet depth

6 mm/¼ in

Side

Side

The finished joint.

The joint pulled apart.

Gauged line

Gauged line

The wood marked up.

Squared line

A through dovetail joint with rabbet

Leave a projection on the shoulder of the tail piece to act as a stop for the rabbet on the socket piece. Cut the rabbets first. Mark in the depth of the rabbet on the edge of the tail piece. Mark 6 mm/¼ in in from the rabbet for the first tail. Mark 6 mm/¼ in in from the bottom edge for the last tail. Set out the tails between these marks and make sawcuts between the tails. Remove the waste between the tails. Chalk the end grain of the socket piece. Mark and chop out the sockets (*see* pages 130–1). Cut the rabbet shoulder and the shoulder on the bottom edge of the tail piece.

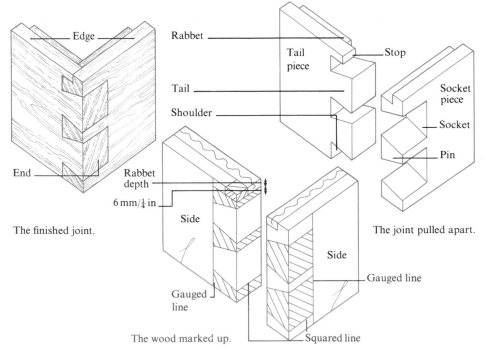

A decorative through dovetail joint

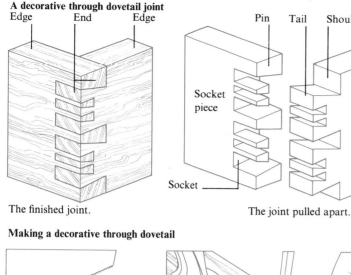

Edge   End   Edge

Pin Tail Shoulder

Socket piece

Tail piece

Socket

The finished joint.

The joint pulled apart.

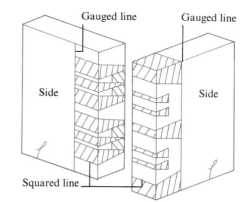

Gauged line   Gauged line

Side   Side

Squared line

The wood marked up.

Making a decorative through dovetail

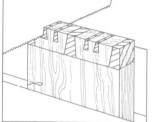

1. Saw the dovetail slope.

2. Chop out the waste.

3. Clean out the corners.

4. Gauge the end of the socket piece.

5. Mark the pins.

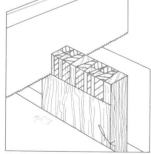

6. Saw between the pins.

7. Remove the waste.

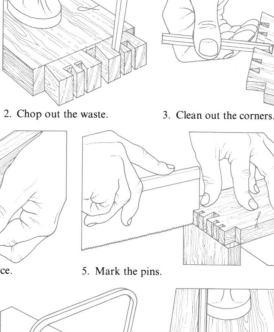

8. Form the small pins.

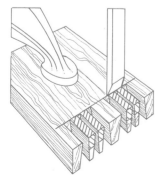

9. Chisel to the cut fibers.

10. Work down to gauged line.

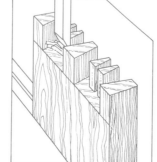

11. Chop two pins at once.

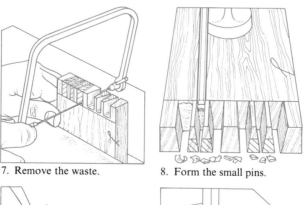

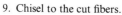

Making a decorative through dovetail

Set the cutting gauge to the thickness of the wood and mark around the tail piece and along both sides of the socket piece. (If both pieces are not of the same thickness, set out the thickness of the tail piece on the socket piece and vice versa.) Mark 6 mm/$\frac{1}{4}$ in in from each edge for the shoulders. Set the gauge to half the thickness of the wood and gauge across the sides of the tail piece to give the length of the small tails. Set out the tails. The small tails need not be spaced equally within the large tails; they should be finely proportioned — more like cabinetmakers' dovetails than carpenters' dovetails. With the wood upright in the vise, saw down the slopes of the tails, making all the cuts on one side first (1). Saw off the shoulders. Remove the waste between the large tails with a coping saw. Chisel out any remaining waste between the tails (2). Clean out the corners (3). Chalk the end of the socket piece. Gauge half the thickness of the wood on the end of the socket piece to mark the depth of the small pins (4). Mark the pins from the tails, using a saw (5). Hatch and crosshatch the waste areas carefully. Make vertical saw-cuts between the pins (6). Remove the waste with a coping saw (7). Form the small pins by chopping across the grain (8). Then chisel as far as the severed fibers (9). Gradually work down to the half-thickness gauged line (10). Using a wide chisel, chop out two pins at once for a cleaner line (11).

Dovetail 4

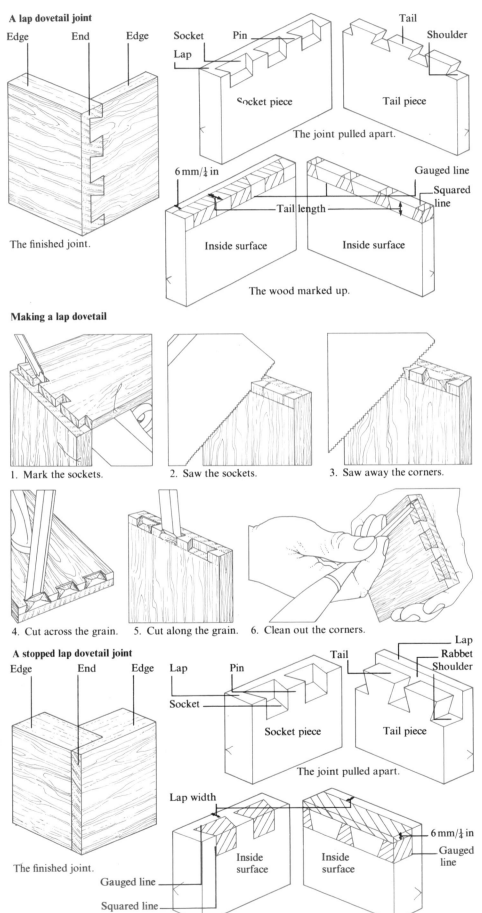

A lap dovetail joint

Edge · End · Edge

The finished joint.

Socket · Pin · Lap

Socket piece

Tail · Shoulder

Tail piece

The joint pulled apart.

6 mm/¼ in

Gauged line

Squared line

Tail length

Inside surface

Inside surface

The wood marked up.

Making a lap dovetail

1. Mark the sockets.

2. Saw the sockets.

3. Saw away the corners.

4. Cut across the grain.

5. Cut along the grain.

6. Clean out the corners.

A stopped lap dovetail joint

Edge · End · Edge

The finished joint.

Lap · Pin

Socket

Socket piece

Tail

Lap · Rabbet · Shoulder

Tail piece

The joint pulled apart.

Lap width

Inside surface

Inside surface

6 mm/¼ in

Gauged line

Gauged line

Squared line

The wood marked up.

Making a lap dovetail

The tails are hidden by being cut shorter than usual and by being overlapped by the socket piece. First decide on the length of the tails. Long tails have a stronger hold, but if they are too long, only a thin lap will be possible on the socket piece. As a general rule, leave about 6 mm/¼ in for the lap. Set the gauge to the tail length and gauge the end of the socket piece from the inside surface. Gauge all around the tail piece. Mark and cut the tails (*see* pages 130–1). Chalk the end grain of the socket piece up to the gauged line. Mark the sockets from the tails (**1**). With the wood upright in the vise, saw the sockets tight against the line (**2**). Stop when the saw touches the gauged lines on the inside surface and the lap. Saw away the corners of the waste to enable it to come away more easily (**3**). Place the socket piece flat on the bench and remove the waste, chopping across the grain (**4**). Then chop along the grain as far as the severed fibers (**5**). Avoid the corners when chopping along the grain, otherwise the wood may split. Clean the corners out carefully, using a 6 mm/¼ in bevel-edged chisel (**6**).

Making a stopped lap dovetail

The tail piece can overlap the socket piece, as shown, or vice versa. (If the socket piece overlaps, the tails must be marked from the sockets.) Decide on the width of the lap (*see* method above). Gauge this width on the end of both pieces from the face sides. From the tail piece end, gauge the rabbet, with the same setting as the lap width, around the inside surface and both edges. Set the gauge to the thickness of the socket piece and gauge the inside surface of the tail piece. To form the rabbet, saw across the grain (**1**). Then chisel down the grain. Clean up the rabbet with a shoulder plane, working from both edges to the center (**2**). Mark 6 mm/¼ in in from each edge of

Making a stopped lap dovetail

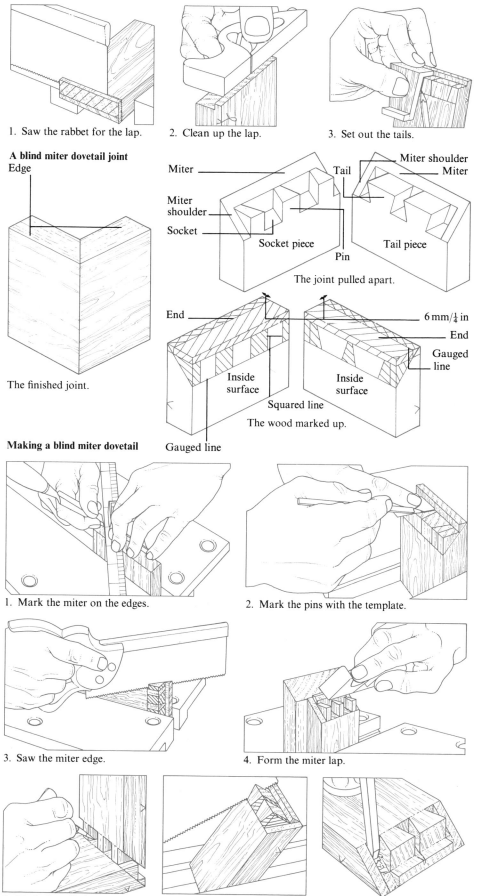

1. Saw the rabbet for the lap.

2. Clean up the lap.

3. Set out the tails.

A blind miter dovetail joint

Edge

The finished joint.

Miter

Miter shoulder

Socket

Socket piece

Pin

Tail

Miter shoulder

Miter

Tail piece

The joint pulled apart.

End

Inside surface

Squared line

Gauged line

6 mm/¼ in

End

Gauged line

Inside surface

The wood marked up.

Making a blind miter dovetail

1. Mark the miter on the edges.

2. Mark the pins with the template.

3. Saw the miter edge.

4. Form the miter lap.

5. Mark the tails from the pins.

6. Saw the miter.

7. Chop out the waste.

the tail piece and set out the tails between these marks (**3**). Remove the waste at the shoulders and between the tails in the same way as chopping out the sockets in the lap dovetail joint. Chalk the end grain of the socket piece and mark the sockets from the tails. Saw and chisel away the waste, cutting the corners away first to ease the work of chiseling the wood.

Making a blind miter, or secret, dovetail

Both pieces must be the same thickness. Set the cutting gauge to the thickness of the wood and gauge the inside surfaces of both pieces from the end. To mark the rabbets, set the gauge to 6 mm/¼ in and, on both pieces, gauge the end from the face side and the inside surface from the end. Mark the miter on the edges of both pieces, using a knife or chisel (**1**). Square the rabbet line down to the miter line on both edges. Cut both rabbets (*see* method above). For this joint it is necessary to form the pins and sockets before the tails. Chalk the end of the socket piece. Mark lines 6 mm/¼ in in from the edge of the rabbet on the socket piece, parallel to the edge. Then mark in a further 6 mm/¼ in. Mark the pins using a cardboard template the width of the widest part of the tails and with the required slope on each side (**2**). For marking in a rabbet, a cardboard template is more convenient than a wooden one, which would be awkward, and the sliding bevel could not be adjusted close enough. Square the marks down the inside surface. Saw the miter edges on each side (**3**). Then saw down to the miter shoulders. Chop out the sockets in the same way as in the lap dovetail joint. Place a backing piece, cut and planed to 45°, behind the socket piece and align it with the top of the miter shoulder. Secure it in position so it will support the shoulder plane as you form the miter lap (**4**). (The backing piece is needed because the miter area is so small.) Mark the tails from the pins on the inside surface of the tail piece (**5**). Square the lines into the rabbet. Saw the miter (**6**). Then chop out the waste between the tails and the waste between the miter shoulders and the tails (**7**). Plane the miter lap.

Dovetail 5

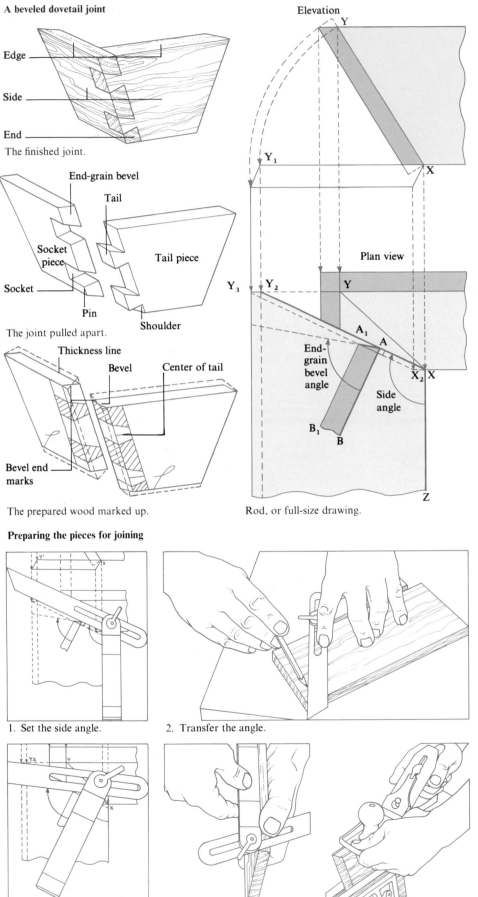

A beveled dovetail joint

Edge

Side

End

The finished joint.

End-grain bevel

Tail

Socket
piece

Socket

Pin

Tail piece

Shoulder

The joint pulled apart.

Thickness line

Bevel

Center of tail

Bevel end
marks

The prepared wood marked up.

Elevation

Y

Y₁

X

Plan view

Y₃ Y₂

Y

A₁

A

End-
grain
bevel
angle

X₂ X

Side
angle

B₁

B

Z

Rod, or full-size drawing.

Making a beveled dovetail
Before this joint can be made, the true shape of the end must be obtained from a rod, or full-size drawing, because the sides as seen in elevation and in plan are sloping and are therefore foreshortened. Draw the elevation showing the slope and the thickness of the wood. Directly underneath, draw the plan view. If the side elevation is pivoted to lie flat, the true shape is seen when looking straight down on it. On paper in the elevation, pivot Y around X to Y₁. Project Y₁ vertically to Y₂, that is, until the line is level with Y in the plan view. Join X in the plan to Y₂. The angle between XY₂ and XZ is the angle of the sides. To work out the end-grain bevel angle, draw the outside edge of the side on the plan (Y₃X₂). From any point on line XY₂, draw a line AB at right angles. Draw a line A₁B₁ parallel to AB, the thickness of the wood away, reaching to line Y₃X₂. Join A to A₁. The angle between AB and AA₁ represents the end-grain bevel angle.

Preparing the pieces for joining

1. Set the side angle.

2. Transfer the angle.

3. Set the end-grain bevel.

4. Mark this on the edges.

5. Plane the end grain.

Preparing the pieces for joining
The pieces of wood must be the same thickness. Set the side angle using a sliding bevel against the rod (**1**). Transfer the angle to both pieces of wood (**2**). Saw and plane the ends. Check the angles with the sliding bevel. Set the sliding bevel to the end-grain bevel angle against the rod (**3**). Mark this angle on the top and bottom edges of each piece (**4**). Connect up the lines along the inner and outer sides of each piece. Plane the end grain to these lines (**5**). Check with the sliding bevel.

Cutting the tails and sockets

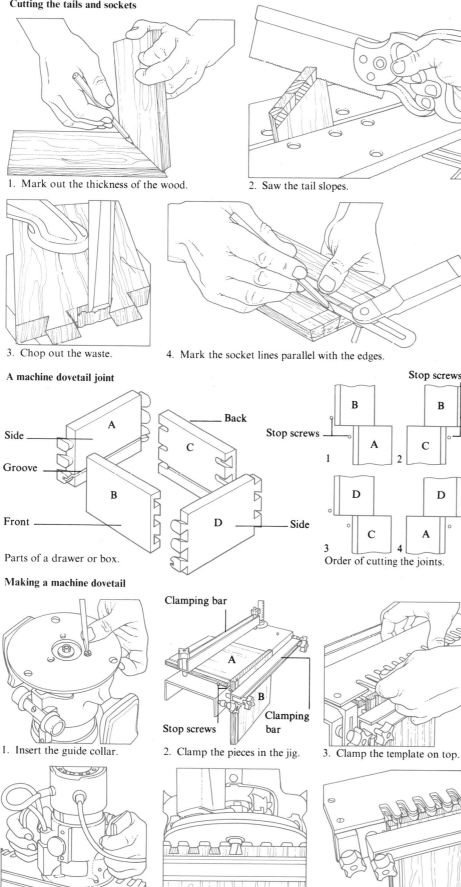

1. Mark out the thickness of the wood.

2. Saw the tail slopes.

3. Chop out the waste.

4. Mark the socket lines parallel with the edges.

A machine dovetail joint

Side

Groove

Front

A

B

Back

C

D — Side

Parts of a drawer or box.

Stop screws

Stop screws

B		B	
A		C	
1		2	

D		D	
C		A	
3		4	

Order of cutting the joints.

Making a machine dovetail

Clamping bar

A

B

Stop screws

Clamping bar

1. Insert the guide collar.

2. Clamp the pieces in the jig.

3. Clamp the template on top.

4. Cut the joint.

5. Let the bit cut each tail.

6. The completed cut.

Cutting the tails and sockets

Mark the thickness of the wood on both sides of each piece (1). Set out the centers of the tails parallel to the top edge. Then set out each tail relative to its center line. Place the wood in the vise so one set of tail slopes is vertical. Cut these (2). Change the angle and complete the cuts. Connect up the two thickness lines across the edges. With the thickness lines vertical, saw off the shoulders. Saw out the waste between the tails. Chop out the remaining waste, working to the center from each side (3). Chalk and mark the socket piece from the tails (see page 131). Mark the socket lines on the side parallel with the edges (4). With socket lines vertical, saw away most of the waste. Chop out the sockets. Plane a bevel on the top and bottom edges so they are horizontal when the joint is put together. The angle is the same as the end-grain bevel.

Making a machine dovetail

Through or lap dovetail joints can be cut with the power router fitted with a dovetail bit guided by a jig and template. Both sides of the joint are cut at the same time. The sides are offset against stops in such a way that, as the router follows the template fingers, it cuts the recess between two tails in the vertical piece at the same time as it cuts a socket in the horizontal piece. For a four-sided box or drawer, mark the pieces and make the cuts in the order shown. Always cut a joint in waste pieces first to determine the depth of cut. If the drawer sides are grooved to take the bottom, cut the groove first. Insert the template guide collar into the router sole plate (1). This will ride against the template. Cut and plane the pieces to size. Clamp them with face sides against the jig and the ends flush and abutting against the stop screws on the jig; this will automatically give the required amount of offset (2). Use the stop screws on the left or the right according to which of the four corners are being joined. When the wood is correctly positioned, clamp the template on top (3). To cut the joint, run the router fully around the contour of the template fingers (4). The guide collar runs against the template, while below it the bit makes the tail cut (5). The completed cut (6). Then unclamp the pieces and cut the next joint.

Housed

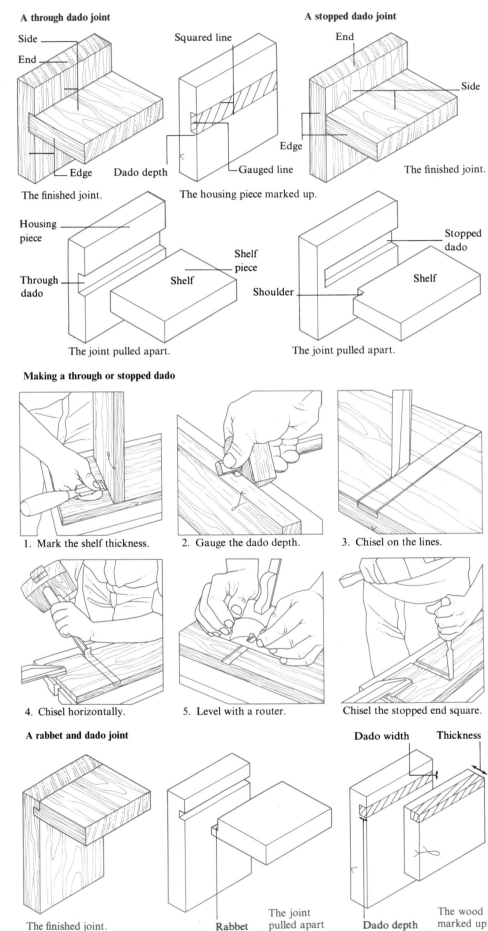

A through dado joint

Side
End
Edge
Dado depth

The finished joint.

Squared line
Gauged line

The housing piece marked up.

A stopped dado joint

End
Side
Edge

The finished joint.

Housing piece
Through dado
Shelf
Shelf piece

The joint pulled apart.

Stopped dado
Shoulder
Shelf

The joint pulled apart.

Making a through or stopped dado

1. Mark the shelf thickness.

2. Gauge the dado depth.

3. Chisel on the lines.

4. Chisel horizontally.

5. Level with a router.

Chisel the stopped end square.

A rabbet and dado joint

The finished joint.

Rabbet

The joint pulled apart.

Dado width   Thickness

Dado depth   The wood marked up

Making a through or stopped dado

Mark the thickness of the shelf piece on the side of the housing piece with a marking knife or a chisel (**1**). Square the lines across the face side and down both edges. Set the marking gauge to one-third the thickness of the housing piece and gauge the dado depth between the squared lines on both edges (**2**). With a bevel-edged chisel and mallet cut across the grain inside the squared lines. Then chisel on the lines (**3**). To provide shoulders for the router to run against, chisel horizontally to a depth of approximately 3 mm/⅛ in (**4**). Work into the center from both edges. Level the dado with a router (**5**). Test the joint for fit; if it is too tight, plane a shaving off the bottom of the shelf piece. (For alternative methods *see* pages 76, 127, 133.)

In a stopped dado joint a shoulder must be cut in the face edge of the shelf piece. Mark the thickness of the shelf on the housing piece in pencil. Square the lines down the rear edges, and gauge the dado depth between these marks. With the gauge at this setting, gauge around the face side, underside and face edge of the shelf. Decide how far in from the edge the dado will be stopped and cut a shoulder at this point on the shelf piece. Re-mark the thickness of the shelf on the housing piece with a marking knife or chisel. Mark the stopped end of the dado. Clear the stopped end by drilling a few overlapping holes with a Forstner bit. Chisel shoulders for the saw to run against. Chisel the stopped end square. Cut the dado in the same way as for the through dado joint.

A rabbet and dado joint

Pencil the thickness of the top on the housing piece. Square the marks down the edges. Mark the dado width to one-third this measurement. Square the lines down the edges. Gauge the dado depth from the inside surface. Cut the dado. Gauge the dado width on the end of the top and gauge the depth on the edges. Saw across the grain down to the gauged marks. Then gradually chisel down to the sawed line. Finish the rabbet with a shoulder plane.

Dowel

A dowel joint

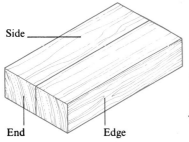

Side

End   Edge

The finished joint.

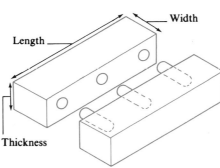

Length   Width

Thickness

The joint pulled apart.

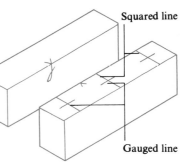

Squared line

Gauged line

The wood marked up.

Making a dowel joint

1. Drive through the plate.

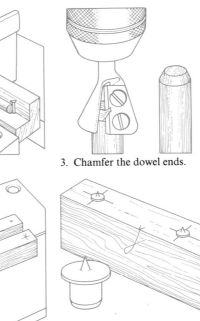

2. Cut equal lengths.

3. Chamfer the dowel ends.

4. Saw a kerf in the dowel.

5. Mark the joint with pins

OR with dowel centers.

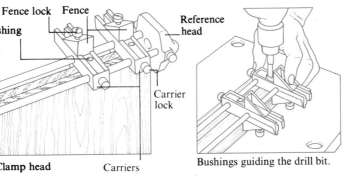

Fence lock   Fence

Drill bushing   Reference head

Adjustable head

Steel rod

Carrier lock

A doweling jig.   Clamp head   Carriers

Bushings guiding the drill bit.

Variations on the dowel joint

Edge-to-end dowel joints are often used as a substitute for mortise and tenons. Dowels can also strengthen miters and join pieces meeting at an awkward angle. Stagger dowels when joining thick pieces of wood. When doweling into end grain, drill the end-grain holes first. When joining at an angle, insert the dowels at right angles to the joint surface for a strong joint. Clamp the wood at an angle so that the holes are drilled vertically wherever possible.

Variations on the dowel joint

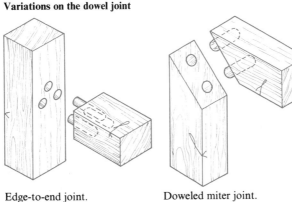

Edge-to-end joint.

Doweled miter joint.

Making a dowel joint

Dowels can be made with a metal dowel plate, which has holes of varying sizes. (Proprietary dowel rods and standard-size dowels are the alternatives.) Roughly chisel a cylindrical shape from square-sectioned beech wood cut to the diameter of the appropriate hole. Cut a lead-in at one end and drive the wood through the hole with a hammer (1). To cut all dowels to the same length first make a jig. Butt the dowel rod against a nail the required distance from a kerf in the back batten. Saw through the kerf (2). Chamfer both ends of the dowel for easy entry into the hole; use a dowel rounder in a brace or power drill (3). Saw a kerf along the dowel to allow air and surplus glue to escape from the hole (4). Dowel holes must be accurately aligned in both pieces. Mark up one piece, squaring the hole positions across and gauging them centrally along the wood. Tap in panel pins and cut off their heads. Align the second piece and tap it down. Remove it, revealing the central impressions (5). Remove the pins. (Alternatively, drill the holes in the first piece, and insert a dowel center into each hole. Align the second piece and tap it down.) All holes in both pieces must be the same depth — the combined depth being slightly greater than the length of the dowel — so use a depth stop. Countersink all dowel holes for easy entry. A doweling jig makes accurate doweling easier. The type shown has two steel rods on which two carriers slide. Select the right sized bushings and insert in the carriers to guide the drill bit. Position and tighten the carriers and fences, with the reference head against one end of the work. Lock the adjustable head against the other end. Drill the holes in one piece. Invert the jig and drill the second piece.

Edge and tongue & groove

A rubbed glue joint

A tongue and groove joint

A slot-screwed joint

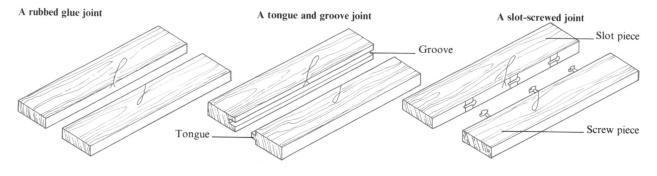

Groove

Tongue

Slot piece

Screw piece

Making a rubbed glue joint

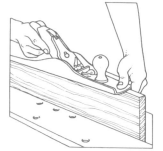

1. Plane the first edge.

2. Check the mating edge.

3. Check the alignment.

4. Glue both edges.

5. Rub the edges together.

6. Check the joint.

Making a tongue and groove joint

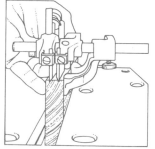

1. Make a cut at one end.

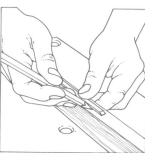

2. Mark the blade position.

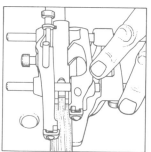

3. Center the tonguing blade.

Making a slot-screwed joint

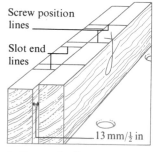

Screw position lines

Slot end lines

13 mm/½ in

1. Square slot lines across.

2. Gauge along the edges.

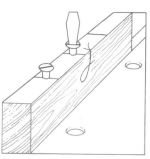

3. Insert the screws.

Making a rubbed glue joint
Hold one piece in the vise and plane the edge with the longest bench plane available (**1**). Try to plane a hollow in the center of the piece. The length of the plane will make this impossible, but the attempt will help to prevent the ends being rounded over. Finish planing by taking one continuous shaving. Check for perfect squareness. Plane the mating edge similarly, checking it continually by moving it against the first edge to feel for points of friction at the ends (**2**). Check the alignment of the sides with the plane edge (**3**). Hold the edges of both pieces in contact, and glue them (**4**). Rub the edges together to remove any air from the joint (**5**). Clamp up the joint. Apply light pressure at first; then tighten, checking the evenness of the joint as the pieces are tapped down flush with a hammer (**6**).

Making a tongue and groove joint
Cut the whole joint with a plow plane. To center the groove, make a cut at one end with the rabbeting blade (**1**). Then reverse the plane and make a second cut. When the cuts coincide, cut the groove. Lay the rabbeting blade centrally on the edge of the tongue piece and mark its position (**2**). Guided by the marks, center the tonguing blade and cut the tongue (**3**).

Making a slot-screwed joint
Select the face sides and mark accordingly. Mark the positions for the screws 44 mm/1¾ in from each end and at intervals of 150 mm/6 in. Place the pieces in the vise with the screw piece offset by 13 mm/½ in. Square screw position lines across both pieces. Square slot lines across the slot piece 13 mm/½ in to the left of the screw position marks (**1**). Gauge half the wood thickness along the marked

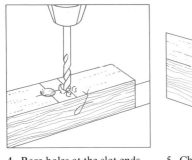

4. Bore holes at the slot ends.

5. Chisel the slots.

6. Tap the ends flush.

edges (2). Insert the screws, leaving the heads projecting by 13 mm/½ in (3). Bore holes to take the screw-heads in the slot piece on the screw position marks, 13 mm/½ in deep. Drill holes to take the shanks at the end of the slot lines, 13 mm/½ in deep (4). With a narrow chisel, cut slots 13 mm/½ in deep (5). Fit the larger holes in the slot piece over the screw-heads and tap the ends flush (6).

A miter joint with spline

A keyed miter joint

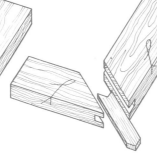

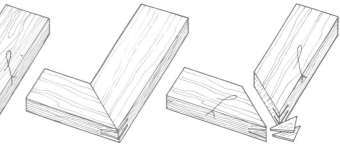

The finished joint.

The joint pulled apart.

The finished joint.

The joint pulled apart.

Making a miter joint with spline

1. Mark the spline width.

2. Measure across the miter.

3. Gauge around the miter.

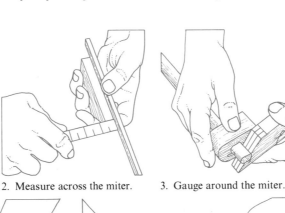

4. Chisel the groove.

5. Mark the waste on the spline.

Making a miter joint with spline
Cut the miter in a miter box. Cut a strip of wood for the spline about one-third the thickness of the mitered pieces and about 16 mm/⅝ in wide. Put the miter pieces together. Lay the spline centrally on top and mark its width (1). Square the marked line parallel to the miter and down both edges. Place the spline on edge on a miter end, flush with one side. Measure across what remains of the miter (2). Divide by two. Set the marking gauge to this measurement and gauge around the miter to the squared lines (3). Saw down the miter on the waste side of the gauged lines to the squared lines. Chisel the groove in the miter end (4). Fit the joint together with the spline in place and mark the waste on the spline (5). Remove the spline; saw it to size. Glue the joint and spline together and leave them to dry.

Making a keyed miter joint

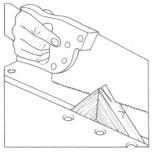

1. Saw two kerfs.

2. Tap the key into the kerfs.

3. Saw the keys off flush.

Making a keyed miter joint
Cut and glue the miter pieces; clamp and leave to dry. Position the miter in the vise so the saw cuts level with the work surface. Cut two kerfs, angling each slightly inwards (1). Cut the key strips from 2 mm/³⁄₃₂ in three-ply or from veneer, depending on the size of the kerf. Glue the key strips and the kerfs and tap the keys into the kerfs (2). Allow to dry. Saw the keys off and plane them flush with the joint (3).

Pivoting

A knuckle joint

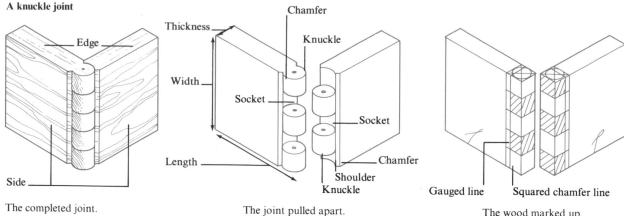

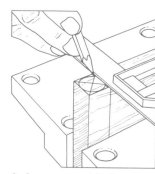
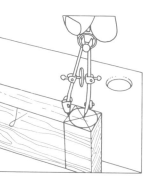

Labels: Edge, Side — The completed joint.

Thickness, Width, Length, Socket, Chamfer, Knuckle, Socket, Chamfer, Shoulder, Knuckle — The joint pulled apart.

Gauged line, Squared chamfer line — The wood marked up.

Making a knuckle joint

1. Draw diagonals.

2. Draw a circle on each edge.

3. Square around both pieces.

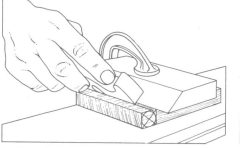

4. Saw the chamfer.

5. Plane the chamfer.

6. Round the knuckle.

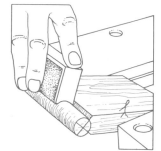

7. Smooth the rounded end.

8. Divide the width equally.

9. Gauge around the knuckles.

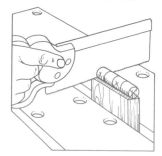
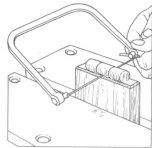
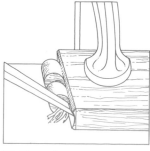

10. Saw between the knuckles.

11. Remove the waste.

12. Shape the curve.

Making a knuckle joint
The pieces must be of equal thickness. Set the cutting gauge to the thickness of the wood; gauge around the sides and edges of both pieces. Draw diagonals on the edges (1). Set a compass to half the wood thickness and draw a circle on each edge (2). Position a try square where the diagonals bisect the circumference of the circle near the gauged line and square around the sides and edges of both pieces to mark the chamfer lines (3). With a dovetail saw, cut along each chamfer line down to the circle (4). Before forming the chamfers, make a 45° backing piece; clamp it behind the chamfer line. With a shoulder plane, form the chamfer flush with the backing piece (5). Partially round the knuckles from the chamfers using the shoulder plane (6). Round the end of the wood to the circle mark using first a smoothing plane and then the shoulder plane. With a gouge make a concave-shaped sanding block to fit the curve. Using the block and a coarse and then a fine abrasive paper, smooth the rounded end of the workpiece (7). Divide the width of the workpiece equally by the total number of pins in the knuckle (in this case, five) (8). Set the gauge for each mark and gauge around the knuckles from the face edge on each piece (9). Hatch in the waste areas. Saw down between the knuckles on the waste side of each gauged line allowing for a filed finish (10). Saw off the shoulders. Remove the waste between the knuckles with a coping saw (11). Clean out the sockets with a chisel. Shape the curve on the shoulders using a chisel with a rocking motion and an in-cannel gouge (12). Finish by paring vertically with an in-cannel gouge up to the knuckle (13). For a smooth action, file

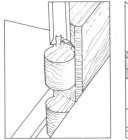

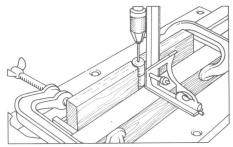

13. Pare the shoulders.　14. File the knuckles.　15. Drill a hole through the joint.

the knuckle surfaces lightly and test the joint for a friction fit (**14**). Complete the rounding of the sockets with a narrow chisel. Clamp the joint. Bore a hole through the center of the joint, working from both sides (**15**). To ensure that the drill remains vertical, position a try square alongside the bit or use a drill stand. Insert a steel pivoting rod through the hole and cut it flush.

A rule joint

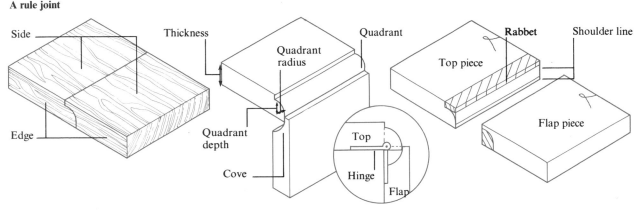
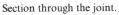

The joint with flap up.　The joint with flap down.　Section through the joint.　The wood marked up.

Making a rule joint

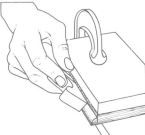
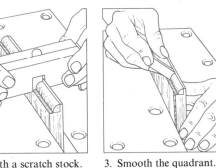

1. Plane to the quadrant mark.　2. Shape with a scratch stock.　3. Smooth the quadrant.

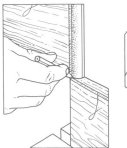

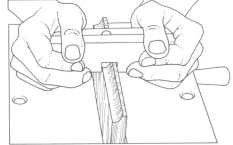

4. Mark the cove.　5. Form the cove.　6. Shape with the scratch stock.

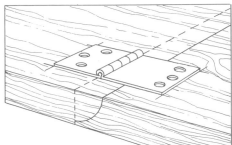
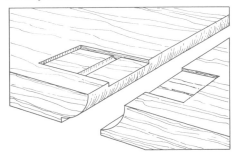

7. Mark around the hinge.　8. Chisel the knuckle recess.

Making a rule joint
Glue waste wood to the edges to prevent splintering. Finger gauge a line 3 mm/⅛ in down from the face side of the top along the end and both edges. Repeat on the underside. The distance between these two lines is the radius of the quadrant. Set a plow plane to this distance and cut rabbets on the face of the top and the underside of the flap. Square the shoulder line down one edge of the top; where it intersects the bottom shoulder line is the circle center. Draw the quadrant with a compass. Clamp a square backing piece flush with the shoulder and plane to the quadrant mark (**1**). Shape the quadrant finally with a scratch stock made to fit (**2**). Smooth using a shaped sanding block (**3**). Mark the cove from the quadrant (**4**). Chamfer the cove edge of the flap against a 45° backing piece. Roughly form the cove profile with an out-cannel gouge (**5**). Check the profiles of the edges together, then finish with the cove cutter in the scratch stock (**6**). Smooth. Chisel and plane off the waste wood. Square the rabbet line around to the underside of the top. Align the hinges, knuckles up and centered on the line. Mark around each hinge (**7**). Chop out flap recesses. Mark and chisel the knuckle recess with a scooping action (**8**).

145

Working manufactured boards 1

The most popular manufactured boards are plywoods (three-ply, multi-ply, battenboard, blockboard and laminboard), chipboard and hardboard. They make maximum use of wood material that might otherwise be wasted, and so they are usually more economical than solid wood. The natural figure in plywoods is generally plain due to the rotary slicing process; chipboard and hardboard have no figure.

The chief advantage of working with manufactured boards is that they do not distort as a result of shrinkage or natural warping. Carcass construction is therefore made considerably simpler because allowance does not have to be made for dimensional change. Distortion in battenboard, blockboard and laminboard is minimal as the plies sandwiching the solid wood strips restrain the tendency of the growth rings to straighten themselves. Any distortion that might occur in three-ply or multi-ply, where each veneer is laid at right angles to the next, is balanced out by its construction. Both chipboard and hardboard have no natural grain direction across which to shrink.

Manufactured boards, however, will warp if they are incorrectly stored. Damp boards usually warp and may even disintegrate. They should be stacked vertically against a dry wall, resting fully on their edges with the top corners supported and protected from damage. Apart from chipboard, manufactured boards can only be used outside when specially treated. Tempered hardboard is impregnated with oil, and marine multiply is resin bonded, making it water resistant.

In addition to their stability, manufactured boards are useful because they are supplied in large sheets.

Cutting

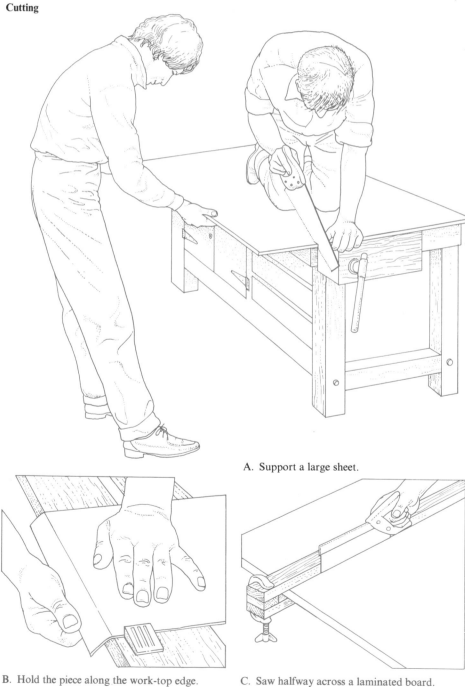

A. Support a large sheet.

B. Hold the piece along the work-top edge.

C. Saw halfway across a laminated board.

Cutting

A power saw is the most effective tool to cut manufactured boards, but, unless laminated, they may also be cut with a panel or a tenon saw in the same way as solid wood. A tenon saw with its fine teeth, held almost horizontally, is preferable for thin boards. Never use a ripsaw. The resin adhesive in marine three-ply and chipboard quickly blunts saw teeth, so for these boards tungsten-carbide-tipped teeth are best. Make sure the saw blade always cuts into the face so any ragged edge that may result will be on the underside. To minimize this ragged edge, stick adhesive tape on the underside along the cutting line. When sawing, support both sides of the board close to the cutting line. To cut a large board, climb onto the workbench and kneel close to the cutting line. Have an assistant support a large overlap to prevent it from snapping off unevenly towards the end (A). Three-ply that is up to 3mm/⅛ in thick can be scored with a cutting gauge and snapped cleanly in two by holding it along the work-top edge against a stop and pressing on the overlapping piece (B). A laminated board can be sawed with a panel or tenon saw, but the leading edges on the teeth should be raked back more than usual. Always saw into the laminate and keep the teeth well clear of the laminate on the backward stroke to prevent it from chipping. To hold a laminated board firmly while sawing, clamp two battens close to the cutting line. Place blocks of the same thickness as the laminated board between the two battens at both ends. Then saw halfway across the board from each side (C).

This makes them particularly suitable for wide, flat components. Hardboard and sheets of three-ply or five-ply are commonly used for paneling; multi-ply is for carcass tops, shelving and doors. For shelving, the solid core strips must run from end to end for greatest support. For doors, the strips should run vertically as screws do not hold in the end grain. Chipboard is most suitable for carcass construction and doors.

The edges of manufactured boards are both unsightly and easily damaged if they are not lipped, either with veneer or solid wood. Nails, screws and pins should be set between 100 mm/4 in and 150 mm/6 in from one another, and should never be fixed closer to the edges than 13 mm/½ in as the edges are liable to break. Use a twist drill for boring small holes in manufactured boards and a flat bit for large holes.

Before drilling plastic laminated surfaces, stick two layers of adhesive tape over the spot to be bored. Then make a pilot hole with a small twist drill.

It is not always necessary to apply a surface finish to manufactured boards because most are ready primed and sealed; many are also laminated or veneered. If not treated, three-ply, multi-ply and blockboard may be finished in any way suitable for solid wood. Both these and chipboard are porous and should be well sealed and may be lightly keyed for veneer.

While three-ply and hardboard are suitable for bending, chipboard and blockboard are rigid. To achieve the maximum curve, three-ply should be steamed and bent across the short grain of the outer veneers. Hardboard is even more flexible when bent with the rough textured side on the outside.

Smoothing

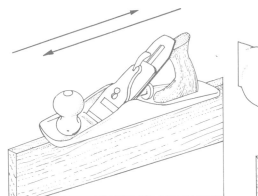

A. Plane the edges from the ends inwards.

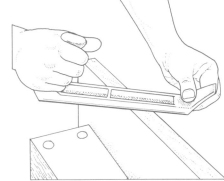

B. Plane laminated board edges at an angle.

Smoothing
The wide surfaces of a manufactured board should not be planed, as they are pre-finished. Plane the edges from the ends inwards (**A**). If the plane runs off the end, chipping will occur. Cut resin-bonded boards to exact size to avoid planing as resin blunts the blades. Smooth chipboard edges with a Surform tool or rasp. Remove the woolly burr left on planed hardboard edges by sandpapering using a block. Smooth the edges of laminated boards away from the laminate and at an angle, with a Surform tool (**B**).

Lipping

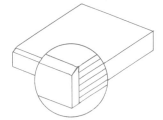

A. Veneer lipping laid second.

B. Veneer lipping laid first.

C. Solid wood lipping.

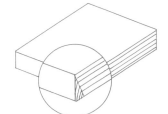

D. Feathered lipping.

E. Grooved multi-ply.

F. Grooved chipboard.

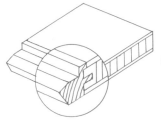

G. Grooved laminboard.

H. Spline lipping.

J. Tongued multi-ply.

Lipping
Where edges have split, reglue them and clamp until dry before lipping. Hardboard is lipped only in combination with solid wood. If veneer lipping is laid after face veneer, the thin lipping edge will show on the face. If the lipping is done first, the face veneer will cover the lipping edge, but there is a risk of the face veneer lifting. To avoid both problems, smooth the top edge of the face veneer to a 45° angle (**A, B**). All solid wood lipping that shows on the face side should be mitered. Lipping should be left proud of wide surfaces until the glue has dried. It can then be smoothed to any shape. Apply self-adhesive strip veneer lipping, using a warm iron. Edges can also be lipped with solid wood (**C**). Feathered lipping does not show on the top surface (**D**). Lipping that is tongued and grooved provides the strongest join. Multi-ply, blockboard and chipboard may be grooved (**E, F, G**). Splines can be used (**H**). It is always best to cut the tongue from solid wood if lipping blockboard or chipboard, though a tongue in multi-ply is quite strong (**J**).

Working manufactured boards 2

Fixing

Three-ply and multi-ply can be fixed with screws or nails against a solid wood frame or another sheet of three-ply or multi-ply. Choose nail and screw sizes to ensure a firm grip in the underneath piece. Nails can be used to secure glue joints. Their heads must be left projecting above the surface so they can easily be withdrawn when the glue is dry (**A**). As well as fixing with nails, veneer pins, panel pins and oval brads can also be used (**B**). When using screws, drill a clearance hole, which can be countersunk (**C**). The most durable method of fixing three-ply and multi-ply is with a combination of glue and nails (**D**) or glue and screws (**E**). Nails and screws must not be fixed into three-ply and multi-ply edges as they will split the plies.

Knock-down fittings

If three-ply or hardboard is used to panel hollow doors, secure a fitting, such as a handle, to the face with a hollow door fitting (**H**). Push the anchor partially through a drilled hole in the face with a hammer; then screw it tight. Once the anchor is in position, its arms will have opened and will be pulled flat against the other side of the face. A chipboard screw (**J**) has a thread profile designed specifically to grip in chipboard without the need for a bushing. The countersunk screw-head will lie flush with the surface. It can be concealed with a cover cap, which is pressed or screwed into a special recess in the head. Use a block joint (**K**) to join chipboard panels at right angles. Screw one part into the chipboard side and screw the other into the side of a perpendicular piece of chipboard. The two dowels in one part fit into the recesses in the other. The parts are then screwed together tightly. Use a modesty block (**L**), which is less obtrusive, for corner joints in light constructions or for strengthening. The knock-down dowel fitting (**M**) is strong and neat. Fit the plastic dowel into a drilled hole in the side of chipboard and fit the bushing into a perpendicular board in the same way. Then tighten a machine screw into the bushing. A strong joint can be made with a cross dowel and guide dowels (**N**). The metal cross dowel has a threaded hole across its axis to accommodate a machine screw.

 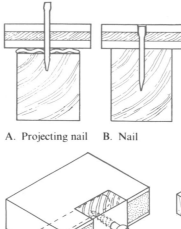  

A. Projecting nail   B. Nail   C. Screw   D. Glue and nail   E. Glue and screw

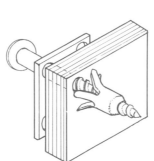 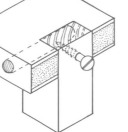

F. Solid wood dowel   G. Plastic bushing

Nails and screws should not be driven between core sections in battenboard, blockboard and laminboard. Nor will nails and screws hold in the edge of chipboard, unless a solid wood dowel is inserted at right angles to the screw or nail, to give it a firm grip (**F**). Alternatively, push a plastic bushing into a chipboard edge to hold a screw firmly (**G**). Hardboard pins and round-head screws give the best hold in hardboard.

Insert the dowel into a housing in edge X. Ensure the threads in the dowel are parallel with edge Y. Bore a hole through the side of the other board and into edge Y and countersink. Pass the machine screw through and into the dowel. Turn it tightly with a screwdriver or an Allen wrench. Use guide dowels to prevent the pieces rotating. Bore holes on edge Y and its mating edge and insert the dowels.

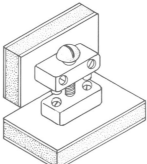 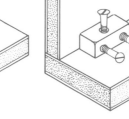

H. Hollow door fitting   J. Chipboard screw

K. Block joint   L. Modesty block

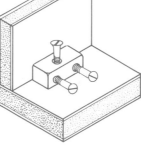

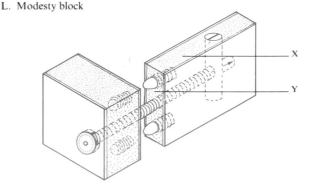

M. Knock-down dowel fitting   N. Cross dowel and guide dowels

Joining multi-ply

A number of solid wood joints can also be cut in multi-ply. These include the rabbet and dado joint (**A**) and all other housed joints; the pin joint (**B**) and the dowel joint (**C**). The halved cross lap joint (**D**) can be cut in the width of the multi-ply. The bare-faced tenon joint (**E**) and bridle joint (**F**) can be cut if the mortise piece is thick enough. The tongue and groove joint (**G**) makes a strong edge joint. Joints such as the through multiple dovetail (**H**), decorative through dovetail (**J**) and stopped lap dovetail (**L**) can be cut, provided the tails and pins are coarse. The comb joint (**K**) can be cut only on a table saw. With the projected and wedged mortise and tenon joint (**M**), the wedge is cut from solid wood and should be four times the thickness of the tenon.

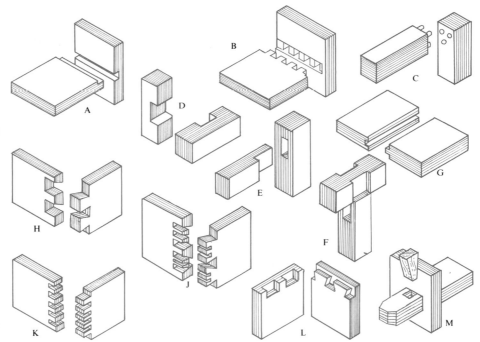

Joining boards other than multi-ply

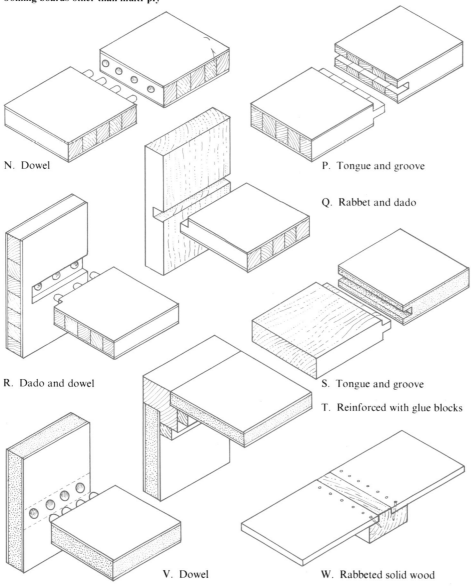

N. Dowel

P. Tongue and groove

Q. Rabbet and dado

R. Dado and dowel

S. Tongue and groove

T. Reinforced with glue blocks

V. Dowel

W. Rabbeted solid wood

Joining boards other than multi-ply

In many respects battenboard, blockboard and laminboard can be joined in the same way as solid wood. Among the most suitable joints are dowel joints (**N**) and dado and dowel joints (**R**). A tongue and groove joint can be cut across solid core strips — never along their length (**P**). Battenboard, blockboard and laminboard can be cut under similar conditions for a tongue but not a groove; a spline can also be used. A rabbet and dado joint cut into solid wood (**Q**) is quite strong. Few of the traditional joints can be cut in chipboard as the fibers tend to crumble. Joints are usually made in conjunction with solid wood, the chipboard being rabbeted into the solid wood and reinforced with glue blocks (**T**). A tongue and groove joint can be used, providing only the groove, and never the tongue, is cut from the chipboard (**S**). The tongue must always be cut from plywoods or solid wood. Grooves in chipboard are best cut with a portable circular saw fitted with wobble washers. A dovetail joint is not strong in chipboard, but dowels give adequate strength (**V**). Since hardboard is generally used for paneling in conjunction with a solid wood frame, hardboard rarely needs joining. Where appearance is not important, a strip of solid wood double rabbeted to take a sheet of hardboard on either side and secured with panel pins can make a strong joint for paneling (**W**).

Carcasses and frames

Every piece of furniture can be classified into one or more of three types of unit — storage, supporting or seating. Each type of unit consists of components assembled into one of three basic constructions — frame, box or stool. From these basic constructions any kind of furniture unit can be made.

Common to all constructions are the allowances for movement of the wood. When a solid board is fixed to another, splitting along their length may occur because wood shrinks across the grain. The development of frame and panel construction allowed for this as the panels were able to move independently of the frame. The direction of movement must be considered when boards have to be joined side edge to side edge, or end to end at right angles as for a corner joint. Wood moves across the grain and the growth rings tend to straighten as the wood dries, causing cupping. Join narrow boards side edge to side edge so that the movement of one board is counteracted by the next. Ideally, join boards with their growth rings at right angles to the wide surface.

Before assembling a carcass, make a preliminary unglued knock-up to ascertain the order of assembly and to check fit. Wherever possible, break the carcass down into subcarcasses each comprising several components, and assemble each unit separately. Assemble and roughly adjust all clamps; prepare any cross-bearers and clamping blocks. Collect the necessary tools: a hammer, a mallet and a striking block of waste wood, a soft pencil, measuring and testing tools, a damp rag for wiping off surplus glue and a quirk stick for cleaning glue out of corners. Mark joints and components with the soft pencil on the face side and number them with the order of assembly. Smooth all inside surfaces and inside all joints with a smoothing plane and abrasive paper. Glue and assemble the components into their various units. Any dowels or splines can be glued into one side only; then check for length before gluing into the other side. Wipe off all surplus glue before it sets. Check for squareness and leave until the glue has set before assembling the units. Remove heavy clamps as soon as possible.

Types of structure

Piece of furniture	Constructional form		
	Frame	Stool	Box
Blanket chest	*	*	
Writing desk		*	*
Sideboard	*	*	*
Table		*	
Chair		*	
Bed		*	

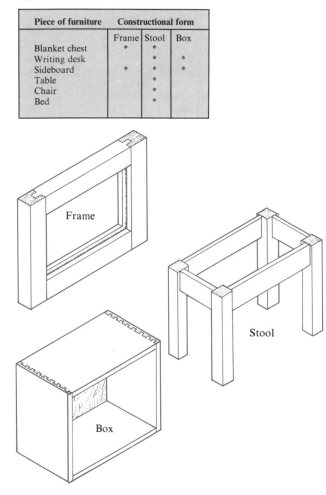

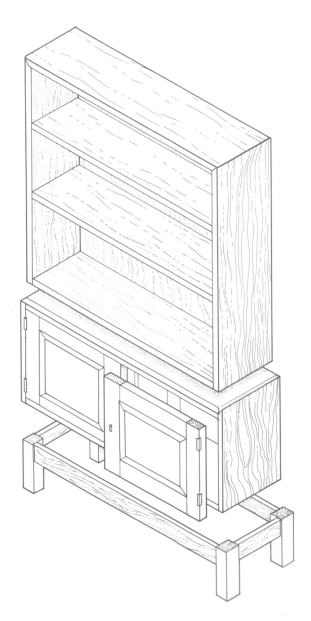

Frame, box and stool are the three basic solid wood constructions from which all units of furniture can be made. They give strength and also allow for the free movement of the wood.

The sideboard is a complex form of storage unit consisting of a stool frame supporting two kinds of storage units — a shelf unit, or large solid-sided box, and a cupboard closed by frame and panel doors.

Why frame and panel construction developed

Early planked oak chests, made from solid boards nailed together, split at the front and back because the vertical grain of the boards at the sides resisted the movement in the wood at the front and back.

In frame and panel construction a solid frame, mortised and tenoned together, is filled in with panels that are free to move in grooves formed along the inside edges of the rigidly fixed frame.

How wood moves

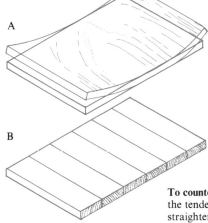

A

B

C

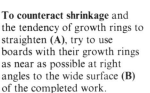

D

To counteract shrinkage and the tendency of growth rings to straighten (**A**), try to use boards with their growth rings as near as possible at right angles to the wide surface (**B**) of the completed work.

A joint made with the heart side inwards as at C will tend to loosen as the outside edges curl away. For a tight, well-fitting joint, arrange the boards with the heart side outwards and the grain as shown at D.

For a solid wood box construction arrange the boards so that the grain in the top, bottom and cheeks runs around the carcass, allowing all the boards to move in a uniform manner.

Measuring and testing carcasses

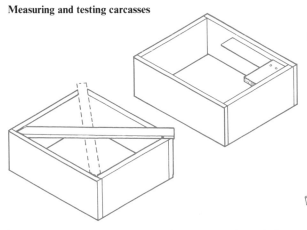

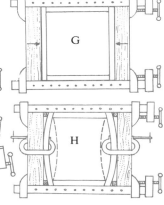

E

F

G

H

To check if a carcass is square measure the diagonals with a squaring rod placed inside one corner. If the diagonals differ, then the carcass is out of square by half that distance; readjust with clamps.

Check the openings of all carcasses, frames and boxes for squareness with a try square. Check all four corners. Put small flat frames on a known level surface and check them for winding.

If a frame is out of square (E), correct it by slanting the clamps to pull the frame square. If the frame is in winding (**F**), apply clamps to pull the higher corners down and the lower corners up.

When clamping across the width of a carcass use crossbearers from front to back (**G**). Any surfaces that bow inwards can sometimes be corrected by clamping in the opposite direction (**H**).

Basic carcass construction 1

Carcass types

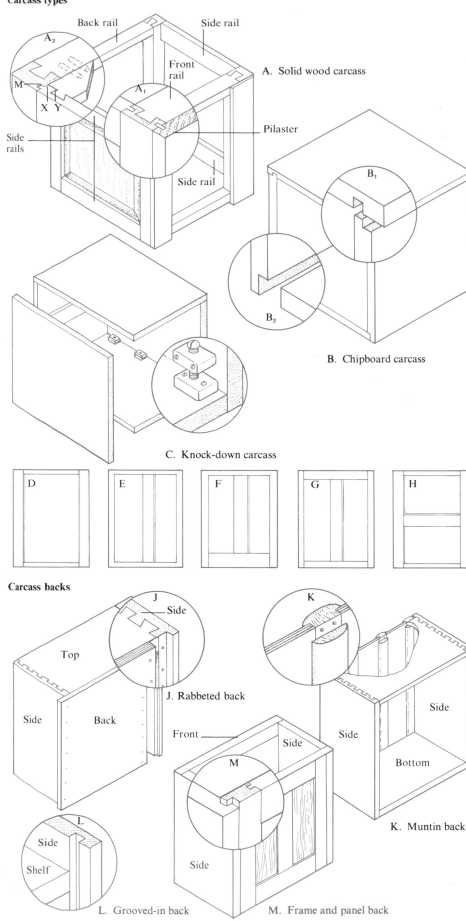

A. Solid wood carcass

B. Chipboard carcass

C. Knock-down carcass

Carcass backs

J. Rabbeted back

K. Muntin back

L. Grooved-in back

M. Frame and panel back

Carcass types

In solid wood carcasses, rails are lap dovetailed to the corner posts, which are mortised to receive the tenons on the side rails (A_1). Posts and side rails are grooved to accept fielded panels. A triangular fillet, doweled to the front or back rail and dovetailed to the side rail, is often added for strength (A_2). Note that the short grain at x is weak and may split, so replace the acute angle at y with a right angle. M can be a tongue in a groove (as shown), or a mortise and tenon. Front pilasters, added for rigidity, may be tongued and grooved or attached with small glue blocks on the inside. Carcass construction using manufactured boards can be simpler than solid wood construction as movement is negligible. Manufactured board tenons and dovetails are structurally weak, so corner joints can be barefaced tongued and grooved (B_1), rabbeted (B_2) or mitered and glue blocked. A lap joint can be reinforced by screwing and pelleting. Most knock-down fittings, such as the block joint (**C**), have been designed for use with chipboard (*see* pages 256–7). If a carcass is not rigid, distortion can lead to doors and drawers jamming, and failure of joints. Rigidity is achieved by fixed shelves and partitions, and by the addition of pilasters secured to the carcass front or back at the sides (**D**), at the center (**E**), attached to a bottom rail (**F**), in T (**G**) or in H (**H**) form.

Carcass backs

A back does more than cover a gap: the carcass depends on it for rigidity and stability. The simplest form is a five-ply panel screwed into rabbets worked in the sides and top of the carcass (**J**). The carcass bottom is flush with or overlapped by the back. Hardboard backs may be screwed or pinned. Five-ply and hardboard can also be fitted in grooves worked in the sides. The carcass bottom and any shelves must be flush with the groove to allow the back panel to be slid in after assembly (**L**). A frame and panel back, suitable for a large carcass, can be screwed into a rabbet, or may be tongued and grooved (**M**). A muntin back (**K**) consists of one or two grooved uprights, or muntins, tenoned into the top and bottom with fill-in panels, usually of three-ply.

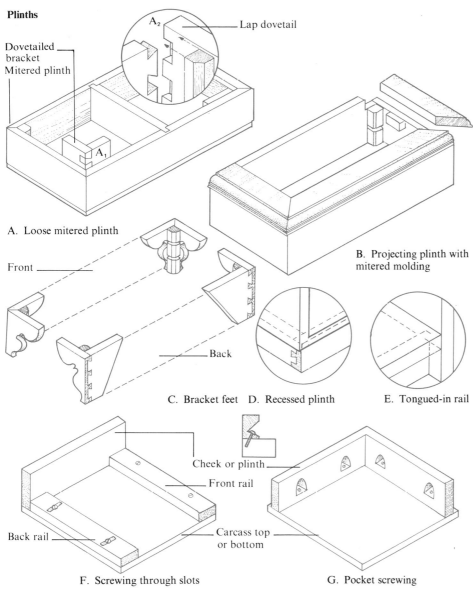

Plinths

Dovetailed bracket
Mitered plinth

A₂ — Lap dovetail

A₁

A. Loose mitered plinth

Front

Back

B. Projecting plinth with mitered molding

C. Bracket feet D. Recessed plinth

E. Tongued-in rail

Cheek or plinth

Front rail

Back rail

Carcass top or bottom

F. Screwing through slots

G. Pocket screwing

Carcass tops

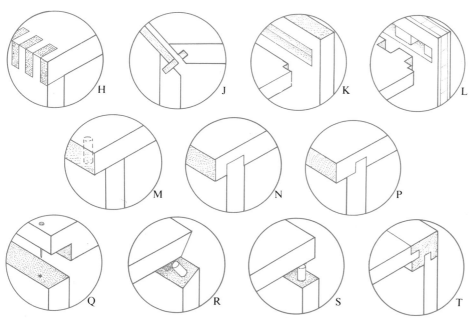

H

J

K

L

M

N

P

Q

R

S

T

Plinths

Blind miter and stopped lap dovetail joints may be used on plinths, but it is more common to make a plain miter joint at the front, reinforced by a glue block or dovetailed bracket (**A₁**). The back may be housed into the sides, tongued and grooved or lap dovetailed (**A₂**). A large plinth will require a crossbearer, which should be housed or slot dovetailed to front and back. A projecting plinth with molding is constructed with mitered corners. The molded board supports the carcass and is mitered at the front and glue-blocked in the corners beneath and along the sides (**B**). Bracket feet are used on antique and reproduction furniture. The front feet are mitered together and glue blocked, while the back feet are lap dovetailed (**C**). Recessed plinths, which do not need precise fitting, allow toe room when one stands facing the furniture. A lap dovetail at the front is typical (**D**). A variation often used with cupboards and wardrobes is to bring the cheeks down to floor level and rest the bottom on an inset plinth rail tongued into the cheeks (**E**). Plinths and tops can be secured in the same way. Solid wood ones should be fixed by gluing and screwing to the front rails and screwing through slots at the back (**F**). Wooden buttons (*see* page 163) or metal shrinkage plates (*see* page 257) can also be used. Plinths and tops of manufactured board can be glued and pocket screwed (**G**).

Carcass tops

Where the top forms an integral part of the carcass and is not screwed on, dovetail or lap dovetail are the most common joints. Stopped lap and blind miter dovetails are used for a hidden joint. The machine-made comb joint can also be worked (**H**). A miter joint should be reinforced with a spline (**J**). Tops that are set below the top ends of the carcass cheeks can be secured in a stopped housing (**K**). However, a pin joint, because of its greater gluing area, is stronger (**L**). With manufactured boards, projecting tops can be fixed to the cheek ends by dowels (**M**), housings (**N**) or tongues (**P**). Flush tops can be housed in a grooved corner block of solid wood (**T**), rabbeted, then pinned and the holes filled (**Q**), mitered and doweled (**R**), or butted and doweled (**S**).

Basic carcass construction 2

A carcass without a back
Strong, well-fitting corner joints are necessary to ensure strength and rigidity. In solid wood, secret dovetails with a single lap joint at the bottom (A_3) and a stopped lap joint at the top (A_1) are usual. Shelves increase strength and prevent the cheeks bowing. All forms of housed joints are suitable, in any materials, although a through dado looks unsightly on the front edge. A stopped dado is neater and preserves the vertical line at the sides of the carcass. The addition of blind tenons, making a pin joint, increases the strength. A shoulder-housed dovetail joint (A_2) can be cut in solid wood only; this joint is strong as it resists any outward pull. The housing is tapered in its length and the shelf is inserted from the rear. Fixed shelves in manufactured boards may be doweled (B_1) or tongued (B_2) to the cheeks. Other methods include screwing through the cheeks into the ends of shelves or using knock-down fittings, such as a cabinet connecting screw (*see* page 256).

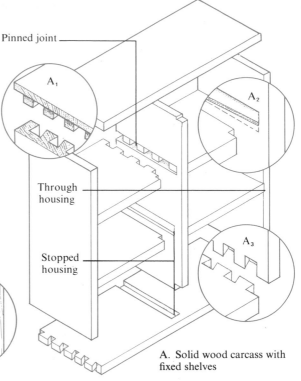

Pinned joint

Through housing

Stopped housing

A. Solid wood carcass with fixed shelves

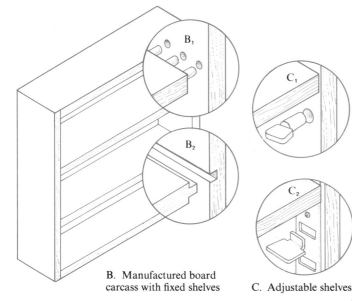

B. Manufactured board carcass with fixed shelves

C. Adjustable shelves

Carcass rails
In the construction of a carcass without a back, such as a bookshelf, that is likely to be subjected to heavy loading and have a tendency to ricking, the addition of full-width rails will increase rigidity considerably. Most large knock-down cupboard or shelving units made of chipboard need rails to hold the side panels vertical. They may be pediment rails, front rails, back rails or plinth rails inset under the base. The rails may be lap dovetailed to the cheeks, doweled (D_2), blind tenoned (D_1), tongued or secured with a knock-down fitting such as a block joint or a bolt and cross dowel (*see* pages 256–7). Rails can be used for fixing the carcass to a wall; this will increase the rigidity in the same way as the addition of a back. Rails may also be used for hanging the carcass.

Pediment rail

Back rail

Front rail

Plinth rail

Adjustable shelves
These are best made from manufactured boards as they are less likely to warp. They may be supported by plastic studs tapped into holes in the cheeks (C_1); ideally, the holes should be fitted with a metal or plastic bushing. Metal bookcase fittings (C_2) should be let flush into the cheeks. Magic wires (C_3) provide a concealed fitting (*see* page 264). When adjustable shelves are included, it is advisable to have a back panel to ensure rigidity. Increase the strength by fixing one of the shelves permanently in position by the addition of a rail to the front or back, or by screwing through the back panel into the back edge of one or more shelves.

Hanging a carcass without a back

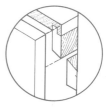

A beveled rail tongued to the carcass or screwed to the carcass back can hook over a corresponding rail attached to the wall. Flushmount fittings and keyhole plates can also be used, as can cabinet hangers (*see* page 257).

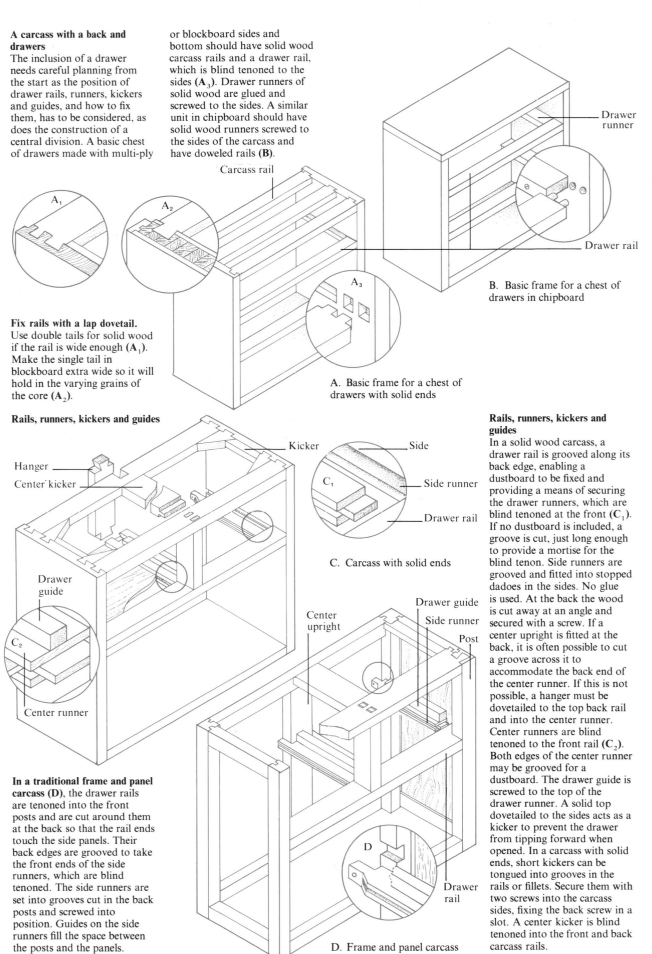

A carcass with a back and drawers

The inclusion of a drawer needs careful planning from the start as the position of drawer rails, runners, kickers and guides, and how to fix them, has to be considered, as does the construction of a central division. A basic chest of drawers made with multi-ply or blockboard sides and bottom should have solid wood carcass rails and a drawer rail, which is blind tenoned to the sides (A_3). Drawer runners of solid wood are glued and screwed to the sides. A similar unit in chipboard should have solid wood runners screwed to the sides of the carcass and have doweled rails (**B**).

Carcass rail

Fix rails with a lap dovetail. Use double tails for solid wood if the rail is wide enough (A_1). Make the single tail in blockboard extra wide so it will hold in the varying grains of the core (A_2).

A. Basic frame for a chest of drawers with solid ends

B. Basic frame for a chest of drawers in chipboard

Drawer runner

Drawer rail

Rails, runners, kickers and guides

Hanger

Center kicker

Kicker

Side

Side runner

Drawer rail

Drawer guide

C_2

Center runner

C. Carcass with solid ends

In a traditional frame and panel carcass (**D**), the drawer rails are tenoned into the front posts and are cut around them at the back so that the rail ends touch the side panels. Their back edges are grooved to take the front ends of the side runners, which are blind tenoned. The side runners are set into grooves cut in the back posts and screwed into position. Guides on the side runners fill the space between the posts and the panels.

Center upright

Drawer guide

Side runner

Post

D

Drawer rail

D. Frame and panel carcass

Rails, runners, kickers and guides

In a solid wood carcass, a drawer rail is grooved along its back edge, enabling a dustboard to be fixed and providing a means of securing the drawer runners, which are blind tenoned at the front (C_1). If no dustboard is included, a groove is cut, just long enough to provide a mortise for the blind tenon. Side runners are grooved and fitted into stopped dadoes in the sides. No glue is used. At the back the wood is cut away at an angle and secured with a screw. If a center upright is fitted at the back, it is often possible to cut a groove across it to accommodate the back end of the center runner. If this is not possible, a hanger must be dovetailed to the top back rail and into the center runner. Center runners are blind tenoned to the front rail (C_2). Both edges of the center runner may be grooved for a dustboard. The drawer guide is screwed to the top of the drawer runner. A solid top dovetailed to the sides acts as a kicker to prevent the drawer from tipping forward when opened. In a carcass with solid ends, short kickers can be tongued into grooves in the rails or fillets. Secure them with two screws into the carcass sides, fixing the back screw in a slot. A center kicker is blind tenoned into the front and back carcass rails.

155

Drawers

The corners of drawers are always dovetailed into solid wood in quality work. The sides are lap dovetailed to the front and through dovetailed to the back. This is the strongest form of construction as the dovetails resist the tendency of the front and back to pull away from the sides while the drawer is being pulled and pushed. A feature of the best handmade drawers is that the corners of the lap dovetails almost meet, giving a neat appearance because of the narrow pins. For large drawers, the front should be approximately 22 mm/$\frac{7}{8}$ in thick. Medium-size drawers have fronts about 19 mm/$\frac{3}{4}$ in thick, and small drawers about 16 mm/$\frac{5}{8}$ in thick. The sides and back should be kept as thin as possible, from 6 mm/$\frac{1}{4}$ in to 10 mm/$\frac{3}{8}$ in thick.

Drawer grooves and slips
Groove the inside front to take the drawer bottom; the lowest front pin must be positioned low enough for the groove to be contained in the socket (**A**), otherwise the grooves will show on the pin. If the sides are thick enough, they too are grooved level with the lowest dovetail. In the best quality work the sides are too thin, so special grooved slips are glued in place to take the bottom (**B**). The simplest slip, commonly used with a three-ply bottom, has a quadrant molding worked on its top edge. The drawer back must be arranged so that its bottom edge and lowest pin rest on the drawer bottom.

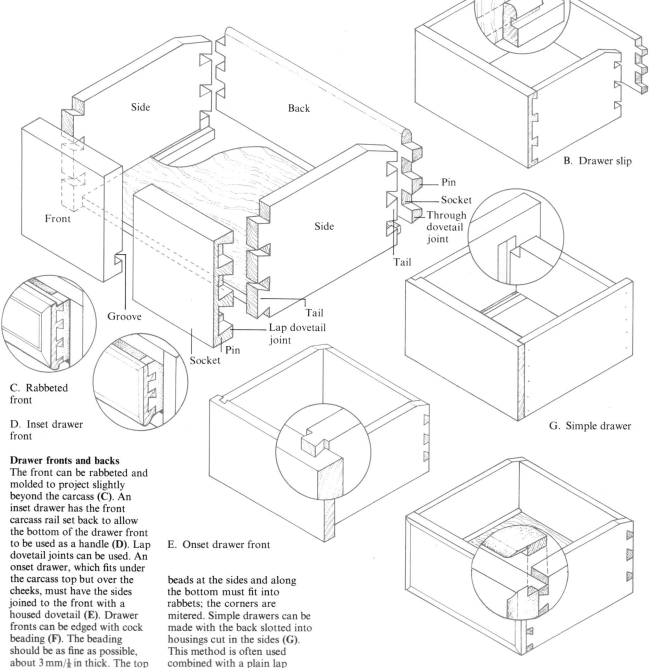

A. Dovetailed drawer

B. Drawer slip

Pin
Socket
Through dovetail joint

Side

Tail

Side

Back

Front

Groove

Pin

Socket

Tail

Lap dovetail joint

C. Rabbeted front

D. Inset drawer front

E. Onset drawer front

G. Simple drawer

F. Cock beading

Drawer fronts and backs
The front can be rabbeted and molded to project slightly beyond the carcass (**C**). An inset drawer has the front carcass rail set back to allow the bottom of the drawer front to be used as a handle (**D**). Lap dovetail joints can be used. An onset drawer, which fits under the carcass top but over the cheeks, must have the sides joined to the front with a housed dovetail (**E**). Drawer fronts can be edged with cock beading (**F**). The beading should be as fine as possible, about 3 mm/$\frac{1}{8}$ in thick. The top length of beading covers the top edge of the drawer, but the beads at the sides and along the bottom must fit into rabbets; the corners are mitered. Simple drawers can be made with the back slotted into housings cut in the sides (**G**). This method is often used combined with a plain lap dovetail joint between the sides and front.

Drawer bottoms

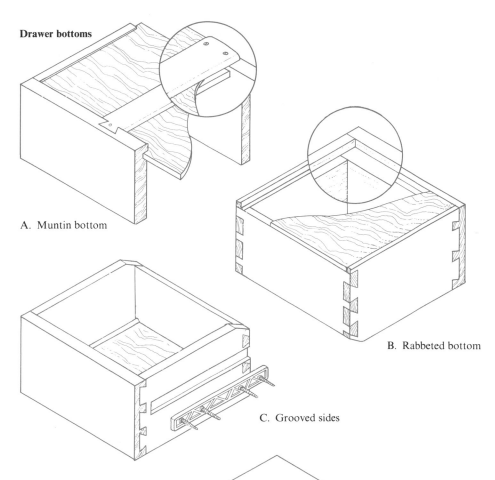

A. Muntin bottom

B. Rabbeted bottom

C. Grooved sides

D. Fitting front and sides

Drawer bottoms

Three-ply is generally used, but make those for small drawers from hardboard. If solid wood is used, the grain must run from side to side, otherwise the bottom may be pulled out of the side grooves. Cut the bottom over-wide to project at the back and secure it with screws in slots; then if it shrinks it can be unscrewed, tapped forward and screwed in place again. Fit a muntin down the center of drawers over 600 mm/24 in wide (**A**). Dovetail it into the drawer front or blind tenon it. Notch and screw it to the drawer back. For a shallower drawer, the bottom can be rabbeted in and secured by gluing and pinning (**B**).

Side-hung drawers

Modern drawers are often side hung, so the drawer sides must be grooved to take the runners (**C**). These are attached to the carcass sides. (No guides or kickers are needed.) The runners can be wooden fillets screwed to the sides, or special plastic strips that are secured by pressing projecting studs into holes bored in the sides of the carcass.

Drawer stops

Stops keep the drawer in the correct position and prevent it from hitting the carcass back (**F**). Gauge the drawer front thickness on the rail. Glue two stops and place them slightly forward of the gauged line. Carefully slide the drawer into place. Remove it, and the stops will be correctly positioned. Fix them with pins.

Making a drawer

Make each drawer to fit its own opening in the carcass. Cut the front and plane the edges to fit, in the correct order (**D**). Follow the same procedure for the back, which should have its top edge set down by 6 mm/¼ in and its bottom edge resting on the drawer bottom. Fit the sides into the carcass, and plane the ends square. Mark each side "Left" and "Right", and check that they are equal in length. Gauge in any grooves for the drawer bottom in the front and sides before marking out the dovetails. Mark mating ends A:A, B:B, C:C and D:D. Mark and cut the dovetails (*see* pages 130–9). Form the grooves; clean up the inside faces. To assemble the drawer, fix the front vertically in the vise, sockets outwards. Glue one side into the front. Remove from the vise. Glue the pins and sockets on the drawer back. Place the back in the vise and partially tap it into the glued side and front. Release from the vise. Place the U frame, front upwards, on the bench and tap the joints fully home. Glue the tails on the second side. Turn the frame upside down and tap the side home. Wipe off surplus glue

and check for squareness. When glue dries, fit the bottom in place but do not secure it yet. Fit the drawer into the carcass, locating tight spots carefully. These are revealed by a shiny surface on the wood when the drawer is moved in and out. Remove them by taking fine shavings with a plane. Clean up the drawer, supporting it with the vise and a board secured with the bench holdfast (**E**). If the drawer is racked about, the joints could fail. Then secure the bottom, using a few drops of glue in the front groove and screws or pins to secure the back.

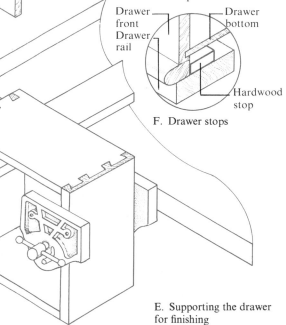

Drawer front
Drawer rail
Drawer bottom
Hardwood stop

F. Drawer stops

E. Supporting the drawer for finishing

Doors 1

A solid wood framework gives a frame-and-panel door its strength; it is made using mortise and tenon joints. Horns are left on the stile ends to prevent the wood from splitting during working. The exact construction of the frame depends on the method used to fit the panels. These are simply fill-in pieces and may be of solid wood or three-ply, fitted into grooves or rabbets. Grooved-in panels are usually made of three-ply, while fielded and raised panels are normally made of solid wood. Flush panels may be of either material. All types

of panels may be beaded into a rabbeted frame. To allow for movement, solid wood panels must never be glued, and care must be taken during assembly so that no glue is squeezed out of the joints thereby accidentally locking the corners of the panel into the frame. As it does not move as much as solid wood, three-ply may be glued to stiffen the whole structure. The panel area may be divided into smaller sections by vertical dividers called muntins. These are often used purely for visual effect, and are blind tenoned and glued.

Frame-and-panel construction
A grooved-in panel made from a manufactured board, which is to have a brushed or sprayed surface finish, should be assembled in its frame before the finish is applied. If the finish is to be applied in any other way, or if the panel is made from solid wood, the surface should be coated before assembly. Framed carcass backs do not need a surface finish. A haunched mortise and tenon is used for the frame (**B**). With a rabbeted frame, the panel can be fitted after assembly and application of a surface finish. A long and short shoulder tenon is used (**E**). The panel is beaded in from the back. Molding may be applied to the front edges of a plain frame to form the rabbet. The panel is beaded in (**D**). In the best molded and rabbeted frames the molding is worked in the solid. It is cut away locally at the mortise and is mitered (**C**).

A. **Door parts**
1. Horn
2. Meeting stile
3. Hinging stile
4. Top rail
5. Middle rail
6. Bottom rail
7. Muntin
8. Top panels
9. Bottom panel

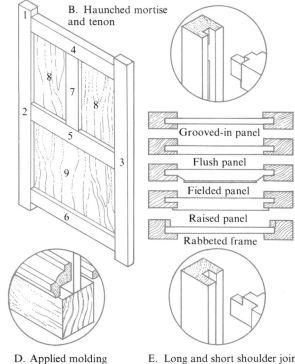

B. Haunched mortise and tenon

Grooved-in panel

Flush panel

Fielded panel

Raised panel

Rabbeted frame

C. Molded door frame

D. Applied molding

E. Long and short shoulder joint

Making a frame-and-panel door

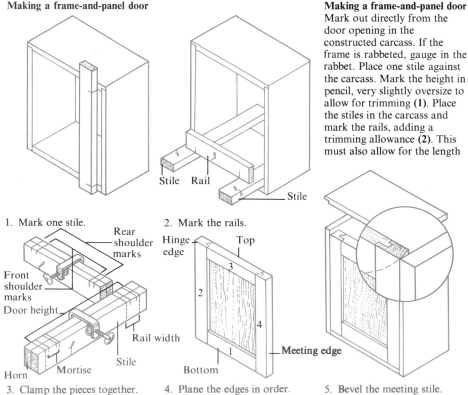

1. Mark one stile.

Rear shoulder marks

Front shoulder marks

Door height

Horn   Mortise   Stile

3. Clamp the pieces together.

2. Mark the rails.

Stile   Rail

Stile

Hinge edge   Top

Bottom

Meeting edge

4. Plane the edges in order.

Making a frame-and-panel door
Mark out directly from the door opening in the constructed carcass. If the frame is rabbeted, gauge in the rabbet. Place one stile against the carcass. Mark the height in pencil, very slightly oversize to allow for trimming (**1**). Place the stiles in the carcass and mark the rails, adding a trimming allowance (**2**). This must also allow for the length

of the tenons and long front shoulders. Clamp the stiles and rails together to mark out the joints (**3**). Cut the joints, then the rabbets. Glue and clamp the frame together. Plane the edges of the frame in the correct order until it fits the door opening (**4**). Plane the bottom edge so that, combined with the hinging edge, the door fits the bottom right corner. To fit the top edge, rest the door on a piece of glasspaper placed on the carcass bottom, to give clearance for the surface finish. Plane the meeting edge, leaving it slightly oversize. Hinge to the carcass and complete the fitting on the meeting edge, giving the door a clearance fit. Bevel the meeting edge (**5**).

5. Bevel the meeting stile.

Increasing use is being made of manufactured boards for flush doors. Multi-ply, laminboard and chipboard must have their edges covered to ease shaping and application of a surface finish, and to conceal the core material. Wide edgings of solid wood are necessary to provide a secure grip for hinge and lock screws because manufactured boards do not give a secure grip on screws inserted into their end grain (*see* pages 148–9). The edging can be butted and glued on, or tongued and glued into a groove for extra strength (*right*).

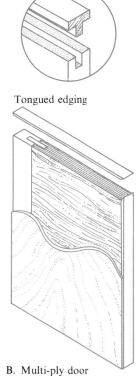

Tongued edging

Flush doors
A traditional flush door **(A)** is made with a core formed of straight-grained, well-seasoned wood strips, 50 mm/2 in wide, glued together. Smooth both sides, then face with constructional veneer laid at right angles to the grain of the core. Veneers of the same thickness must be used on both sides. Over the constructional veneers, lay decorative veneers with their grain parallel to the core strips. Multi-ply **(B)** and laminboard doors should have their grain at right angles to that of the face veneer. A framed and covered door **(C)** has a solid wood frame covered with three-ply. The covering is supported by cross rails, which should be drilled to allow the internal air pressure to equalize with that of the surrounding atmosphere, otherwise sinking or bulging may occur in the three-ply. Bore holes in the bottom rail, not in the top rail.

A. Traditional flush door

B. Multi-ply door

C. Framed and covered door

Internal supports for framed and covered doors can vary. Lighter methods of supporting panels on covered doors are particularly useful for large doors. These include slats of insulation board (D_1) arranged as shown and cardboard honeycombing (D_2).

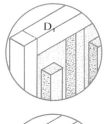

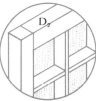

D. Internal supports

Fitting hinges

E. Inset door

F. Lay-on door

G. Inset door

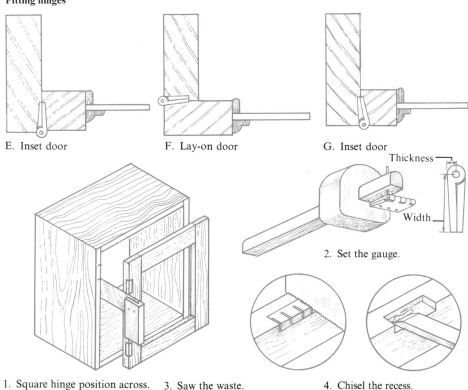

1. Square hinge position across.

2. Set the gauge.

Thickness

Width

3. Saw the waste.

4. Chisel the recess.

Fitting hinges
The butt hinge is the most common in cabinetmaking. The hinge flaps can be recessed equally into the carcass and the door **(E)** or the whole knuckle can be let into the door with one flap taper recessed. The other flap is recessed from the extreme edge of the carcass and tapered to the flap thickness in both lay-on **(F)** and inset doors **(G)**. As a general rule, position a hinge its own length from the end of the stile. On a framed door, the top end of the hinge is usually lined up with the inner edge of the rail. To position a hinge, square penciled position marks across the door frame and carcass **(1)**. Set a marking gauge to the hinge **(2)**. Gauge its width and thickness from the knuckle to the edge on the door frame. Make vertical sawcuts in the waste and on the waste side of the gauged lines **(3)**. Chisel across the ends **(4)**. Then pare the recess flat. Form the recesses in the carcass and screw the door in position, center screws first.

Doors 2

Sliding doors

These are usually made of chipboard or laminboard, edged all around with hardwood. Lightweight doors have a softwood frame, faced with three-ply. The simplest way to make doors slide is to cut grooves in the carcass top and bottom (A) or to attach grooved hardwood, fiber or stainless steel guides to the carcass. Doors may be tongued to slide in the grooves (B). Grooves must be slightly wider than the thickness of the door or tongue, for easy running. Allow sufficient extra depth in the top grooves for the doors to be inserted by pushing them up and then dropping them into the bottom groove. Fiber track cuts down friction. Glue it into grooves worked in the carcass top and bottom. Let mating plastic gliders into a groove in the bottom of the door, which should fit close to the carcass bottom (D). At the bottom, fit the fiber track in two lengths. Glue the first

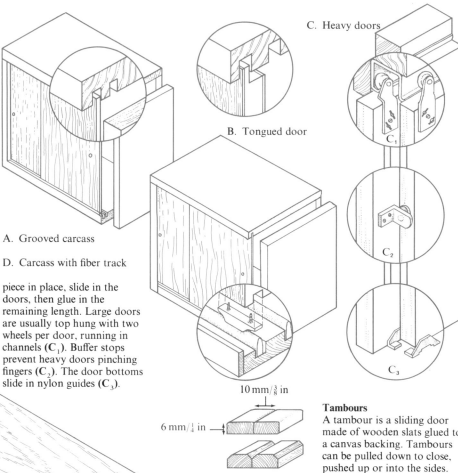

A. Grooved carcass

D. Carcass with fiber track

B. Tongued door

C. Heavy doors

piece in place, slide in the doors, then glue in the remaining length. Large doors are usually top hung with two wheels per door, running in channels (C_1). Buffer stops prevent heavy doors pinching fingers (C_2). The door bottoms slide in nylon guides (C_3).

Tambours

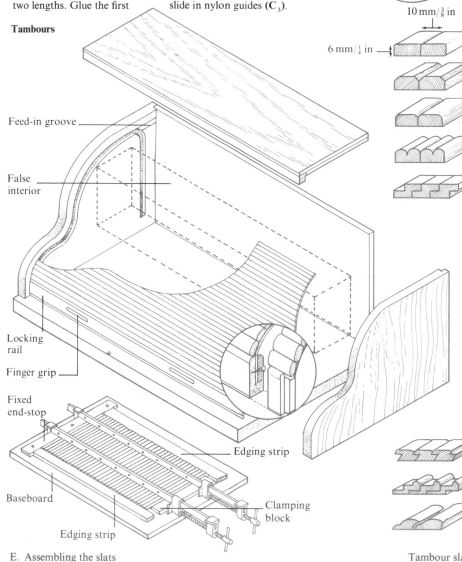

Feed-in groove

False interior

Locking rail

Finger grip

Fixed end-stop

Baseboard

Edging strip

Edging strip

Clamping block

E. Assembling the slats

10 mm/$\frac{3}{8}$ in

6 mm/$\frac{1}{4}$ in

Tambour slats

Tambours

A tambour is a sliding door made of wooden slats glued to a canvas backing. Tambours can be pulled down to close, pushed up or into the sides. Surface finish the slats before assembling them face down on a board between edging strips and a fixed strip at one end, all slightly thinner than the slats. With a clamping block at the other end, lightly clamp up the slats (E). Ensure absolute squareness. Then nail or screw the clamping block to the board and remove the clamps. Glue fine canvas over the slats, stretching it well. Leave sufficient canvas at one end for it to be attached to the locking rail. The tambour runs in grooves in the carcass sides. Widen the grooves around the curves so that the slats can negotiate the curves. Run out the side grooves at the carcass back, which must be detachable, so the tambour can be slid in after the carcass has been assembled and finished. Then fill in the feed-in grooves. The locking rail, which is thicker than the slats and is rabbeted at the ends, is then sprung into the grooves from the front after the tambour is in place. The canvas is then beaded into a rabbet cut in the back of the rail (see inset). To prevent the contents of the carcass from interfering with the travel of the tambour, and for appearance, fit a false top, sides and back.

Fall flaps

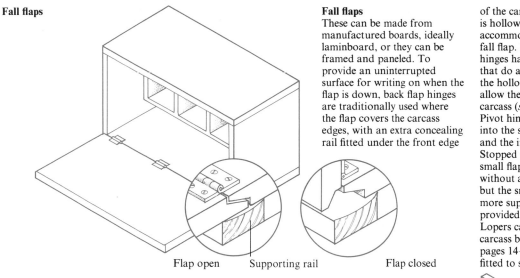

Flap open   Supporting rail   Flap closed

Fall flaps
These can be made from manufactured boards, ideally laminboard, or they can be framed and paneled. To provide an uninterrupted surface for writing on when the flap is down, back flap hinges are traditionally used where the flap covers the carcass edges, with an extra concealing rail fitted under the front edge of the carcass bottom. This rail is hollowed out so that it can accommodate the swing of the fall flap. However, modern hinges have been developed that do away with the need for the hollowed-out rails and allow the flap to sit within the carcass (*see* pages 258–9). Pivot hinges can be recessed into the side edges of the flap and the inside of the carcass. Stopped pivot hinges hold small flaps in the open position without additional support. All but the smallest flaps need more support than can be provided by the hinges alone. Lopers can be built into the carcass beneath the flap (*see* pages 14–15), or a stay must be fitted to support it.

Barred doors
The tracery in barred doors comprises a face molding with a separate stiffening rib grooved into the back to form rabbets for the glass, which is beaded or puttied in (**B**). Lay out the ribs first on a three-ply panel that has been cut to fit the exact size of the door frame rabbets. Set out the pattern, working to the center of the molding. Cut the ribs and hold them in place with wooden blocks pinned to the three-ply panel (**C**). Join the ribs together with halved cross lap joints or simple butt joints reinforced with glued canvas (**D₁**). Two curved ribs meeting an upright can be spliced together and slotted into the end of the upright (**D₂**).

A. Barred door patterns

Blind tenon the rib assembly into the frame. Place the frame over the assembly to mark the tenon lengths and the mortises in the rabbets. Cut the joints. Glue the tenons and spring the whole assembly into the frame. The face moldings are mitered where they meet each other and where they meet the frame (**A₁**), so they must be the same section as the frame molding. The miter always bisects the angle between adjacent parts. Cut the face moldings and glue them onto the ribs. Cut the glass panes with a slight clearance all around; then bead or putty them into the ribs.

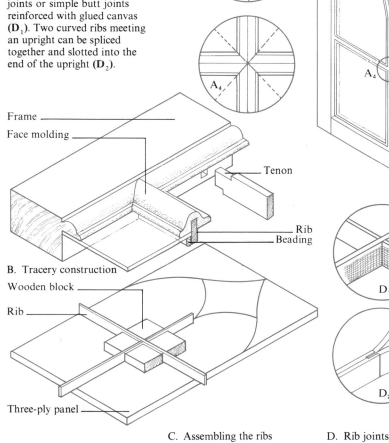

Frame
Face molding
Tenon
Rib
Beading
B. Tracery construction

Wooden block
Rib
Three-ply panel
C. Assembling the ribs

D. Rib joints

Tables

Leg and rail joints

The mortise and tenon is the most common joint. Wide underframe rails should have haunched tenons as the haunch supports the upper third of the rail without it pulling free (**C**). Miter tenons where they meet within the leg (**D**). Leave a small gap between the mitered faces to allow for the contraction of the legs and so the shoulders can be pulled tight. To avoid the haunch showing on the top of the leg, use a tapered haunch resting in a correspondingly tapered mortise (**E**). A dowel joint between rail and leg is simpler but not so strong (**F**). Stagger the dowels. Reinforce leg and rail joints with blocks glued to the inside (**A, B**). To join rails to round legs, either scribe the shoulder of the rail so that it fits the leg, in which case a dowel joint (**H**) is easiest, or plane a flat section on the leg (**G**) and work a mortise and tenon joint in the usual way.

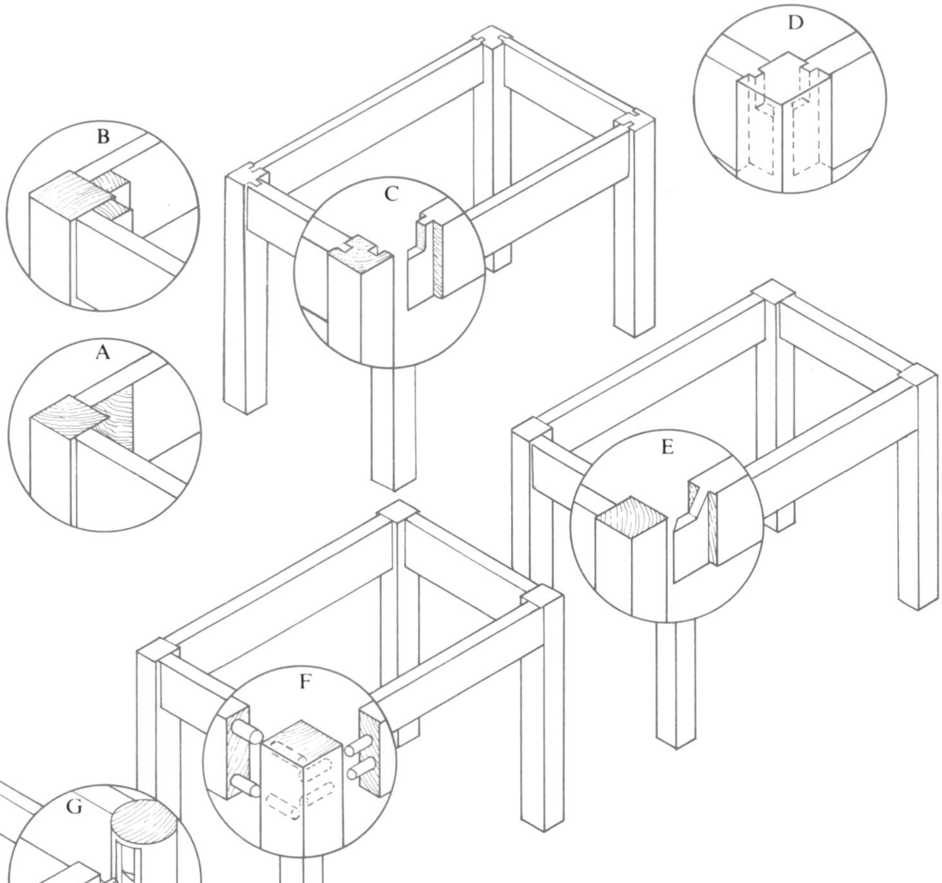

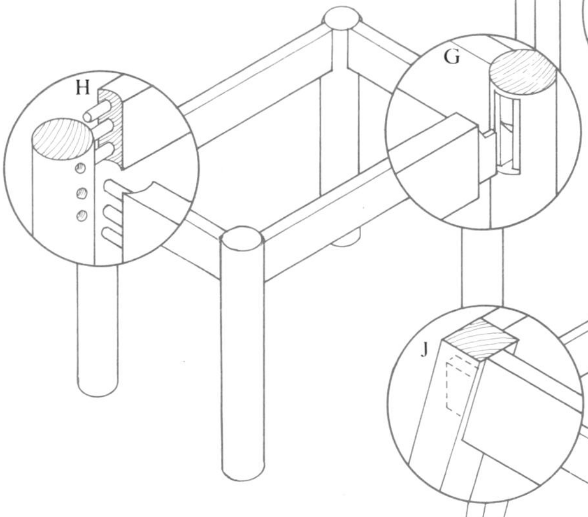

Cabriole legs

Tables with cabriole legs are constructed differently from other styles. Make the leg and rail frame with mortise and tenon joints. Glue the joints and allow them to dry. Then fit the ear pieces to each leg. These should be doweled for the strongest fixing (**N**). They can also be glued, have screws driven into counterbored holes and then be pelleted (**P**).

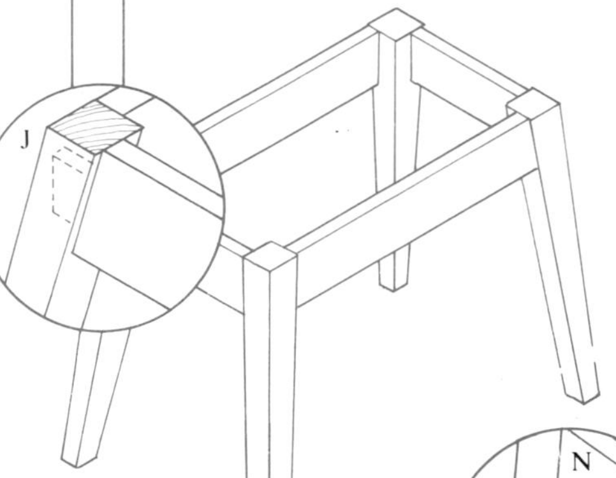

Canted legs

When a table has canted legs, cut the tenon shoulder to slant at the same angle (**J**) or plane an area flat at the top of the leg to take a standard tenon (**K**). For a strong frame, first join the four rails together with through dovetails. Then saw and plane off the corners of the frame. Hold a leg in the vise. Place the frame corner on top and scribe its outline on the end of the leg (**L**). Saw and chisel out a recess in the leg to the frame depth. Slide the frame into the recess (**M**). Reinforce the joint with glue blocks and with a screw that is long enough to run through the leg and the joint and bite into the leg on the other side.

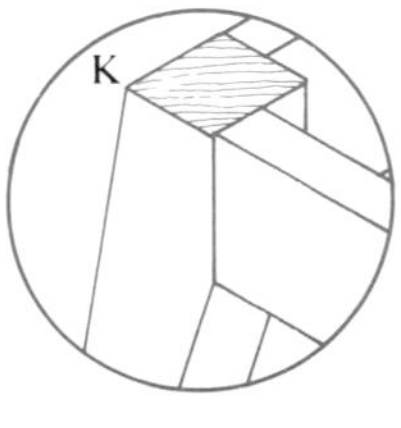

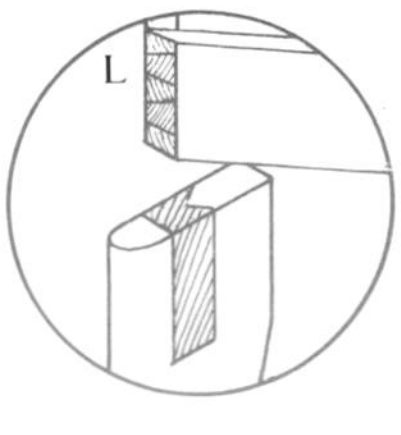

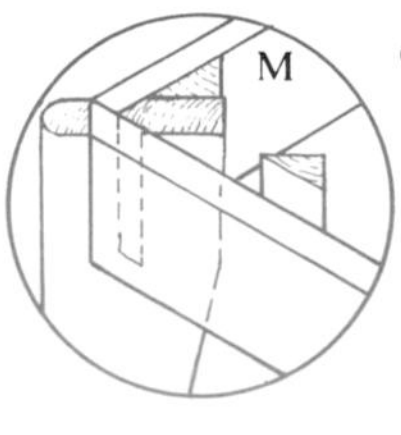

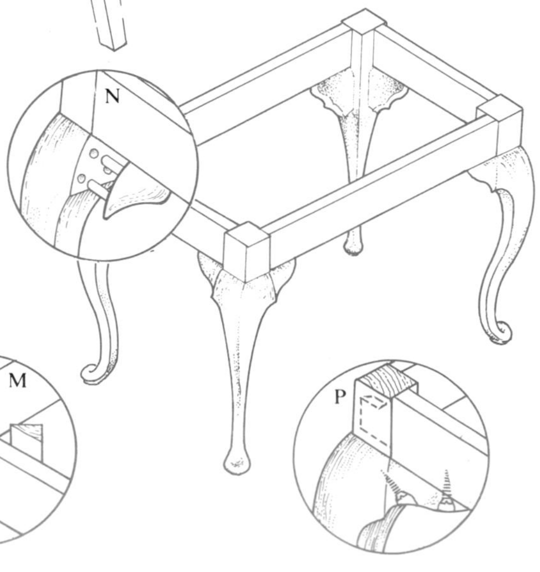

Knock-down fittings

These can be particularly useful on tables. Make firm secure fittings with a bolt and cross dowel (**A**). Insert the steel cross dowel in the rail at right angles to the threaded bolt, which passes through the leg and is screwed into threads in the dowel using an Allen wrench. The table plate is a simple fitting (**B**). Screw each side to a rail. Screw a threaded bolt into the corner of the leg and secure the table plate with a wing nut. Use two short locating dowels with knock-down fittings to prevent the components twisting. Do not glue the dowels if the table is to be dismantled.

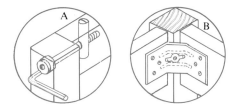

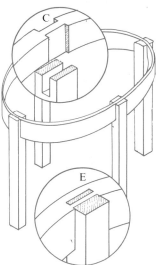

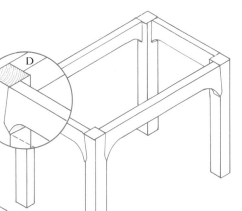

Curved rails

Dovetail curved rails on a round or D-shaped table into grooves on the legs. If a leg is in the middle of the curved rail and is flush with it, join it to the rail with a combined bridle and tenon (**E**). If the rail is inset, use a bridle joint (**C**). When the lower edge of a rail is curved where it meets the leg, a weak feather edge will be formed if the rail is simply planed to a curve. To prevent this, cut a small dado in the edge of the leg around the tenon, and shape the rail (**D**).

Stretcher rails

These add strength and rigidity to the whole frame. They are usually thinner and narrower than the underframe rails and are joined to the leg in the same way, though without a haunch. If the rail is very thin, it can be fitted completely into the mortise; the tenon, however, should have a top shoulder to give it a definite length and to enable the top edge of the rail to be molded (**F**). If stretchers run diagonally, work a flat section on each leg (**G**). Where diagonal stretchers join a central stretcher to form a Y configuration, tenon them into the central stretcher (**H**), or mortise them to receive the tenon on the central stretcher (**J**). Flat stretchers are sometimes required. Make up the stretcher frame with mortise and tenon joints or tongued miters. Secure turned feet with dowels passing through holes bored in the stretcher frame and into the bottom of the legs (**K**).

Solid wood tops

Fix solid wood tops in such a way that the wood is allowed to shrink or expand. Wooden buttons screwed to the table top and sliding in grooves on the inside of the rails are the traditional method (**L**). Metal shrinkage brackets are a more modern method of holding the top down (**M**). These have slots on one face for the screws to slide along.

Chairs

The chair is the most used and abused piece of furniture in the average household. It has to withstand being rocked backwards and sideways as well as having to stand up to the torsional stresses imposed by the continual readjustment of the sitter's position. All chair joints must fit well and be strong. Make up front and back frames first and then join them to the side rails. Always make a full-sized drawing of the projected chair to enable angles to be measured correctly. Any design more elaborate than a simple dining chair should be made up in softwood as a full-sized working model to test the design and construction.

The modern dining chair can be quite simple to construct. Use mortise and double blind tenon joints to fix the front and back rails to their legs. Attach the side rails with dowels. Increase the strength and rigidity of the frame by gluing and screwing stout corner blocks in place. Screw a fillet to the back rail to support the three-ply base of the seat. Tenon any stretcher rails if they are used. Position them so as not to catch the back of the sitter's legs and so they cannot conveniently be used as footrests. Make the chair back from a doweled frame; pad and screw it to the back posts from the inside. Take the cover all around to hide the frame.

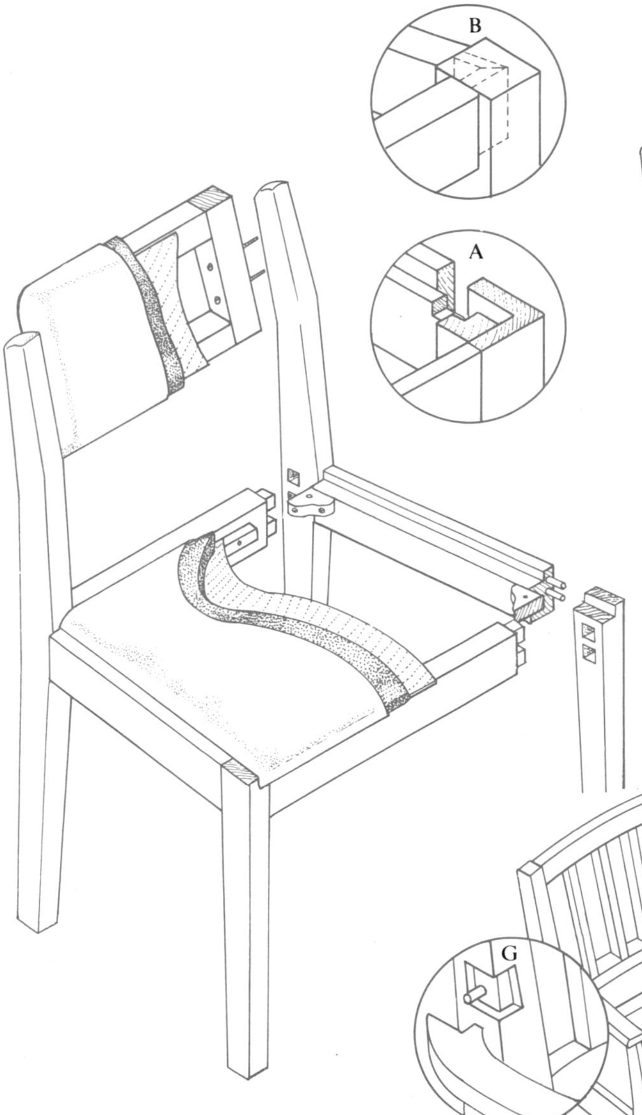

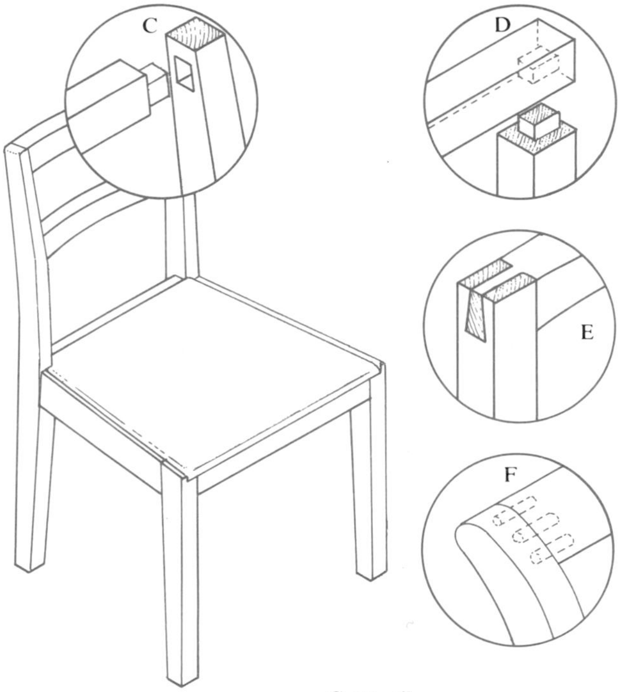

Crest rails

These may be tenoned into the back post (C) or be mortised to receive the tenon on the back post (D). Alternatively, use a dovetail bridle joint (E). Where these joints would be difficult because of shaping or the awkward joint line, use dowels to fit the rail (F).

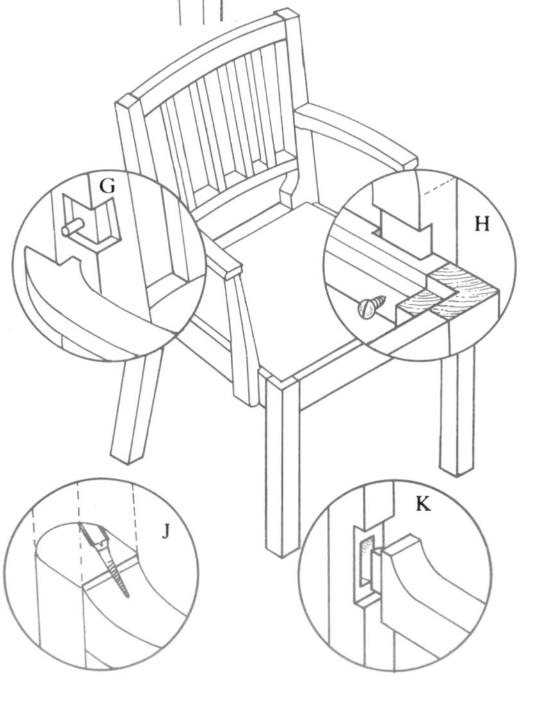

Seat frames

Mortise and tenon joints are traditionally used, but modern chairs rely more on dowel joints or on a combination of dowel and mortise and tenon joints. Set down the front rail flush with the rabbets in the side rails. A drop-in seat requires rabbets to be cut around the inside edges of the rails (A). Use a haunched tenon to accommodate the rabbet. Angle the side rails and connect them to the front and back frames with dowel joints, a canted tenon (B) or a canted mortise. If the leg is to be flush with the rail, plane off the outer part of the leg after the frame has been assembled.

Armrests

Blind tenons, dovetail bridle joints or dowels can be used to join armrests to front arm posts. At the back, notch an armrest and blind dowel it into the back leg (G), or tenon it. The armrests can be housed in, screwed from the back and pelleted (J). A down-swept arm must be tenoned and housed (K). House the front arm post into the seat rail, and glue and screw it (H).

164

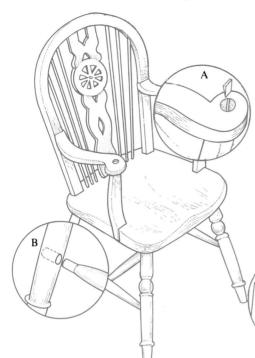

Turned chairs
A turned spindle or leg should ideally fit into a tapered hole, bored with a tapered bit. If the hole has parallel sides, however, a spindle fitted into it requires a shoulder on it to limit the depth of entry (**B**). A small flat area must also be formed on the leg. When fitted to a flat arm, a turned leg must have a spigot formed on the end (**A**). The spigot is taken through and wedged.

Front to back folding chairs
Folding chair frames may be doweled or mortised and tenoned. Hinge the back frame to the top rail of the front frame (**C**). Pivot the back of the seat on steel pins running in stopped grooves in the back leg. Rest the seat frame on dowels or steel pins with the ends riveted over washers (**D**).

Beds

H. Halved bed frame

J. Divan bed

Side to side folding chairs
Mortise and tenon the side frames. Tenon the folding X frames into the seat rail on one side (**F**) and bore them to accept spigots turned on the other seat rail. Use steel pins to connect the folding frames to the side frames (**G**) at the bottom. Fix the straps at the top with screws (**E**).

Beds
The divan type of bed (**J**) has a box framework on stump legs. Its slatted base is simple to make. The side and end rails are of softwood about 150 mm/6 in wide and 22 mm/$\frac{7}{8}$ in thick. The slats are 100 mm × 22 mm/4 in × $\frac{7}{8}$ in. These may be let in flush with the side rail, where they are glued and screwed, or screwed into slots. Make a flush headboard or use frame-and-panel construction. Screw the wooden uprights directly to the back of the bed, slotting them for height adjustment (**L**). Make a bed frame with the top and bottom rails halved to the side rails (**H**). Bolt the side rails to the head- and footboards. Lengthen the uprights to act as legs for the bed. Side frames can also be tenoned into head- and footboards (**K**). Dowels through the bored tenons lock the joints; they can be knocked out to dismantle the bed.

K. Tenoned bed frame

L. Divan headboard

Solid wood bending

Solid wood can be bent for structural purposes. The degree of effective curvature depends on the species of wood. Gradual curves can be made by bending dry wood around a framework, such as a former, and clamping it, but the wood has a tendency to spring back, at least partially, to its original shape when released. Cutting curves from solid wood results in short-grain weakness around the curve or at the ends. By making sawcuts across the grain on a piece of wood its effective thickness will be reduced, allowing a bend of quite small radius, which will be retained when the wood is glued.

The most satisfactory method for bending solid wood, however, is to steam it first to make it supple. Using this method solid wood can be bent without kerfing and its shape will be retained when the wood has dried. As the wood is subjected to the steam, the wood cells will soften and consequently become more amenable to compression. The cells are not, however, equally amenable to expansion and are liable to fracture. The outer surface of the curve must therefore be supported.

The wood to be bent should be straight grained and free from defects. It should be longer than will ultimately be necessary and must be planed smooth. Steaming is done in a rectangular or cylindrical chest, made from wood, metal or plastic, that is large enough to hold the wood to be bent. The chest should be set at a slight incline, so that the condensed water will run to one end. Holes at the bottom of the chest will allow condensation moisture to escape. Place wooden slats inside to support the timber to be bent and to allow the steam to circulate around it. Bore one end of the chest to allow for the insertion of the tube supplying the steam. Loosely block the other end with a cloth to prevent steam loss. Exterior lagging will help reduce condensation and heat loss.

An adequate supply of steam is essential. A 23 liter/5 gallon drum, filled two-thirds with water and heated by a gas burner, will give a sufficient quantity of steam. Bore a hole in the top of the drum to take the width of the rubber tubing, and connect the tubing to the chest. An average temperature of 100°C/212°F at normal atmospheric pressure should be maintained

Cross-kerfing

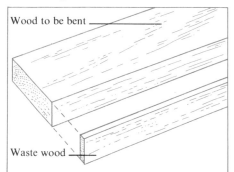

1. Select a piece of waste wood.

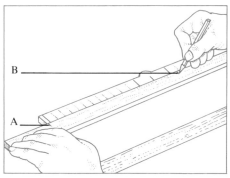

2. Mark the radius.

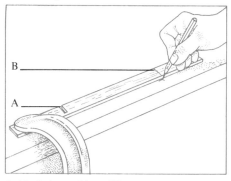

3. Mark the position of line B.

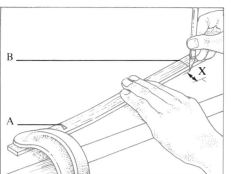

4. Mark the surface again.

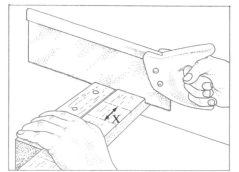

5. Cut a series of kerfs.

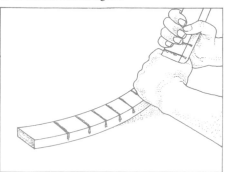

6. Bend the glued wood.

Cross-kerfing

Before dry wood can be cut and bent to the required radius, the distance between kerfs must be calculated. To do this take a thin piece of waste wood of the same width as the thickness of the wood to be bent (1). Mark the required radius measurement along its length, 50 mm/2 in from one end (2). Square the first line (A) across the waste wood. Gauge two-thirds the width of the waste wood. On line A make a single kerf down to the gauged line. Clamp the end to a flat surface. On this surface mark the position of line B (3). Push that end until the kerf at A closes at the top. Holding that position, mark the surface again at line B (4). The distance (X) between the two marks on the surface is the distance that should be left between one kerf and the next in order to make a curve of the required radius. From the face side of the actual wood to be bent, gauge a line on the edges two-thirds the thickness of the wood. With the same saw, cut a series of kerfs down to the gauged lines, allowing the calculated distance between each (5). Insert glue into every kerf. Then bend the wood (6). Hold the curve with a loop of cord, thus closing all the kerfs at the top. Clamp the wood in position when all the kerfs are closed. When the glue has dried, release the clamps and cord. Smooth the bent wood; then veneer its edges to conceal the kerf marks and apply a surface coating to the edges and sides.

within the chest for approximately 3 hours for every 25 mm/1 in of wood thickness.

Steamed wood is bent around a former, which may be constructed from solid wood and/or manufactured boards and should, where necessary, allow for use of clamps. The radius of the former should be slightly smaller than that ultimately required of the wood to be bent as the wood will straighten a little when un-clamped. Each former is usually a one-use con-struction as its curves are difficult to adapt to a different shape. Smooth the former frame so that the wood to be bent can lie directly against it and protect it with a surface coating. A metal strap is used to support the wood on the outside of the bend. Usually the strap is made of steel, but where the wood is particularly acid it must be of aluminum. The strap must be flexible, wider than the wood to be bent and longer, so that some form of handle can be fixed to each end. Each handle is comprised of the backplate, the metal strap end and an end-stop, bolted together. Backplates are essential; they extend along the outside of the strap and prevent the wood from twisting out of plane. End-stops prevent the wood from stretching longitudinally during the bending process. Wedges may be inserted between the end-stops and the bent wood to tighten the fit. A sash clamp should be used to hold the bent wood ends tightly to the former. Concentrated pressure on any one part of the metal strap will dent the wood to be bent, so place a block of waste wood between the clamp and the strap to prevent any damage. The wood may take at least 5 hours to dry in the former in a room at 15°C/60°F.

Some good timbers for steam bending	
Acer pseudoplatanus — Sycamore	*Quercus rubra* — Red oak
Carya spp. — Hickory	*Quercus mongolica* — Japanese oak
Fagus sylvatica — European beech	*Robinia pseudoacacia* — Robinia
Fraxinus excelsior — Ash	*Ulmus americana* — White elm
Juglans regia — European walnut	*Ulmus glabra* — Wych elm
Prunus avium — European cherry	*Ulmus hollandica* — Dutch elm
	Ulmus procera — English elm
	Ulmus thomasii — Rock elm

Steam bending

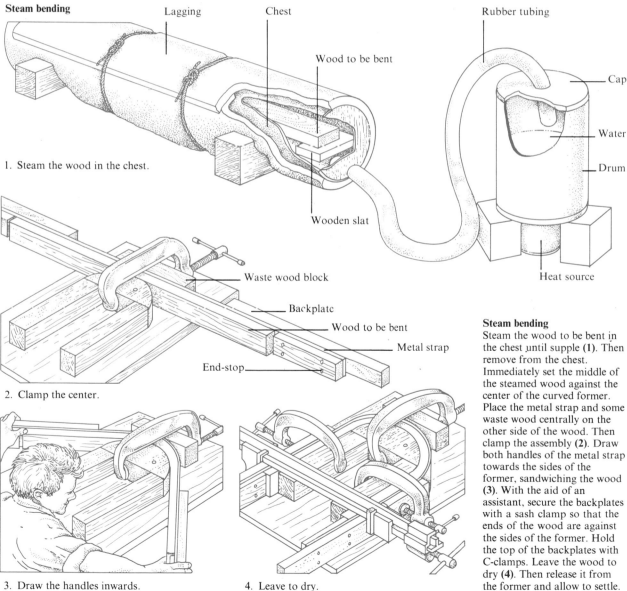

Lagging Chest Wood to be bent Rubber tubing Cap Water Drum

1. Steam the wood in the chest.

Wooden slat Heat source

Waste wood block Backplate Wood to be bent Metal strap End-stop

2. Clamp the center.

3. Draw the handles inwards.   4. Leave to dry.

Steam bending
Steam the wood to be bent in the chest until supple (**1**). Then remove from the chest. Immediately set the middle of the steamed wood against the center of the curved former. Place the metal strap and some waste wood centrally on the other side of the wood. Then clamp the assembly (**2**). Draw both handles of the metal strap towards the sides of the former, sandwiching the wood (**3**). With the aid of an assistant, secure the backplates with a sash clamp so that the ends of the wood are against the sides of the former. Hold the top of the backplates with C-clamps. Leave the wood to dry (**4**). Then release it from the former and allow to settle.

Laminated wood bending

A strong, durable, curved component may be constructed from a number of glued veneers bent around a former. Some complex shapes can be formed by this process, which is known as laminating, but it does have aesthetic limitations as the edge of any laminated component will be parallel layers of veneer and glue.

All veneers must be of the same thickness, which should be no more than 2 mm/$\frac{3}{32}$ in each. Their width should be at least half as wide again as the finished component, to allow for trimming and smoothing of the edges when the glue has dried. Allow an extra 50 mm/2 in length for each veneer. The number of veneers glued together can range from two to 18, depending on the strength and pliability.

The veneers are glued together with their grain running lengthwise in the same direction. Urea formaldehyde is the most suitable type of glue for laminating. It will take 8 hours or less to dry at a room temperature of 15°C/60°F, provided the atmosphere is dry and the room is well ventilated. Once the glue has dried, the veneers are unlikely to move. It is then that the component can be cut to its final length before its edges are smoothed.

The veneers are bent to shape in formers, which are purpose-built for each project from solid wood or manufactured boards. The former must accommodate the excess length and width of the veneers. The area around which the veneers are bent should be smoothed and surface coated before use. Protect the former surface from the glue with a plastic sheet or newspaper. Where small radii are to be bent, veneers can be made more pliable by dampening them and leaving them clamped and unglued between formers to dry out.

Depending on the size and shape of the component required, there are various types of former and different ways of holding the veneers in position while the

Laminating with a pair of male and female formers

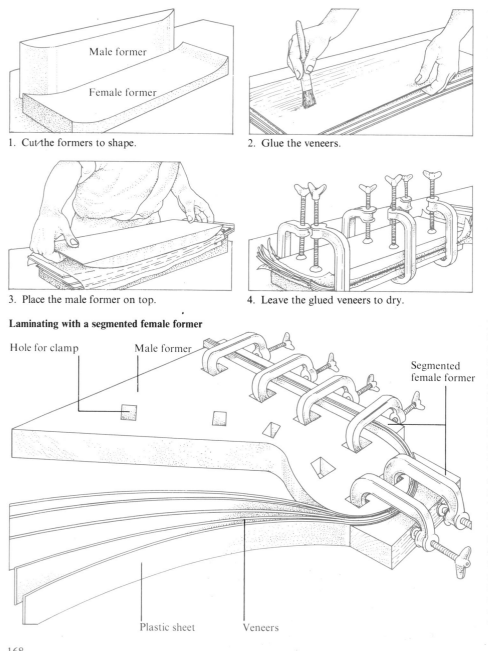

1. Cut the formers to shape.

2. Glue the veneers.

3. Place the male former on top.

4. Leave the glued veneers to dry.

Laminating with a segmented female former

Hole for clamp   Male former

Segmented female former

Plastic sheet   Veneers

Laminating with a pair of male and female formers
Cut the formers and shape them to the required component shape (1). Smooth and coat them. Stack the veneer leaves, with any decorative veneers at the top and bottom. Then with a brush apply glue to one side of each leaf, restacking them in reverse order (2). Place a plastic sheet over the female former. Then lay the glued veneers on the sheeting. Put another plastic sheet on top of the veneers. Place the male former on top (3). Clamp the formers together, using three or more pairs of clamps placed opposite each other. Apply pressure first at the center and work outwards in order to squeeze the glue to the side edges and ends. Then tighten all the clamps and leave until glue is dry (4).

Laminating with a segmented female former
Cut the formers to the desired shape. Smooth and seal each former section. Glue one side of each veneer leaf, using a brush. Then stack them in the required order. Place a plastic sheet around the male former. Set the center of the veneers against the center of the male former. Put a plastic sheet on top of the veneers. Place the appropriate segment of the female former against the veneers at this point and clamp the whole together. Bend the veneers around the male former progressing from the center to the ends, clamping the female segments in position. Each segment must fit the curve perfectly. Retighten the clamps from the center outwards. Leave the glued veneers to dry.

glue dries. If the shape is comparatively flat or has a reverse curve, a pair of formers would be most suitable.

Where a more complex shape is required, a single former is less likely to slide easily into place over the other of the pair. The female former can then be made in a number of segments of different shapes and sizes designed to fit closely around the unbroken edges of the male former. Provision must be made within the male former for the insertion of clamps wherever segments of the female former need securing.

Shallow, wide shapes can be made from a male former alone by creating a vacuum around the former and veneers placed in an airtight bag, usually made from thick rubber or plastic. A vacuum cleaner can be used to remove the air from the bag. The pressure is sufficient to make the veneers wrap around the former. This method can only be used for lightweight components. To test their shape use a paper template cut to the required curve. Wrap the template around the male former; when it fits, cut the veneers accordingly.

If a curve is required only in the end of a piece of solid wood, glued veneers can be inset into kerfs cut into the end to be bent.

Some good timbers for laminating	Mansonia altissima — Mansonia
Acer pseudoplatanus — Sycamore	*Quercus cerris* — Turkish oak
Cassipourea malosana — Pillarwood	*Quercus rubra* — Red oak
Fagus sylvatica — European beech	*Sterculia rhinopetala* — Brown sterculia
Fraxinus excelsior — Ash	*Ulmus americana* — White elm
Gossweilerodendron balsamiferum — Agba	*Ulmus glabra* — Wych elm
Juglans regia — European walnut	*Ulmus hollandica* — Dutch elm
	Ulmus thomasii — Rock elm

Vacuum bending

Plastic sheet — Veneers — Adhesive tape — Airtight bag — Batten — Baseboard — Wooden blocks — Male former — Air pump nozzle

Vacuum bending

Cut a male former to the desired shape. Smooth and coat it. Where necessary, make a template and cut the veneers accordingly. Place a baseboard in an airtight bag. Rest at least four wooden blocks on the baseboard; then set the male former, covered with a sheet of plastic, on the wooden blocks. (The blocks and baseboard will ensure efficient air extraction.) Glue the veneers and stack on the former. Place a plastic sheet on top of the veneers. Check the assembly carefully. Secure in position over the former with adhesive tape. Insert the nozzle of a vacuum pump well into the bag. Seal the ends with battens and clamps. Tie string around the bag where the nozzle is inserted. Turn the pump on. Leave the veneers in the vacuum while the glue dries.

Veneer inserts

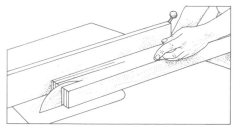
1. Cut stepped kerfs.

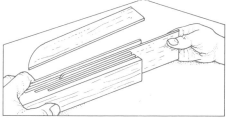
2. Insert each veneer.

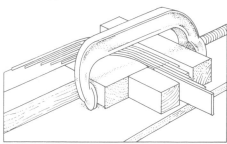
3. Clamp between formers.

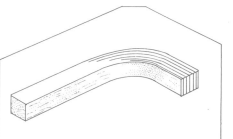
4. The final product.

Veneer inserts

Cut stepped kerfs into the end of the wood, so each tongue of wood and each kerf is approximately the same thickness (**1**). The longest kerf must be that closest to the side that will become the outside curve and should be cut so that it reaches right around the curve; the shortest kerf must be on the inside of the curve. Cut leaves of constructional veneer slightly wider than the component and slightly longer than the kerfs. Shape one end to fit the circular saw curve. Glue them and insert well into the kerfs (**2**). Place plastic sheets on either side of the component. Clamp the component between formers until the glue has dried (**3**). Then saw off the excess veneer and smooth the end. The final product (**4**).

The lathe

Turning wood on a lathe is a particularly absorbing skill for the woodworker, who will work more closely with the lathe than with most other power tools. As the lathe revolves, the cutting tools will make symmetrical cuts in the wood.

When selecting a lathe it is best to limit the choice to manufacturers that are well known in the field, since they will have overcome design problems and will be able to supply additional equipment and spares if necessary. First decide on the type of work to be turned; the size of the work should dictate the size of the lathe. For major woodturning operations, such as table legs and large diameter bowls, the lathe illustrated is best. When purchasing a lathe with its own stand remember to check that it is at a comfortable working height. An approximate guide is for the elbow to be level with the center of the tailstock, so no bending or stooping is necessary. Ensure that a bench-mounted lathe is supported on a solid base.

Whatever its size, a good quality lathe should have the following features. The motor, usually a $\frac{3}{4}$- or 1-horsepower induction motor, should be totally enclosed. Motors that are cooled by air passing through vents in the casing are to be avoided as they will draw in wood dust and chippings. The headstock should be sturdy and free from vibration and nipples should be provided for easy lubrication of the bearings. A plain phospher bronze bearing running on a taper is an ideal arrangement. The lathe bed must be strong and should be of heavy casting. In service, a few drops of oil on both parts of the bed will allow easy movement of the tailstock. The more expensive tailstocks eject the centers when wound fully back — a useful luxury.

Most lathes have the capacity for three or four speed changes. The changes of speed are made using a system of stepped pulleys in the headstock. Easy access to the pulley system is therefore important. In general, cleaner cuts can be made with the lathe turning at its highest speed. Any speed adjustment is dependent on both the size of the stock and the type of wood being turned. Thus heavy stock, such as uncut wood for a bowl, requires a slow initial speed. As the wood is removed the speed can be increased.

The type of wood to be used should be one of the first considerations when planning a wood turning session. Coarse textured timbers such as oak as well as the majority of the softwoods will not hold sharp corners or minutely detailed profiles; if attempted the fibers will crumble. These timbers are best suited to designs incorporating rounds, slow curves and straight sections. For intricate profiles, choose fine textured timbers such as mahogany or beech.

As a general rule, wood should be well seasoned and free from knots and shakes. Knots can cause tools to jump, possibly resulting in injuries to the operator or damage to the stock. It is possible to rough turn the stock first and then leave it for a while in a room of similar humidity and temperature to the place for which it is ultimately intended. Especially when rough turning bowls, fresh surfaces are exposed, which may not be thoroughly seasoned and so, later on, the finished article may distort under changes of temperature and humidity.

Headstock
The headstock contains the pulleys, which provide the driving mechanism from the motor. The lathe speed is altered by slipping the driving belt from one pulley to another after first releasing the tension. The pulleys are mounted on a spindle that projects beyond the headstock on both sides and is threaded so that a faceplate can be screwed onto it. The spindle facing the lathe bed is also fitted with a morse tapered hole for holding attachments that are not threaded.

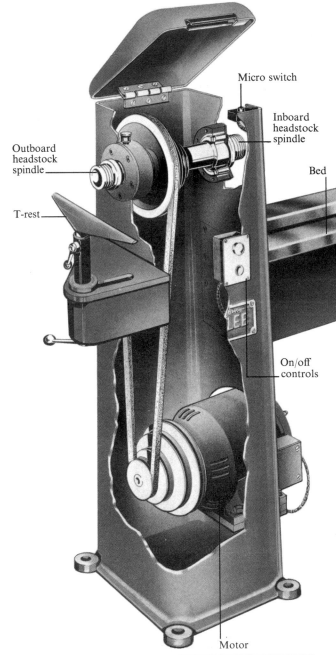

Micro switch

Inboard headstock spindle

Bed

Outboard headstock spindle

T-rest

On/off controls

Motor

Safety
○ Before switching on check that all levers are tight.
○ Before switching on turn the stock by hand for one full revolution to check that there are no obstructions.
○ Always turn the lathe off before making any adjustments.
○ Never leave the lathe running unattended.
○ Wear goggles to protect the eyes from flying wood chips.
○ Avoid any clothing that may become caught in the stock.
○ When sandpapering wear a facemask to protect the lungs.
○ When polishing never wrap the rag around the hand.

T-rest

A T-rest provides support for the lathe tools when turning. Its height is fully adjustable and can be fixed in position by a clamping lever; its head can also be moved around and clamped at any angle. Always stop the lathe while adjusting a T-rest. The top edge of a T-rest is rounded to give the tools greater ease of movement and support. T-rests are available in different lengths and are interchangeable between the headstock side and the lathe bed. They can be slid along the lathe bed.

Tailstock

The tailstock is used when turning between centers. It is fitted with a spindle, which has a morse tapered hole to accommodate a tailstock center. The spindle feed hand wheel is for moving the spindle and center against the piece of wood to be turned until the wood is held firmly. The hand wheel can be locked in position, usually with a screw at the side of the tailstock. The whole tailstock can be slid along the lathe bed and held in position by the cam operated lever at the side.

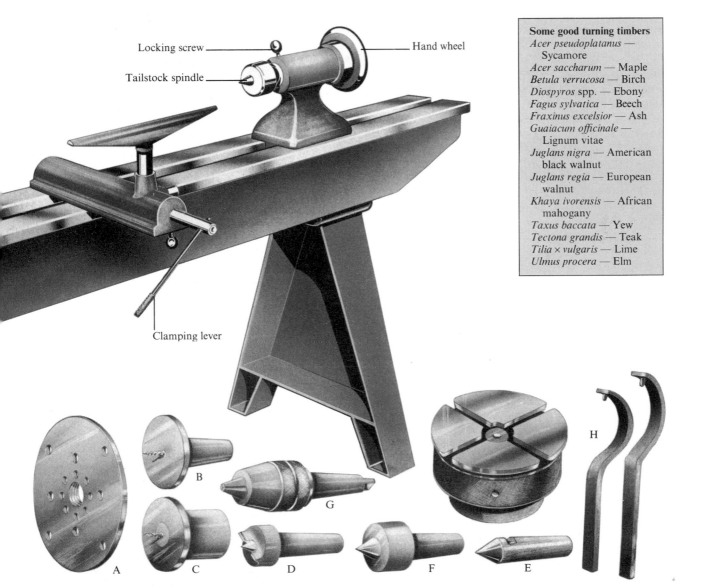

Locking screw

Tailstock spindle

Hand wheel

Clamping lever

A   B   C   D   G   F   E   H

> **Some good turning timbers**
> *Acer pseudoplatanus* — Sycamore
> *Acer saccharum* — Maple
> *Betula verrucosa* — Birch
> *Diospyros* spp. — Ebony
> *Fagus sylvatica* — Beech
> *Fraxinus excelsior* — Ash
> *Guaiacum officinale* — Lignum vitae
> *Juglans nigra* — American black walnut
> *Juglans regia* — European walnut
> *Khaya ivorensis* — African mahogany
> *Taxus baccata* — Yew
> *Tectona grandis* — Teak
> *Tilia × vulgaris* — Lime
> *Ulmus procera* — Elm

Accessories

A A faceplate is a metal disk with screw-holes and a central threaded hub that is fastened onto the headstock spindle. Each faceplate is threaded to tighten when the lathe is switched on. Each faceplate is threaded to fit either the outboard or the inboard spindle; faceplates are therefore not interchangeable. A faceplate is used when bowl turning to hold the stock onto the inside or outside of the lathe. A leather or fiber washer can be fitted between the faceplate and the spindle to prevent the faceplate jamming.

B A screw flange chuck is a small faceplate with a central wood screw and a tapered shank or threaded hub. When stock is screwed to this chuck all surfaces, both inside and out, can be shaped.
C An adjustable screw flange chuck is similar to the screw flange chuck except that the length of its screw can be altered to suit various thicknesses of wood.
D A drive center is attached to the inboard side of the headstock spindle when turning between centers. At one end it usually has a morse tapered shank, which slots into

the spindle, and at the other it has a central point and two spurs, which are fitted into the stock. A four-spurred drive center is also available and is particularly suitable for small-diameter and light work as it gives a very positive drive.
E A tailstock center supports the other end of the stock when turning between centers. It has a tapered shank, which fits into the tailstock, and a point at the other end, which is often called a dead center as it does not turn with the stock. The center is usually removed by tapping a dowel rod against the shank base.

F Alternatively the other end of a tailstock center may rotate on bearings with the stock. It is then called a revolving center. Its advantage is that heat and friction are avoided.
G A drill chuck holds drill bits for boring operations and usually has a maximum shank size of 13 mm/½ in.
H A universal chuck embodies the four traditional chucks — screw, ring, internal screw and faceplate — as well as the recently developed split ring chuck and expanding collet chuck. It has a wide range of uses and is a good investment for any wood turner.

171

Wood turning tools

The hand tools for wood turning are divided into those that cut and those that scrape. Gouges, chisels and parting tools cut into the wood and give a clean finish. Scraping tools are easier to use; but as the name implies they scrape and this tends to give a rougher finish, so considerable sandpapering is necessary. Sparing use of scrapers is a useful adjunct to the craft of wood turning, but the turner should resist the temptation to follow the initially easier scraping method most of the time. Once the art of using cutting tools has been mastered, the turner will produce much better results.

Gouges have channeled or fluted blades with curved or flat cutting edges. They are used for roughing down and reducing stock to a cylindrical shape as well as for cutting coves, beads and sweeping curves. Gouges are beveled on the convex side at an angle of 30° to 45°, depending on use. A large angle is often required when turning bowls and a smaller angle when turning work between centers.

Chisels have flat blades and straight cutting edges either at right angles or skewed at an angle of 70° to 80° to the edge of the blade. Skew chisels are used to make V cuts, beads and tapers, to cut down cylindrical stock and to smooth shoulders. They are beveled on both sides and are sharpened to form a combined beveled edge of 20° to 30°.

Gouges and chisels are held with the bevel of the tool rubbing the stock behind the cutting edge. This also gives the tool support when cutting.

Parting tools are used to cut through the stock and to make narrow recesses or grooves to a desired depth, forming a cut with parallel sides and a square bottom. The tool is beveled on both sides to give a combined angle of 20° to 30°.

Scrapers are flat bladed with bluntly angled bevels and variously shaped cutting edges. They are correctly used for light finishing cuts when turning on the faceplate or between centers and for end-grain work. The cutting edge is ground at an angle of 70° to 80°. Never use old files as scrapers as they are brittle and therefore dangerous.

The tools illustrated form a good basic kit for wood turning and they will enable the turner to tackle a variety of interesting work from chair legs to large-diameter bowls. Always buy good-quality turning tools from a reputable manufacturer. Cheap tools will only prove a continual disappointment and they will also make it difficult to acquire the basic wood turning skills.

Tools must be sharp when wood turning so they will give a clean crisp cut that will require the minimum of effort by the turner. During a lengthy turning session the tools will need sharpening periodically because the cutting edges will have removed a considerable amount of wood very rapidly and will thus have become dull. It is not necessary to regrind the tools completely every time but, as flat bevels are essential in wood turning, they must not be allowed to become too round. Careful use of the oilstone will be sufficient to renew the cutting edges many times between grinding sessions. While grinding, the blade must be kept cool. Dip it frequently in cold water when using an emery wheel; where possible a sandstone running in water is preferable.

Deep fluted bowl gouge
10 mm/$\frac{3}{8}$ in for general shaping inside and out when working on the faceplate.
Spindle gouge
6 mm/$\frac{1}{4}$ in and 13 mm/$\frac{1}{2}$ in for coving, shaping, hollowing and rounding over.
Roughing gouge
25 mm/1 in for roughing down to cylinders, waste removal and long curve shaping.

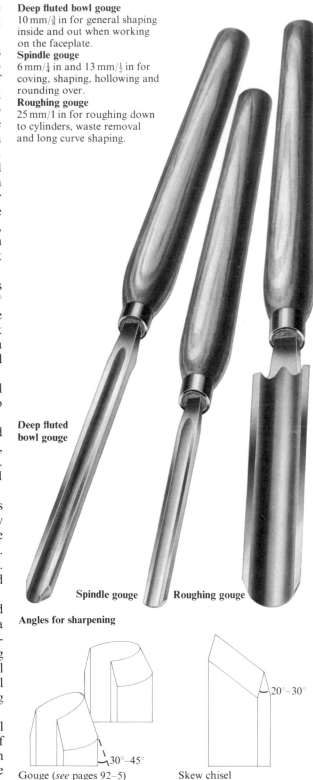

Deep fluted bowl gouge

Spindle gouge   Roughing gouge

Angles for sharpening

30°–45°   20°–30°

Gouge (*see* pages 92–5)   Skew chisel

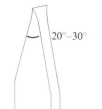

20°–30°

Parting tool

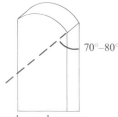

70°–80°

Round-nosed scraper

Skew chisel
19 mm/¾ in and 25 mm/1 in for
V grooves, beading and
smoothing.
Round-nosed scraper
19 mm/¾ in for final end-grain
shaping and smoothing
concave surfaces in faceplate
work.

Square-nosed scraper
25 mm/1 in for smoothing
convex and flat surfaces.
Parting tool
6 mm/¼ in for cutting work to
length and quickly forming
narrow grooves.
Sizing tool
Used with the parting tool for
cutting grooves or for setting
out diameters.

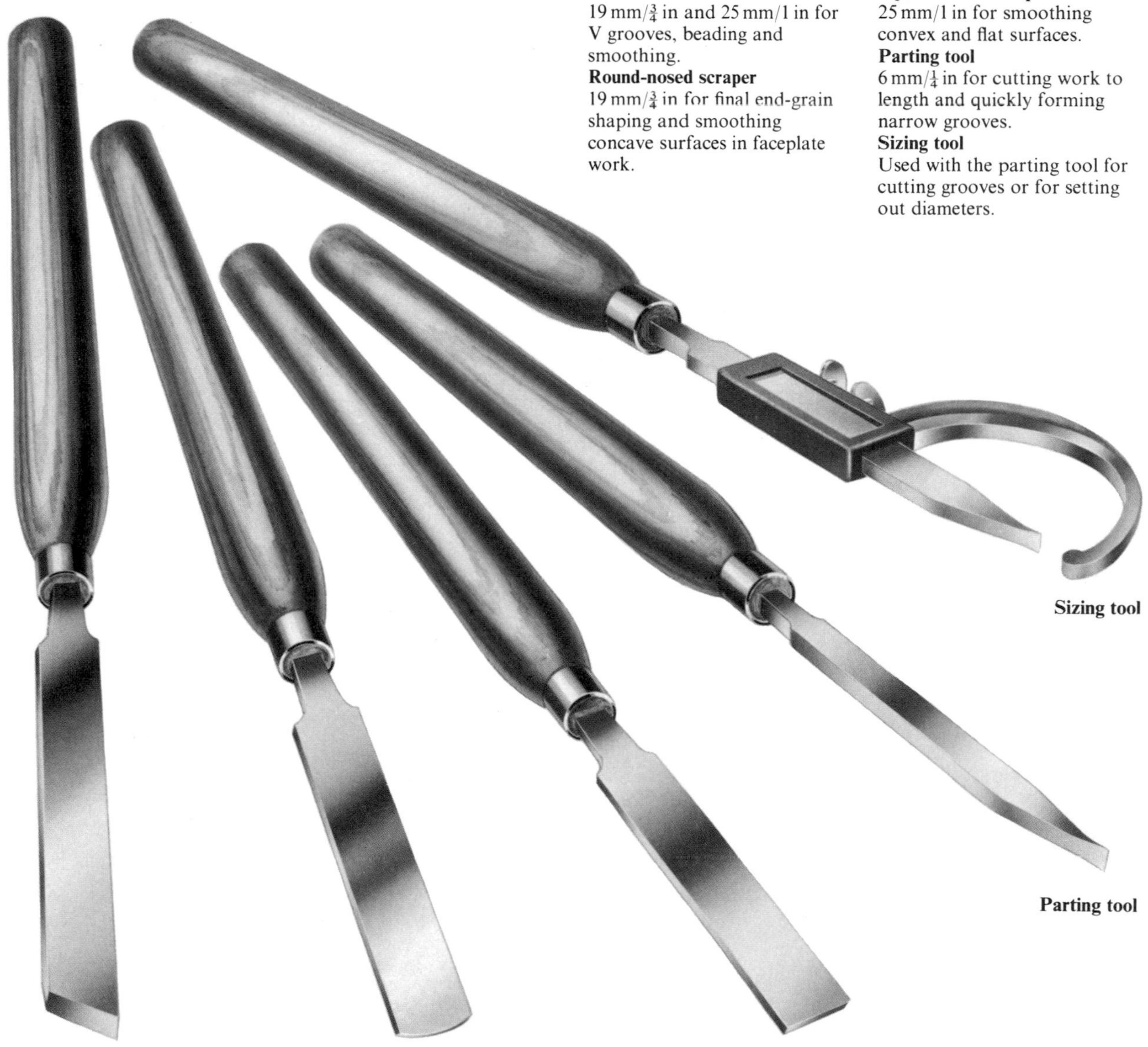

Sizing tool

Parting tool

Skew chisel

Round-nosed scraper

Square-nosed scraper

Methods of sharpening

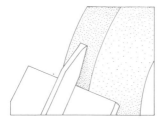
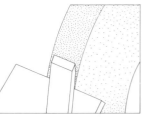
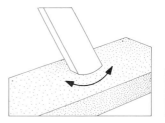
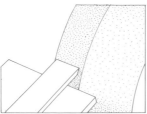

A parting tool needs frequent
sharpening. Using a tool-rest,
which can be adjusted to the
bevel required, grind the blade
to a 20° to 30° angle between
bevels. Move the blade across
the face of the wheel while
exerting light pressure. Hone
with the bevels laid flat against
the oilstone and work it back
and forth. When a wire edge
appears, hone each side
alternately until the edge
disappears.

Sharpen a skew chisel on a
grinding wheel to an angle of
20° to 30° between bevels.
Make the angle of the skew
between 70° and 80°. Place the
center of the bevel against the
rotating wheel. Press until the
complete bevel is ground to the
shape of the wheel. Grind the
other bevel in the same way.
Then hone each bevel
alternately on an oilstone,
working it back and forth to
remove the wire edge.

A round-nosed scraper may be
sharpened to any convex
tailor-made shape. Hold its
single bevel on the tool-rest at
a 70° to 80° angle to the
grinding wheel. Swing the tool
from left to right to keep the
shape even until a wire edge is
produced. At the grinding
angle hone the bevel on the
oilstone, with a rocking
motion. Then lay the blade flat
and move it back and forth to
remove the wire edge.

Sharpen a square-nosed scraper
on its single bevel at a 70° to
80° angle. Using an even
rhythm move the tool from
side to side across the grinding
wheel until a wire edge is
produced. Dip the blade in
cold water at frequent intervals
to stop the steel overheating.
Then hone the bevel, back and
forth, at an angle of 70° to 80°
to the oilstone. Finally rub the
blade flat on the stone to
remove the wire edge.

Turning between centers 1

When turning between centers, the wood to be turned is supported at one end by the tailstock center and at the other by the headstock driving center. The grain of the wood always runs parallel with the lathe bed, which means that the turning tools will always be cutting across the grain of the wood. It is from this technique that long slender items and cylindrical work with shaped profiles can be produced.

Until some confidence and skill have been acquired, it is advisable to plane the stock to an octagon before turning it on the lathe, because turning from a square can present problems. When roughing down to a cylinder or sandpapering, the stock must be revolved at a low speed. Continually adjust the T-rest so that the gap between the stock and the T-rest is 3 mm/⅛ in.

When turning between centers, try to develop the habit of keeping an eye on the general shape of the stock, especially the rear profile. Initially the temptation will be to look at the cutting edge of the tool, but this should be resisted as it will be difficult to assess whether the stock is cylindrical.

Both a scraping and a cutting method can be used in turning between centers. Although the scraping method may appeal to the beginner as being a more cautious approach, the finished surface will always be rougher than a surface that has been cut, and practice using the cutting method will in the long run be rewarded with more professional results. All cutting tools can scrape but not all scraping tools can cut, and so sandpapering will be required after scraping.

Preparing the stock and mounting it on a lathe

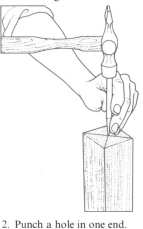

1. Draw diagonals.

2. Punch a hole in one end.

3. Make a cut in the other end.

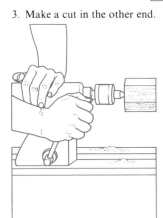

4. Plane to an octagon.

5. Mount the stock on the lathe.

6. Adjust the T-rest.

Preparing the stock and mounting it on a lathe
To locate the center of the stock draw diagonals from corner to corner at both ends (1). Then, with dividers, mark out circles on both ends of the stock. Make a location hole at the center of one end, using a nail-set or bradawl (2).
At the other end make a saw cut about 3 mm/⅛ in deep across one of the diagonals (3). This should stop the wood splitting. To relieve some of the initial stress of cutting a cylinder from a square, it is advisable to plane the stock to an octagonal section before mounting it on the lathe (4). Fasten a revolving or dead center onto the tailstock spindle. If using a dead center on the tailstock apply a spot of grease to the central hole in the stock to avoid friction when the work is turning. (This is unnecessary when a revolving center is used.) Then fit a drive center into the headstock spindle and press the center of the kerfed end of the stock into the spurs on the drive center. Push the tailstock up to the center of the stock and clamp into position with the lever. Then turn the spindle feed hand wheel to drive the tailstock center fully into the stock center (5). Retract the center a fraction to ease friction, then fix in position with the tightening screw. Adjust the T-rest parallel and central to the stock, 3 mm/⅛ in from the edge (6). Before turning, spin the stock by hand to check that there are no obstructions.

Cutting down to a cylinder

1. Lay the bevel on the stock.

2. Rough down the stock.

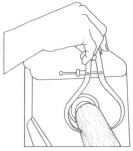

3. Check the diameter.

4. Finish with a skew chisel.

Cutting down to a cylinder
Place a 25 mm/1 in roughing gouge on the T-rest and move the bevel lightly against the stock so that the middle of the blade tip will cut away the waste (1). For quick roughing, place a hand over the gouge (2). For more control, place the hand under the gouge. Begin to cut at one end; gradually move the gouge along to the other end of the stock before bringing it back. The cylinder is then cut down to rough size using a smooth, even rhythm. Continually stop the lathe to adjust the T-rest and check the diameter of the stock, using calipers (3). Finally raise the T-rest 3 mm/⅛ in above center and smooth off the stock with a skew chisel (4). It is important for the cut to come from the center of the blade and not from the blade corners as this would result in the wood tearing or the chisel digging into the stock. To avoid the chisel digging into the shoulder start the cut a little way in from the tailstock and work towards the headstock. Then reverse the chisel and work towards the tailstock. If the skew chisel is used correctly the stock will not need smoothing.

Scraping down to a cylinder

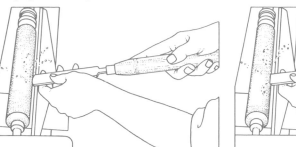

1. Rough down with a round-nosed scraper.

2. Finish with a square-nosed scraper.

Scraping down to a cylinder
Adjust the T-rest to just below the stock center. Tilt the handle of a round-nosed scraper slightly above horizontal (1). Work the tool from one end of the stock to the other. When the stock has been reduced to the required diameter use a square-nosed scraper to remove any unevenness, using a similar motion to that of the round-nosed scraper (2). Sandpaper the stock smooth.

Using a sizing tool

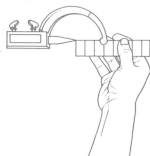

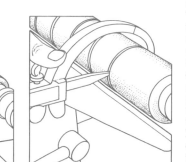

1. Set to the required diameter.

2. Place over the stock.

3. Lift the handle gently.

Using the sizing tool
The sizing tool is a parting tool fitted with an arched attachment and is frequently used for repetition work. Set the sizing tool to the required diameter and screw it in position (1). With the T-rest positioned parallel and central to the stock, make a cut with the handle held slightly below horizontal (2). Hold the position of the blade while gently lifting the handle until the required diameter is cut in the stock (3).

Turning between centers 2

Making a groove

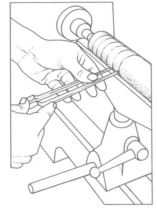

1. Scribe lines on the stock.

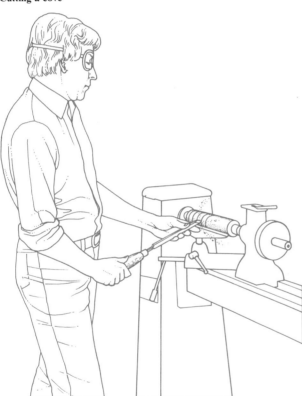

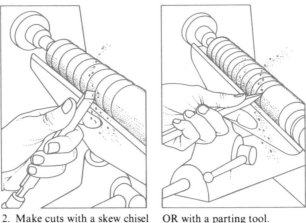

2. Make cuts with a skew chisel OR with a parting tool.

Making a groove

Grooves are commonly cut to shape profiles on cylindrical work. Clamp the T-rest centrally. With a wing compass, score the required widths along the stock (1). Set the toe of a skew chisel against each scribed line and make a vertical cut 3 mm/⅛ in deep. Hold the handle well down (2). For wider grooves use a parting tool, starting with the tool held straight onto the stock and gently lifting the handle from below horizontal until the point is cutting 3 mm/⅛ in into the stock.

Cutting a cove

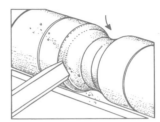

2. Start at one shoulder.

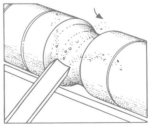

3. Cut down to the cove center.

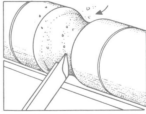

4. Cut from the other shoulder.

1. Make angled grooves.

Cutting a cove

Mark the required cove width with a wing compass. With the T-rest central to the stock make grooves at alternate angles, using a skew chisel (1). Then take out the waste from the center of the cove with a spindle gouge, inclining the handle of the gouge downwards. Twisting the wrist, cut from left down to the center and then from right down to the center, using a rhythmic movement (2, 3, 4). Only cut on the down stroke. On the up stroke the gouge should not touch the wood but the action helps to create an even rolling motion. If a cut is made from the bottom to the top of the cove it will tear the wood. Gradually cut closer to the grooves each time, twisting the wrist from the one cove shoulder to the other. This shaping process is carried out bit by bit on each side of the cove until the required depth is achieved and the cove width stretches from angled groove to angled groove.

Scraping a cove

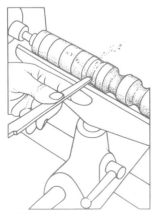

1. Scrape the cove center.

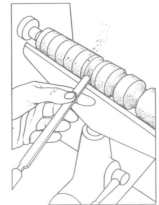

2. Work from each shoulder.

3. Sandpaper to finish.

Scraping a cove

Scribe the stock and cut out grooves that are angled alternately. Adjust the T-rest to just below the center. Using a round-nosed scraper that is slightly narrower than the intended width of the cove, feed the tool straight into the center of the stock until the desired depth is reached (1). Then work the tool from either shoulder down to the center of the cove (2). Remove the T-rest. Switch on the lathe at a low speed. Smooth the coves without removing any corners, using sandpaper wrapped tightly around a purpose-shaped sanding block (3).

Cutting a bead

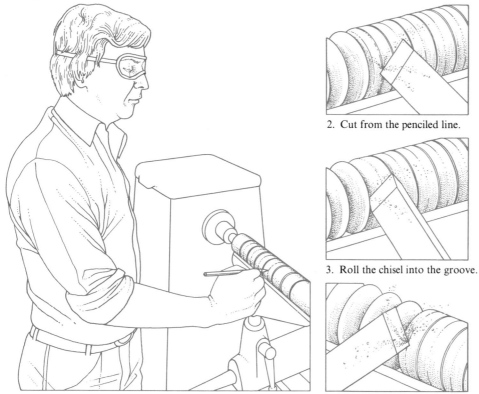

1. Draw lines on the stock.

2. Cut from the penciled line.

3. Roll the chisel into the groove.

4. Cut the other side.

Cutting a bead

Clamp the T-rest centrally. Scribe the stock, using a wing compass. Then make a vertical groove 3 mm/⅛ in deep on each scribed line. Mark the centers of the beads by holding a pencil against the revolving stock (1). Hold the skew chisel against the stock at 25° to 30° with the handle dropped to just below the T-rest. Lay the cutting edge of the chisel heel close to one of the penciled lines. Push the tool forwards so that the cutting edge of the heel is in contact with the surface of the stock, close to the bead center (2). The bevel should be flat against the stock without the heel digging into it. Raise the handle slightly as the chisel is rolled slowly up and over in the direction of the groove. Push the chisel forwards in one continuous motion so that the tool finishes in an almost vertical position (3). Cut the other half of the bead in the same way with the angles reversed (4).

Scraping a bead

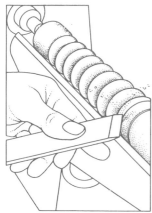

1. Scrape from the pencil line.

2. Move the toe into the groove.

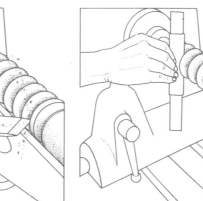

3. Sandpaper to finish.

Scraping a bead
Prepare the stock as if cutting a bead. Then hold a skew chisel flat against the T-rest with the handle slightly below horizontal. Place the cutting edge of the toe close to the central bead line. Work it round towards the groove and back (1). As the bead shape forms the tool should rock, causing the toe of the tool to point deeper and deeper into the groove (2). Never allow the toe tip to plow into the stock. Work each side separately before building up a completely balanced and rounded bead, again from each side. Sandpaper to finish (3).

Making duplicate turnings

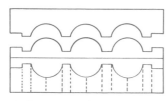

1. Draw the important points.

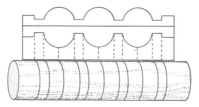

2. Transfer the points.

3. Set a sizing tool.

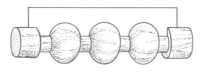

4. Check the finished work.

Making duplicate turnings
First sketch the design on card either side of a center line. Draw up the design accurately on one side, cut along the outline, fold the card flat on the center line and draw along the outline to give the exact symmetrical shape. Draw all the important points, at right angles to the center line, down to the bottom edge (1). With this edge against the stock, transfer the points (2). Set a sizing tool to the diameters on the symmetrical drawing for exact depth of cut (3). Use the "negative" piece of card from the cut outline to check the finished work (4).

Turning on the faceplate 1

Bowls, trays and many other circular objects can be turned on a faceplate, which is fixed either inside the headstock or outside when working on large-diameter stock. Successful turning requires much skill, practice and a certain amount of luck because of the inherent structure of wood. As with turning between centers, the stock must be well seasoned and generally free of knots, although knots can sometimes enhance the work if they are in the base. The stock must be cut with the grain running across its diameter; otherwise the edges of the stock will split and crack. There will however be two distinct end-grain patches on the stock where the chisel or gouge is cutting into the grain and it is these areas that will cause trouble as they tend to tear out under the action of the chisel or gouge. Cutting must be done between the center and the right-hand edge. If the tool is held elsewhere it will be lifted away. As always tool sharpness is most important.

There should be minimum delay between turning the outside and inside of the stock because the wood will shrink slightly across the grain, causing the rim of the stock to become unevenly thick. Ideally, the outside should be turned and finished one day and the inside either that same day or the next.

Preparing the stock and mounting it on a faceplate

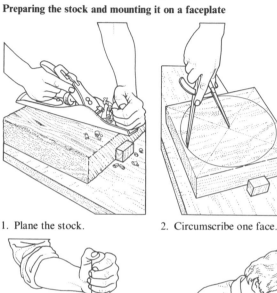

1. Plane the stock.

2. Circumscribe one face.

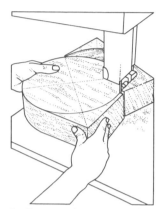

3. Cut away the waste wood.

4. Drill the screw-holes.

5. Secure the screws tightly.

6. Mount the faceplate.

7. Position the T-rest.

Preparing the stock and mounting it on a faceplate

With a jack plane, flatten one face of the stock against a bench-stop so that there are no gaps when it is placed on a faceplate (1). This will avoid vibration during cutting. Locate the center by drawing diagonals across the stock, using a pencil and rule. With a wing compass, mark out a circle that is slightly larger than the intended diameter of the bowl (2). Then cut away the waste wood, using either a bow saw or a band saw (3). Ensure the stock is as near round as possible to minimize the forces during roughing down. It is highly dangerous to turn a stock on the faceplate with the corners left on. Select a faceplate that is of smaller diameter than the stock. When the tools are unlikely to come into contact with the faceplate while turning, a faceplate that is larger than the stock can be used. (Alternatively take some waste wood of larger diameter than the stock; plane both sides flat and screw it to the stock.) Place the planed side of the stock on the faceplate and visually center it. Select at least four screws that will penetrate through the 6 mm/¼ in-thick faceplate so that it cannot judder or vibrate during the cutting process. Drill the screw-holes with a hand drill, ensuring the holes are not aligned along the grain (4). Lubricate the screws with tallow or soft soap before securing them as tightly as possible in the holes (5). Check that there is a leather washer on the hub of the outer faceplate spindle, giving a resilient cushion to prevent jamming. Secure the faceplate firmly to the lathe by turning the faceplate on to the spindle with one hand and pulling the pulley cone gently in the opposite direction with the other (6). Finally position the T-rest so that it is central and parallel with the stock 3 mm/⅛ in away (7).

Turning the underside

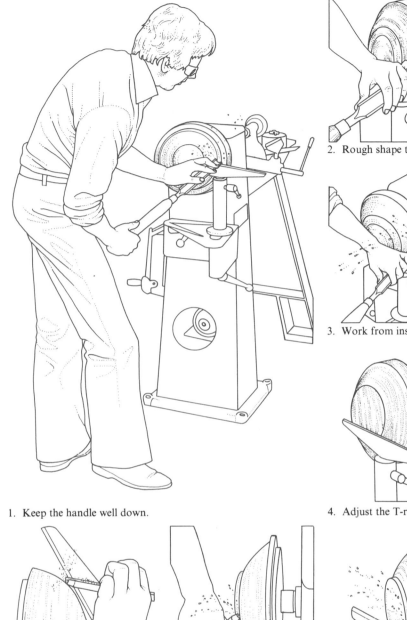

1. Keep the handle well down.

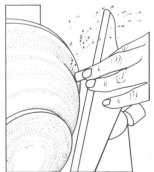

5. Neaten the sides.

8. Mold the lip with a scraper.

2. Rough shape the stock.

3. Work from inside to outside.

4. Adjust the T-rest continually.

6. Cut a plinth in the base.

9. Sandpaper the outside.

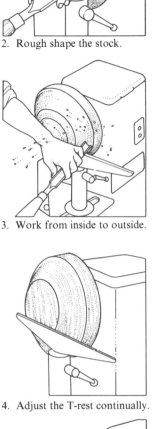

7. Smooth with a scraper.

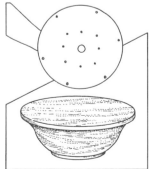

10. Remove the faceplate.

Turning the underside

Spin the stock to ensure there are no hindrances. Then switch on the lathe. Initial roughing down speed should be low. Mark the required diameter of the base with a pencil and steel rule while the stock is spinning. The work of truing up the edge of the stock and rough shaping now begins, using a 25 mm/1 in roughing gouge. With forearm resting lightly on the leg, keep the handle well down and the bevel touching the stock (1). The other hand should steady the tool as it lies on the T-rest. Rough shape the stock by cutting a chamfer at the edge of the stock (2). Gradually increase it towards both the rim of the stock and the diameter of the base (3). Continually adjust the T-rest so it stays parallel with the edge of the stock 3 mm/⅛ in away (4). Once the basic shape has been cut, speed up the lathe. A 13 mm/½ in deep fluted bowl gouge is then used to cut a more accurate shape (5). In order to form the plinth at the base, cut the stock with the gouge turned round (6). Adjust the T-rest to just below the center and use a very sharp square-nosed scraper to smooth the outside of the stock. Point the scraper slightly downwards, steadying it on the T-rest (7). Work outwards and inwards rhythmically. Then use a round-nosed scraper to clean the lip, pivoting it around on the T-rest to make the lip more pronounced (8). Remove the T-rest. Then sandpaper (9). Never use coarse sandpaper as it will cause radial scratches. Set a slow speed and allow the paper to trail in the direction of the stock's rotation. Mold across the shape of the stock, paying special attention to corners and edges. Stop the lathe. The stock is ready for its surface finish. Olive oil, for example, is ideal for a salad bowl and wax suits a fruit bowl. Many other finishes, however, can be applied to the stock. Once the oil or wax finish has dried it should be lightly buffed into the grain. Switch the lathe to the slowest speed and with a clean cloth work out from the center, supporting the wrist with the other hand. Apply a couple of coats within 2 to 3 hours. Leave the stock overnight to dry. Then sandpaper the surface carefully. Remove the stock from the lathe and take out the four screws holding it to the faceplate. Then lift away the faceplate (10).

Turning on the faceplate 2

Mounting the stock base

1. Screw on the faceplate.

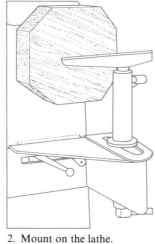

2. Mount on the lathe.

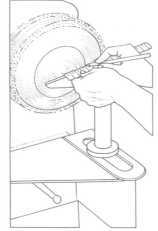

3. Draw a circle.

Mounting the stock base

Before turning the inside of the stock take a 19 mm/¾ in thick piece of waste softwood that is larger than the diameter of the stock base. Plane the waste wood flat so that the faceplate will fit flush against it and remove its corners, which could be dangerous when turning at speed. Screw the waste wood to the faceplate **(1)**. Mount the faceplate on the spindle and position the T-rest parallel to the waste wood just below its center **(2)**. Turn the waste wood by hand to ensure there are no hindrances. With a steel rule and pencil mark out a circle that is the same diameter as the stock base **(3)**. Remove the waste within the circle to a depth of 3 mm/⅛ in, using a square-nosed scraper **(4)**. Stop the lathe frequently to check that the base of the stock fits closely into the recess. Remove the faceplate from the lathe. (For wide work, measure and make a note of the stock depth. Use the measurement at a later stage of turning as a check against the inside depth.) Cut out a piece of thick brown paper to fit the recess exactly. Using a sparing amount of PVA or animal glue stick the paper into the recess; then glue the stock base onto the paper so that the grain of the stock is at right angles to that of the waste wood base. Place a weight on top of the stock and leave to dry **(5)**.

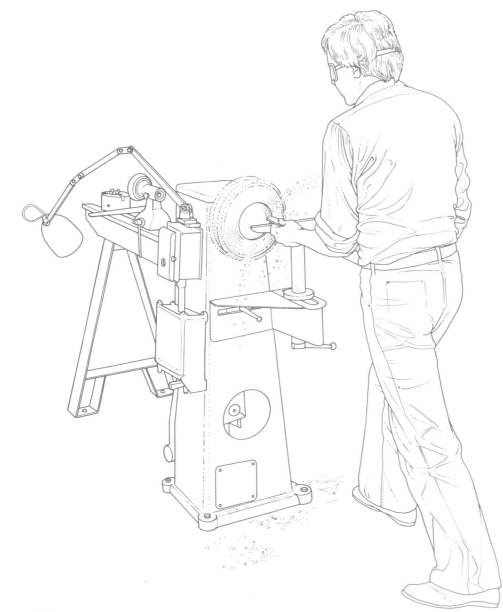

4. Scrape out a recess.

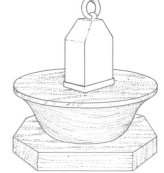

5. Leave the glued base to dry.

Turning the inside

1. Clamp the T-rest.

2. Make a cut into the middle.

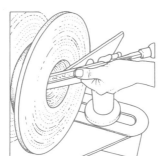
3. Twist the gouge round.

4. Twist the gouge into the stock.

5. Measure the inside depth.

6. Sight the depth cut.

7. Finish off with a scraper.

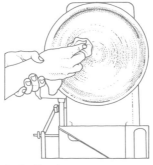
8. Buff with a clean cloth.

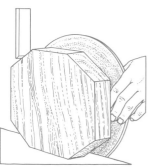
9. Release the stock.

10. The final product.

Turning the inside

Mount the faceplate, together with the waste wood and stock, on the lathe. Clamp the T-rest slightly lower than the stock center and spin the stock manually to check clearance (1). Hold the handle of a deep fluted bowl gouge slightly below horizontal and start cutting into the middle, placing the gouge at right angles to the stock (2). Keep the bevel touching the stock. When cutting for depth, the handle should be lifted up slightly so that the edge of the blade cuts into the stock. As the cut develops, twist the handle around to about a 45° angle to the stock (3). As the hollow becomes wider and deeper the twisting movement becomes more pronounced. Continually adjust the T-rest so that it remains close to the work. Once the hollowing procedure is well under way a roughing gouge may be used to cut away additional waste quickly. Using body pressure twist the gouge into the stock (4). Stop the lathe periodically to dust down the stock and to check its depth. To do this, place a steel rule across the rim and rest another at right angles to it (5). With the second rule held at the inside depth, place the rule against the outside and sight the depth cut (6). At the required depth raise the T-rest to the center of the stock and use a round-nosed scraper to finish off the inside and to smooth the rim. Hold its handle up and the blade down, thus lifting the bevel off the stock (7). Then switch the lathe to its lowest speed and sandpaper the stock, working from the center outwards. Stop the lathe and apply the chosen surface finish. Where necessary buff it into the grain with the lathe on its slowest speed (8). After the surface finish has been applied, remove the stock from the lathe and unscrew the faceplate. Position the waste wood upright on the edge of the bench, with its grain running vertically. With an assistant holding the stock, place a chisel on the edge of the waste wood, with its blade aligned along the grain. Split the grain by giving the chisel a sharp tap with a mallet (9). Work the chisel along from each edge until the stock is released. Remove any surplus glue and paper from the base of the stock and lightly sandpaper it. Touch up the surface finish. The final product (10).

Wood carving tools and basic techniques 1

Wood carving provides the opportunity of combining woodworking skill with creative ability, whether used purely to embellish a piece of wood or to produce a complete work in itself. When choosing wood for carving, ensure that it is free from knots, shakes and splits and is, preferably, fine textured for clean cutting. Lime is thus particularly suited to carving, and the attractive grains of yew and various species of pine are ideal for abstract shapes.

Wood carvers have several different names for each implement. Those illustrated below make up the basic carvers' tool kit; a selection for the more advanced wood carver is shown opposite. All carving tools are beveled on both sides to permit easy entry into and through the wood. Their blades are made of tempered steel to give maximum strength and are bought with unsharpened edges and without inner bevels. Therefore, before a session, the woodworker must always sharpen his carving tools. Bevels should be made to an approximate length of 13 mm/½ in. The harder the wood the shorter the bevel required. The inner bevel on a gouge must be honed on a slipstone (*see* page 94).

With tools that have sharp edges, accidents can all too easily occur unless certain safety precautions are taken: always check the stability of the work surface and the wood itself before commencing work. One good light source is essential. The tools should be laid out on the bench with their handles towards the carver.

Skew chisel
6mm/¼ in. Useful for undercutting. The skew blade is beveled on both sides.
Parting tool
60° 6 mm/¼ in and 45° 13 mm/½ in. For incised lettering, pattern detail and outlining. It has two straight edges to form a V shape of varying angles.

Basic cutting tools

Skew chisel

Parting tool

Straight gouge

Fluter

Straight gouge
3 mm/⅛ in, 6 mm/¼ in, 10 mm/⅜ in, 13 mm/½ in and 16 mm/⅝ in. For general carving. It must be sharpened to a bevel on the inside as well as the outside. Its depth ranges from almost straight to semicircular.
Fluter
10 mm/⅜ in and 13 mm/½ in. A gouge with a sweep greater than a semicircle.

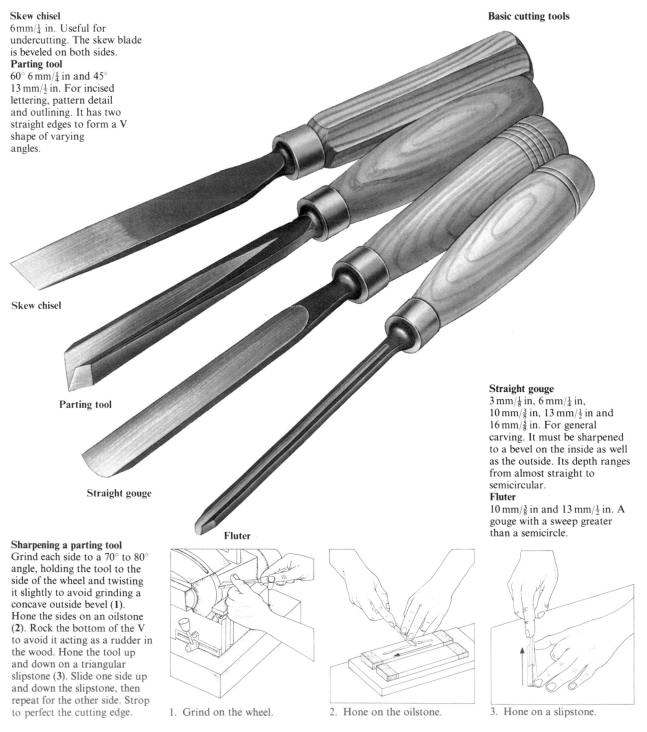

Sharpening a parting tool
Grind each side to a 70° to 80° angle, holding the tool to the side of the wheel and twisting it slightly to avoid grinding a concave outside bevel (**1**). Hone the sides on an oilstone (**2**). Rock the bottom of the V to avoid it acting as a rudder in the wood. Hone the tool up and down on a triangular slipstone (**3**). Slide one side up and down the slipstone, then repeat for the other side. Strop to perfect the cutting edge.

1. Grind on the wheel.

2. Hone on the oilstone.

3. Hone on a slipstone.

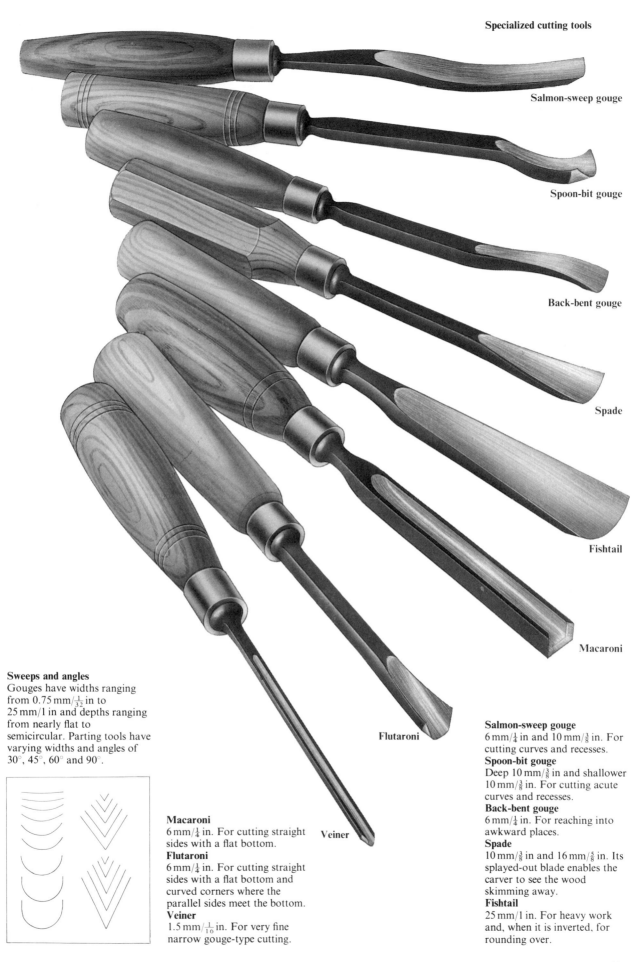

Specialized cutting tools

Salmon-sweep gouge

Spoon-bit gouge

Back-bent gouge

Spade

Fishtail

Macaroni

Flutaroni

Veiner

Sweeps and angles
Gouges have widths ranging
from 0.75 mm/$\frac{1}{32}$ in to
25 mm/1 in and depths ranging
from nearly flat to
semicircular. Parting tools have
varying widths and angles of
30°, 45°, 60° and 90°.

Macaroni
6 mm/$\frac{1}{4}$ in. For cutting straight
sides with a flat bottom.
Flutaroni
6 mm/$\frac{1}{4}$ in. For cutting straight
sides with a flat bottom and
curved corners where the
parallel sides meet the bottom.
Veiner
1.5 mm/$\frac{1}{16}$ in. For very fine
narrow gouge-type cutting.

Salmon-sweep gouge
6 mm/$\frac{1}{4}$ in and 10 mm/$\frac{3}{8}$ in. For
cutting curves and recesses.
Spoon-bit gouge
Deep 10 mm/$\frac{3}{8}$ in and shallower
10 mm/$\frac{3}{8}$ in. For cutting acute
curves and recesses.
Back-bent gouge
6 mm/$\frac{1}{4}$ in. For reaching into
awkward places.
Spade
10 mm/$\frac{3}{8}$ in and 16 mm/$\frac{5}{8}$ in. Its
splayed-out blade enables the
carver to see the wood
skimming away.
Fishtail
25 mm/1 in. For heavy work
and, when it is inverted, for
rounding over.

Wood carving tools and basic techniques 2

Carving methods

Carving stance.

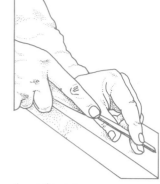

A running cut.

A reverse running cut.

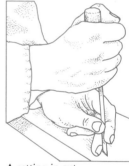

A setting-in cut.

Driving a tool with a fist.

Carving methods

For greatest control when carving, always stand up. Ensure the workbench is at a level that is comfortable so it is not necessary to stoop over the work, which should be held securely. To keep control of the tool, always rest an arm on the work or workbench. Never allow the body to block the path of any cut. Move the work around as each part is carved so the wood being worked is always close to hand.

For a running cut, rest the thumb and the middle finger on the tool while the index finger exerts a backwards pressure to prevent the tool running through the wood too quickly. Push the tool forwards with the other hand and point the index finger down the shank. Gauge the depth of an incision by how much blade is showing out of the wood.

Where it is not always possible to cut away from the carver, move the body away from the direct path of the tool and let the thumb provide the backwards pressure.

For a setting-in cut, stab the tool vertically into the wood. Clench the fist and place the fingers of the other hand close to the wood to ensure accurate positioning.

Where a mallet would prove too forceful and an ordinary running cut not forceful enough, hit the tool handle instead with the fleshiest part of the palm.

For hard wood and large work wrap the fingers around the tool to provide a backwards pressure and rest the wrist against the work. Hit the tool handle with a mallet.

Always cut in the direction of the wood grain where possible. When cutting a circle make four separate cuts in the directions indicated in the diagram.

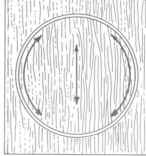

Carving a circle.

Driving a tool with a mallet.

Mallets

A round wooden mallet is traditionally used for driving cutting tools. Mallets with malleable-iron heads are a modern improvement. They require less exertion from a carver and do not bounce on the wooden handles of the cutting tools so much as a wooden mallet.

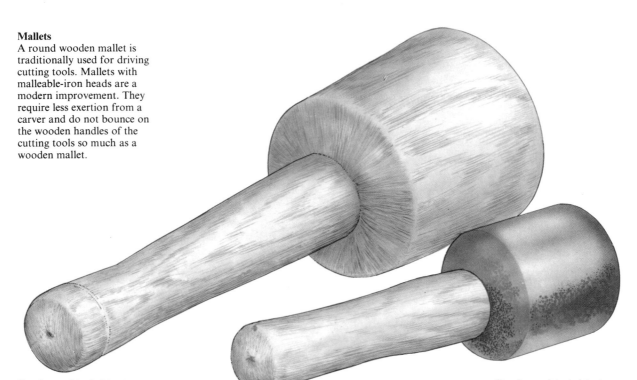

Punches and tooled textures

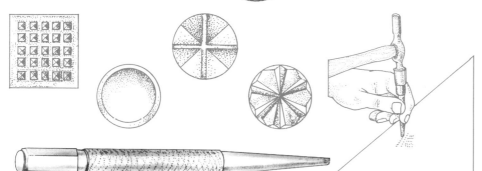

Punch and punch tips.

Tap the punch head gently.

Punches and tooled textures

Wooden surfaces, especially backgrounds, may be punched or tooled. To make a small impression, hit the punch head with a hammer. The frosting punch impressions must overlap; the impressions of a ring punch just touch. A wide variety of patterned punch impressions are available.

A tooled surface is often preferred to a smooth or punched surface. The surface of the wood is carved so as to give it a particular texture. The texture may be chosen at random for, say, a background or it may have been chosen specifically for the purpose of enhancing or emphasizing the character of the carving.

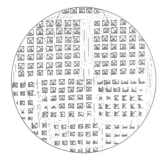

Frosting punch.

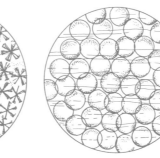

Star punch.

Ring punch.

Textures that can be tooled.

Wood carving tools and basic techniques 3

Rasps and rifflers

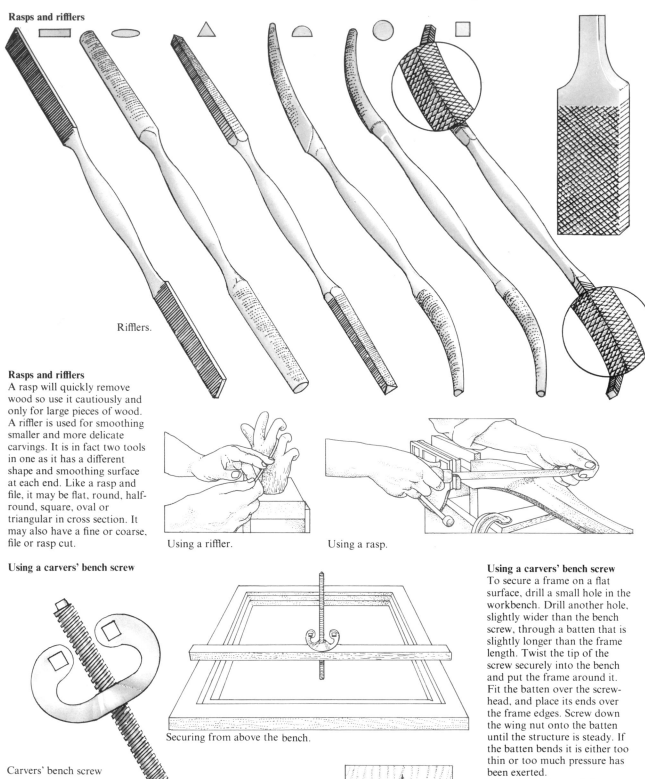

Rifflers.

Rasps and rifflers
A rasp will quickly remove wood so use it cautiously and only for large pieces of wood. A riffler is used for smoothing smaller and more delicate carvings. It is in fact two tools in one as it has a different shape and smoothing surface at each end. Like a rasp and file, it may be flat, round, half-round, square, oval or triangular in cross section. It may also have a fine or coarse, file or rasp cut.

Using a riffler.

Using a rasp.

Using a carvers' bench screw

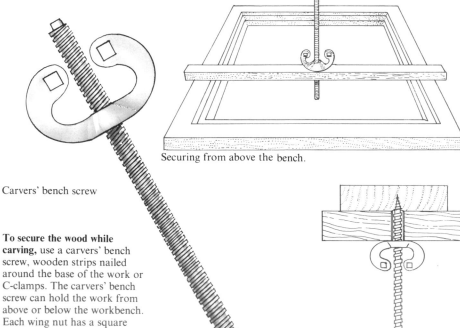

Carvers' bench screw

Securing from above the bench.

To secure the wood while carving, use a carvers' bench screw, wooden strips nailed around the base of the work or C-clamps. The carvers' bench screw can hold the work from above or below the workbench. Each wing nut has a square hole that, when locked on the screw-head, acts as a wrench.

Securing from below the bench.

Using a carvers' bench screw
To secure a frame on a flat surface, drill a small hole in the workbench. Drill another hole, slightly wider than the bench screw, through a batten that is slightly longer than the frame length. Twist the tip of the screw securely into the bench and put the frame around it. Fit the batten over the screw-head, and place its ends over the frame edges. Screw down the wing nut onto the batten until the structure is steady. If the batten bends it is either too thin or too much pressure has been exerted.

To secure the work from below the bench, drill a small hole in the underside of the wood to be carved. Twist the screw tip securely into the hole. Then drill a clearance hole in the workbench. Put the screw-head through the bench hole. From beneath the bench, screw up the wing nut until the wood is held tightly and securely against the bench with the nut tight against the underside.

Marking up with carbon paper

1. Secure the design firmly.

2. Trace on the design.

3. The design transferred.

Marking up with a stencil

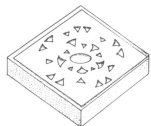

1. Secure the stencil to the wood.

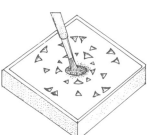

2. Dab on the paint.

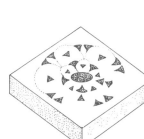

3. Touch up the design.

Marking up with tools

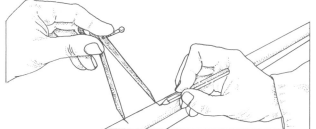

1. Transfer the measurements.

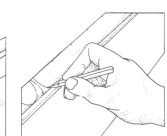

2. Fill in the design.

Checking work in progress

1. Measure the model.

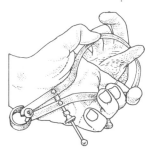

2. Measure the carving.

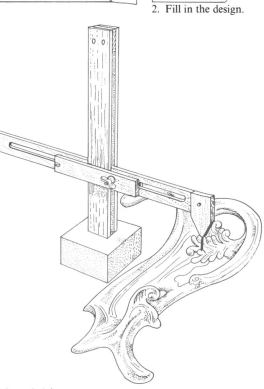

Using a height gauge.

Marking up with carbon paper

Draw the design on a piece of paper. Lay some carbon paper face down on the selected piece of wood and, on top, place the design paper (**1**). Check that no part of the design extends beyond either the edges of the wood or the carbon paper. Secure both papers with cellophane tape. Using a wooden stick with a very thin point, trace around the design (**2**). When complete, lift away both papers. Where necessary, touch up the design. The design transferred (**3**).

Marking up with a stencil

Draw a design on some stiff paper. Then cut away the major features, using a sharp, pointed knife, to create a stencil. Secure the stencil to the wood to be carved (**1**). With a brush, dab some paint wash through the holes (**2**). Carefully remove the stencil. When the paint is dry, draw in the complete design using the paint marks as a guide (**3**).

Marking up with tools

Set dividers against salient points on the design. Move them onto the wood to be carved and transfer the measurement, using a pencil (**1**). Using these datum marks, complete the rest of the design freehand (**2**).

Checking work in progress

Once a carving is under way, use dividers to check the work against the design, and calipers to take measurements when high-relief carving or carving in the round (**1**). The curved legs of the calipers allow them to reach around any projection on the original model. Then check the measurement against the work (**2**). Make a height gauge to measure relief depths. The arm should slide vertically and horizontally and should swing through almost 180° so that depths can be checked against any model. Like the calipers and dividers, the height gauge is secured with a threaded nut.

Chip and incised carving

Chip carving
Technique is based mainly on
cutting inverted pyramids. To
mark up an inverted pyramid
first draw a triangle. Then
bisect each of the angles to give
the center **(1)**. Any design will
consist of a number of
pyramids. Transfer the design
onto the wood, then secure the
wood firmly **(2)**. With a parting
tool make a pilot cut from
each corner of the triangle to
its center **(3)**. Set in along these
cuts with a cabinetmakers'
chisel, increasing the depth
towards the center **(4)**. Pare
back towards the lines of the
triangle, gradually removing
the waste **(5)**. Sharpen up all
sides ensuring that they slope
evenly. Cut out each inverted
pyramid in the design similarly.
The final chip carving **(6)**.

Chip carving

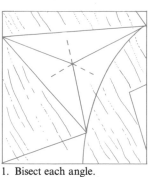

1. Bisect each angle.

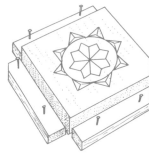

2. Secure the wood firmly.

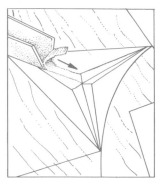

3. Cut from each corner.

4. Set in from each corner.

5. Cut along the outside lines.

6. The final chip carving.

Incised lettering

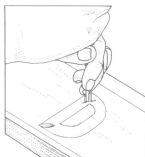

1. Intersect the angle.

2. Cut down the center.

3. Make an angled inside cut.

4. Make an angled outside cut.

5. Cut cleanly around the edges.

Incised lettering
Transfer a design onto the
wood. Secure the wood to the
bench. Where a right angle
occurs in a letter, use a parting
tool to intersect the angle from
the edge to the center of the
letter **(1)**. Cut down the center
of the letter between the design
lines with the parting tool **(2)**.
Using a shallow gouge or
chisel, cut along the inner
design line into the center of
the letter at an angle of
approximately 55° **(3)**. If the
tool is held at a steeper angle it
is likely to jam in the wood.
Cut cautiously to achieve a
uniform depth. If too shallow,
the cut can always be made
deeper. Cut the other side,
using a chisel or a gouge, from
the outer edges of the lettering
(4). Take out the waste as the
cuts progress so that the design
can be seen clearly. Cut cleanly
around the edges with a wide
gouge so that they are sharp
and perfectly shaped **(5)**.
Finally, any open right-angled
corner should be cut away as
an inverted pyramid, as in chip
carving. Cut at an angle of 55°
to the edge and at a similar
depth to the incisions already
cut at the center.

Repeat carving

Repeat carving

1. Make a tapered downward cut. 2. Cut down from the semicircle. 3. Set in around the egg.

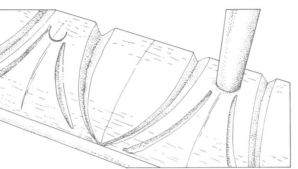

4. Set in the curved line. 5. Cut the semicircle.

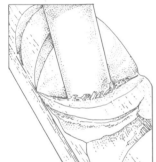

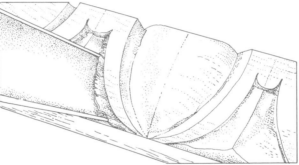

6. Round over the egg. 7. Carve the waste wood from either side of the dart.

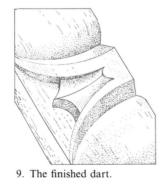

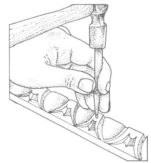

8. Cut the dart. 9. The finished dart. 10. Punch in the decoration.

Egg and dart molding parts

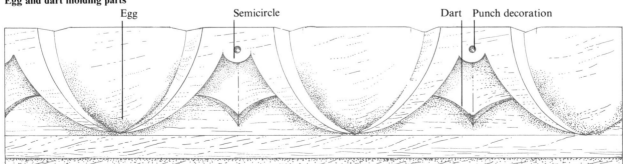

Egg   Semicircle   Dart   Punch decoration

Repeat carving

Having mastered the techniques of chip carving and incised lettering, repeat carving should seem a natural progression. It most often occurs on moldings. Make each cut throughout the length of the molding before moving on to the next. This is to reduce the time spent picking up and putting down tools. With the parting tool make a tapered downward cut on each side of the egg (**1**). Using the same tool, cut a curved downward line from the semicircle towards the bottom center of the egg (**2**). Remove the waste wood. With a wide gouge that fits the contour of the molding, set in around the egg (**3**). Set in the curved line as well (**4**). With a suitably shaped gouge held vertically, cut the semicircle between the eggs (**5**). Round over the egg with an inverted gouge (**6**). Carve out the waste wood from either side of the dart leaving a straight vertical line down the center (**7**). Set in the point of the dart with a spade (**8**). Then finish the point of the dart. Clean out the waste wood, using a spade (**9**). Decorate the center of the semicircle with a punch or nail to complete the design (**10**).

Relief carving

Unlike incised carving where the design is sunk into the wood, relief carving has the design raised above a cut-away surface. Low relief carving can be very simple. The outline is cut to a uniform depth and the waste wood is removed from the background to the depth of the relief line. The design is then ready for setting in and modeling.

What is termed "high relief" is in fact deep relief and the deeper the relief, the greater the number of planes on which the design must be carved. The techniques in high relief work are similar to those required for carving in the round and demand considerable skill and confidence. The main technique of undercutting gives a more three-dimensional effect. Undercutting is done by carving inwards beyond the vertical to give each feature a greater shape and to cast shadows, thus enhancing the three-dimensional effect. Bent gouges and chisels are used to form the corners and crevices.

As with carving moldings, each stage of relief carving must be completed throughout the design before moving on to the next. The background can be either punched or given a tooled texture. Check the finished work for areas that may appear too deep or shallow under fixed viewing light. The Tudor rose design used to illustrate the technique is carved on a number of different planes.

Relief carving

1. Draw a relief line.

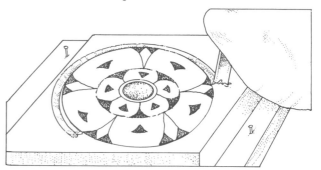

2. Cut the outer circle.

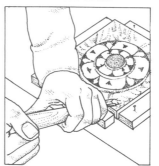

3. Remove the groundwork.

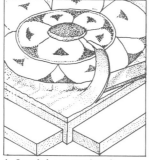

4. Level the groundwork.

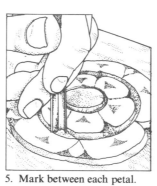

5. Mark between each petal.

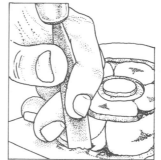

6. Set in the petals.

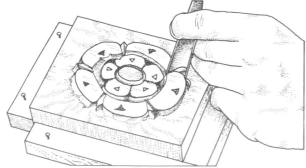

7. Curve around each petal.

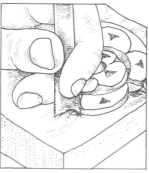

8. Make two vertical cuts.

Relief carving
Before commencing relief carving, transfer a design onto the wood. Then draw a line around the edges of the block to indicate the depth of the groundwork (**1**). The position of this line depends on what depth of relief is intended. Secure the block on four sides with wooden strips and then begin by cutting the outer, middle and inner circles around the rose with a fluter (**2**). Cut closely with the grain where possible. With a larger gouge, remove the groundwork to within 3 mm/⅛ in of the relief line, pressing backwards with the other hand to prevent the tool running into the design (**3**). With a shallow tool, flatten away the groundwork and smooth down to the relief line (**4**). Using a parting tool, mark the division between each petal with a vertical cut (**5**). Set in with a wide gouge (**6**). Curve around each petal to make individual shapes (**7**). Keep referring to the original design to check that no part, such as where an outer petal or leaf is to be carved, is being cut away. It is easy to make a mistake when the design is being carved on many different levels, so always work with great care. Remove the waste wood continually throughout the process. To carve the leaves emerging from under the outer petals, make two vertical cuts down to the leveled groundwork (**8**). Remove the

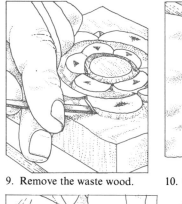

9. Remove the waste wood.

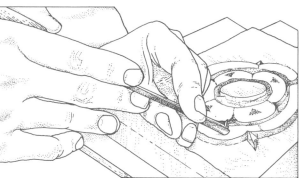

10. Cut the turnover shape.

waste **(9)**. Cut the turnover shape on the inner and outer petals with a fluter **(10)**. Carve gently down into the center on both inner and outer circle of petals, using a shallow gouge **(11)**. With a pencil, draw in the outer petal stems. Then, using a small fluter, take away the wood on either side of each of the stems **(12)**. Set in with a shallow gouge or spade **(13)**. With an inverted gouge, round over the petal edges and the central button of the flower **(14)**. Soften the inner and outer turnovers **(15)**. Undercut the outer petals and the little pointed leaves **(16)**. For a final effect use a punch and hammer to texture the groundwork **(17)**. The Tudor rose carved in relief **(18)**.

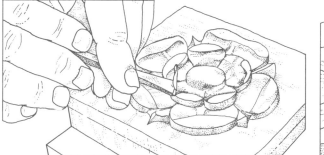

11. Carve into the petal center.

12. Remove the waste wood.

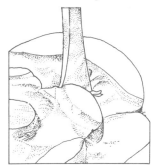

13. Set in each stem.   14. Round over the petal edges.

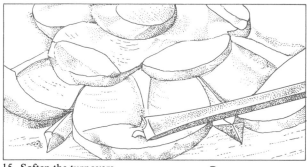

15. Soften the turnovers.

16. Undercut the outer leaves.   17. Texture the groundwork.   18. The final carving.

Veneering

Veneering is an invaluable part of woodworking. Many woods are very expensive or have grains that are unsuitable to work in solid form; thin slices of wood, known as veneers, offer the opportunity of using some of these rare woods and their figures. Beautiful decorative patterns can be achieved with veneers, especially by incorporating banding and borders of contrasting color to the central veneer panel. Leaves of veneer stacked as they were cut from the log are particularly useful for matching veneers.

There are two methods of laying veneers: by using cauls and by hand. Cauls are used for laying thick or difficult pieces and for built-up patterns, as in marquetry and parquetry. The veneers are all taped together before being laid with either animal or resin glue. Hand veneering requires fast and organized working because each veneer is laid separately with hot animal glue, which must remain hot until the veneer is correctly positioned. Contact adhesives do not give good results as positioning the veneer on the ground is difficult, bumps in the surface are almost unavoidable and blisters are difficult to deal with.

Veneers should always cover the base or ground onto which they are to be glued, with a little extra width and length. Avoid veneering over surface joints or end grain wherever possible, especially on face veneer, as sinkage usually results. When veneering both sides of the ground, or substrate, always lay the underside veneer before the face veneer.

To cut veneers use a knife and a metal straightedge, except oak or sycamore when a hardwood straightedge should be substituted. A trimming knife or scalpel is ideal. Cut the fibers with several scoring strokes to avoid splitting the veneer.

Any splits in veneer must be taped with gummed paper tape once the veneer is laid on the ground. Always tape veneer joins with gummed paper tape. This holds the joins together so the veneers will shrink from the sides. Do not use cellophane tape, which is very sticky and can easily tear the veneer on removal.

Veneer grounds (substrates)
The ground is the base for veneers and can be of any thickness. It can be of solid wood provided the growth rings run at right angles to the surface and it is flat, true, unsplit and free from knots and any other defect. Manufactured boards are also suitable as grounds. On all manufactured boards and all solid wood of a thickness less than 13 mm/$\frac{1}{2}$ in lay a plain inexpensive counter veneer on the reverse side of the ground to prevent the ground curling as the glue dries and the veneer contracts.

Preparing the veneers
Unpack the batch of veneers carefully. Keeping the leaves in order, number them consecutively with a soft pencil (1). The adjoining leaves will then match almost perfectly. Gently rub the veneers on both sides with a warm damp cloth to flatten them (2). Lay no more than ten leaves at a time between flat boards, and clamp for at least three hours or overnight until they are almost, but not quite, dry (3). On a flat board cut the sheets singly to the required size, allowing 6 mm/$\frac{1}{4}$ in extra all round. To facilitate a closer join tilt the knife blade slightly away from the straightedge (4). Cut the veneer with several light strokes. Too much pressure, especially across the grain, will split the veneer. It might be helpful to cut the last bit away from the uncut end.

Veneer grounds

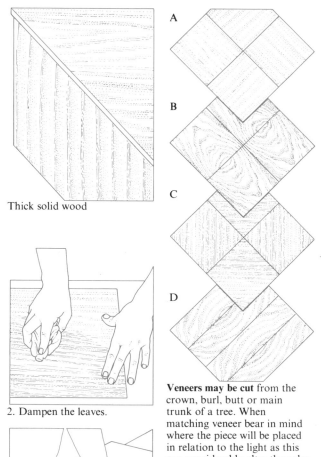

Manufactured boards

Thick solid wood

Preparing the veneers

1. Number the veneer leaves.

2. Dampen the leaves.

3. Place in a clamp.

4. Cut to the required size.

Matching veneers

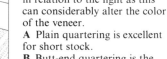

A

B

C

D

Veneers may be cut from the crown, burl, butt or main trunk of a tree. When matching veneer bear in mind where the piece will be placed in relation to the light as this can considerably alter the color of the veneer.
A Plain quartering is excellent for short stock.
B Butt-end quartering is the most commonly used quarter matching pattern.
C Diamond matching can be used when the veneer is straight grained and has a plain figure.
D Book matching has every other sheet turned over.

Using cauls

Pressing veneers using cauls

When using cauls, veneers and ground are glued together, placed between bearers and clamped. The caul itself is a stout piece of laminboard or blockboard, at least 19 mm/¾ in thick, that is clean, flat and free from defects. Its length and width should exceed that of the ground. Use of a joined board is inadvisable. The caul is placed against the veneered side of the ground with newspaper inserted between the veneer and the caul to absorb the excess glue. When the ground is counter veneered, newspaper and a caul are placed against each veneered side. Bearers in pairs are clamped over and under the caul and veneered ground, no farther than 250 mm/10 in away from each other. Bearers are made from sound solid wood, cut on the band saw. For a single caul the bottom bearers are flat while the top bearers have a slight curve. For a double caul one edge of each bearer is curved. The width of the bearer should be twice its thickness and its length should extend 100 mm/4 in beyond the caul's width.

Joining veneers

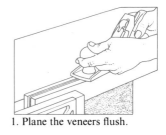

1. Plane the veneers flush.

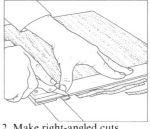

2. Make right-angled cuts.

Using cauls with resin glue

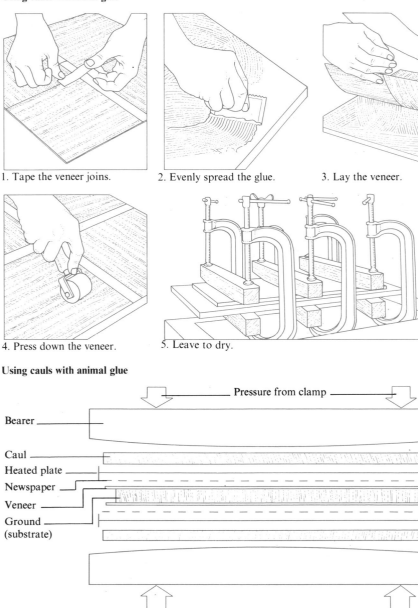

1. Tape the veneer joins.

2. Evenly spread the glue.

3. Lay the veneer.

4. Press down the veneer.

5. Leave to dry.

Using cauls with animal glue

Pressure from clamp

Bearer

Caul

Heated plate

Newspaper

Veneer

Ground (substrate)

Joining veneers

When using cauls, veneers must be joined, pinned and taped together before being glued and laid as one piece. To join two veneers place them between two straight-edged battens with the veneer grain running parallel with the battens. Allow a little of the veneers to extend above the battens. Grip one end of the assembly in a bench vise and hold the other in a C-clamp. Plane the veneers flush (1). Remove from the vise and lay the veneers on a flat surface. With a knife and try square, cut the ends of the veneers at right angles to the planed sides, using several light strokes (2). Cut the unplaned side of the veneers using a sharp knife and straightedge.

Using cauls with resin glue

Pin and tape the joined veneers closely together in the required pattern with short pieces of tape (1). Remedy any defects on the ground to be veneered (see page 223). Roughen the ground diagonally with a wood rasp. Dust off. Assemble the press. Pour the glue onto the ground. Spread it evenly over the surface with a spatula (2). Lay the veneer on the ground, leaving an extra 6 mm/¼ in of veneer around the sides (3). Smooth it down by hand. With a roller, press the glue from the center to the edges (4). If counter veneering, first lay the veneer on the underside of the ground in the same way. Cover the veneered ground with newspaper and place in the press. With the aid of an assistant, clamp the pairs of bearers together, gradually exerting pressure from the center outwards. Leave to dry (5). Then remedy any blisters (see page 200).

Using cauls with animal glue

Prepare the ground as for hand veneering (see page 194). Pin and tape the joined veneers together. Brush the glue onto the ground. Lay on the veneer, leaving 6 mm/¼ in extra veneer around the edges. If counter veneering, first lay veneer on the underside in the same way. Cover the veneered ground with newspaper and a preheated zinc or alloy plate and place in a press. Clamp the pairs of bearers together from the center outwards. When dry, check that the veneer has stuck to every part of the ground. Remedy any blisters that may have occurred (see page 222).

Hand veneering

Preparing the ground

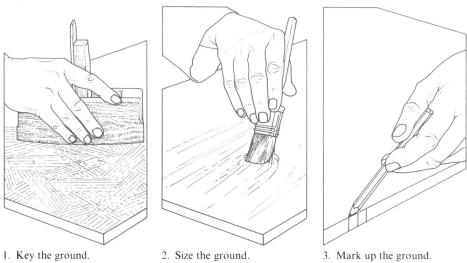

1. Key the ground.

2. Size the ground.

3. Mark up the ground.

Laying veneer by hand

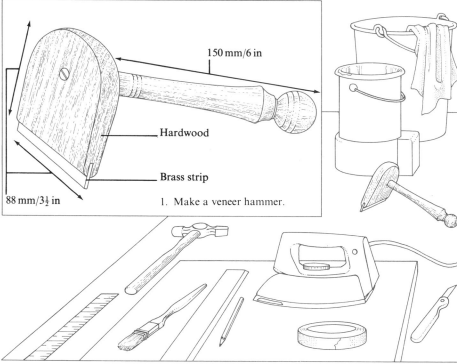

150 mm/6 in

Hardwood

Brass strip

88 mm/3½ in

1. Make a veneer hammer.

2. Assemble the equipment.

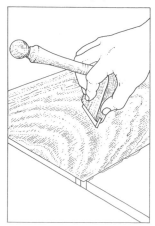

3. Press down the veneer.

4. Use two hands.

Preparing the ground

Cut out any defects and patch (*see* page 223). Using a wood rasp key the ground to be veneered diagonally across the grain in both directions (**1**). Dust off. Size all the ground to seal the pores (**2**). Size the end grain repeatedly until the pores are filled. Leave each coat for at least three hours to dry. Smooth the sized ground diagonally with a medium abrasive paper around a sanding block. Wipe the surface clean. Mark both sides and edges of the ground with a pencil and rule to indicate where all veneer pieces, including borders, will lie. Mark a 6 mm/¼ in overlap on each side on any joins between the veneer pieces (**3**).

Laying veneer by hand

To lay veneer by hand a veneering hammer must be made (**1**). The handle and flat or tapered head of the hammer are of hardwood and are screwed or doweled together. The grain of the head should run at right angles to the brass or aluminum strip inserted at its bottom end. Always hand veneer in a room that is at least 21 C/70 F and have all the required tools and materials at hand so the hot glue does not chill and thus lose its adhesive properties. Collect together the following: freshly prepared hot animal glue, a straightedge, a rule, a knife, a pencil, a brush, an electric iron (a domestic one will do), clean hot water, a clean damp cloth, gummed paper tape, a cross-peen hammer and a veneer hammer (**2**). Dip the veneer hammer in the hot water to heat its metal edge. Warm the ground over a gentle flame or against a radiator. Brush a thin even coat of hot glue onto the ground. Position the veneer against the ground marks, allowing a 6 mm/¼ in overlap for any edge and join. Brush the top side of the veneer lightly with glue. Then press the veneer against the ground with the veneer hammer, working from the center outwards with a zigzag motion and pushing the warm glue and any air bubbles to the edges (**3**). Avoid bringing too much pressure across the grain as this will stretch the veneer and will cause more shrinkage when the veneer is drying. For greater pressure, use two hands on the hammer (**4**). If the glue needs reheating, lay a damp cloth on the veneer and lightly run a warm iron over the

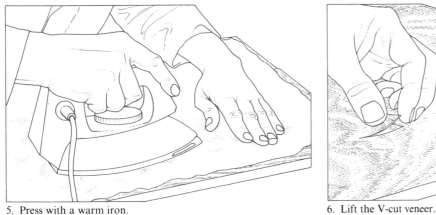

5. Press with a warm iron.

6. Lift the V-cut veneer.

7. Position the second veneer.

8. Apply glue to the top side.

9. Cut through the two veneers.

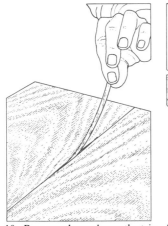

10. Remove the underneath strip.

11. Reglue the veneer join.

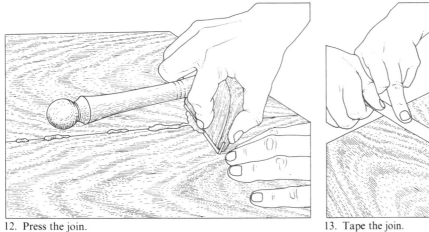

12. Press the join.

13. Tape the join.

surface (5). Then press the veneer against the ground again. When the veneer has been evenly pressed in position, clean away any excess glue. It is important to keep the veneer clean, but take care to avoid unnecessary cleaning as too much water in the glue will cause it to lose its adhesion. Test for blisters or foreign bodies under the veneer by brushing the surface with the palm of the hand and tapping suspect areas with a finger nail. Remedy any blisters (see page 222). To take out a foreign body make a V-cut, following the grain and figure as closely as possible. Lift the veneer, remove the defect and insert a little more glue (6). Then press the veneer down. To lay the second veneer, glue the ground. Then lay the veneer on the glued ground, overlapping the first veneer as marked (7). Brush the top side with glue (8). Press the veneer to the ground. Place the straightedge against the marked center of the overlap and cut through the two veneers (9). Peel away the top waste strip. Lift the veneer slightly to remove the second waste strip and any dirt particles from underneath (10). Reglue the veneer edges (11). Press the center join (12). Clean the area and tape the join so the veneer shrinks from the edges as it dries (13). Repeat this process on any further pieces of veneer.

Banding and finishing

Banding

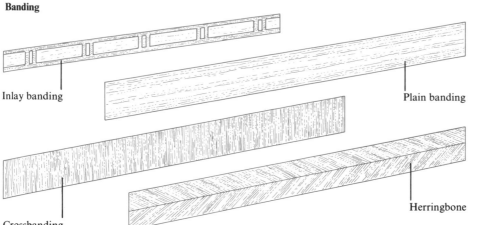

Inlay banding

Plain banding

Crossbanding

Herringbone

Laying banding by hand

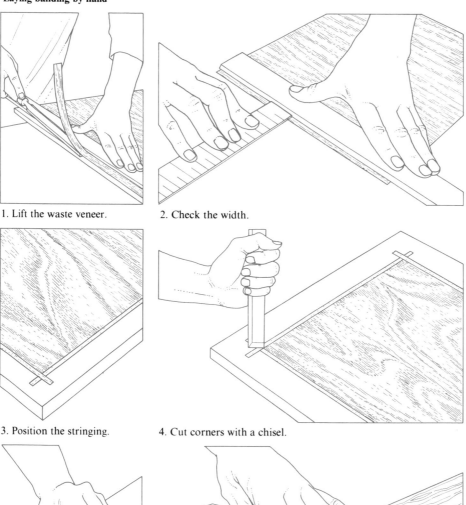

1. Lift the waste veneer.

2. Check the width.

3. Position the stringing.

4. Cut corners with a chisel.

5. Remove the waste veneer.

6. Tape each corner.

Banding

Banding is the overall term for any veneer decoration, of whatever width or pattern, surrounding a veneer panel. The inner decoration is known as stringing and is usually a very fine line of black or cream veneer. The outer decoration is known as a border. Plain borders are cut along the grain as one continuous strip. Crossbanding is cut at right angles to the grain as one or more strips. Herringbone banding is made from two strips cut at 45° to the grain and laid together with one strip turned over.

Laying banding by hand

Both stringing and borders are laid in the same way. The process of cutting and laying stringing is illustrated. Set a straightedge or cutting gauge on the ground against the border marks and cut away the waste veneer from the central panels while the glue is still at the jelly stage. Lift the waste with a chisel (1). Clean the area well. Measure the stringing, allowing 6 mm/¼ in overlaps. Check the evenness of width with a rule set against a straightedge (2). Cut the stringing and apply glue to both sides. Place it on the ground, tight to the central veneer panels. Press the stringing to the ground with a veneer hammer or cross-peen hammer. Glue both sides of a second piece of stringing and lay it at right angles to the first, overlapping at the corner (3). Miter the corners with a wide chisel cutting vertically through both veneer layers (4). (For borders use a knife and straightedge, making a short cut from the outer corner before cutting from the inside corner outwards.) Remove the top waste veneer and lift the mitered corner to extract the underneath waste (5). Reglue the corner. Rub it down and wipe clean. Lay all stringing and then the borders similarly. Tape the mitered corners (6). Then tape the join between the banding and the central panels.

Laying banding prior to using cauls

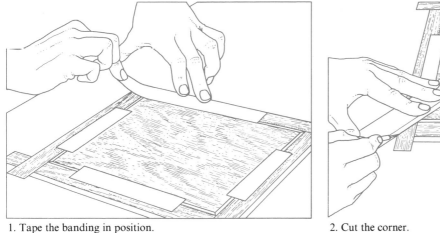

1. Tape the banding in position.

2. Cut the corner.

Laying banding prior to using cauls

Measure and cut each strip of banding so it overlaps 6 mm/¼ in at the ground corners and over the edge. Tape each strip tight against the central veneer panel, leaving the corners free (1). Either miter each corner with a wide chisel, cutting vertically through both layers of veneer, or use a knife and straight-edge to make a short cut from the outer corner before cutting from the inside corner outwards (2). Remove the waste. Tape the corner. Repeat for all corners. Then lay the veneers as a single leaf (*see* page 193).

Trimming excess veneer

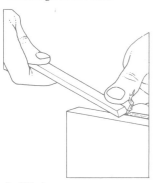

1. Chisel away excess veneer.

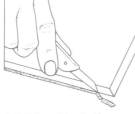

OR Trim with a knife.

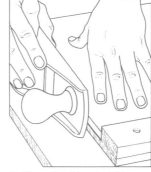

2. Plane the sides straight.

Trimming excess veneer

When the glue has dried, place the veneered ground in a bench vise, protecting the veneer with cork. Chisel the veneer sides flush with the ground sides, working inwards towards the ground (1). Pay careful attention to the direction of the grain. Alternatively tilt the ground slightly on a piece of waste wood. Cut the bottom excess veneer away with a knife. Repeat all around the ground. Using a jack plane smooth the sides straight and square (2).

Edges

Whether the veneer has been laid by hand or using cauls, ground edges must be sized with animal glue. When dry, smooth with abrasive paper. Cut veneer strips of the same color as the borders, allowing 6 mm/¼ in extra width and length. Glue both sides of a strip with hot animal glue and press it against the ground edge (1). Trim both ends with either a chisel or a knife. Lay and trim the other strips in the same way. Tape all the corners. When the glue has dried, trim the edges flush with the sides of the veneered ground. (2)

1. Hammer the veneer down.

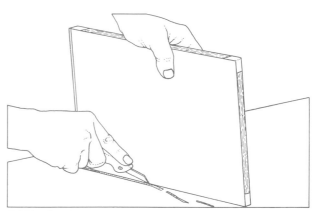

2. Trim the edges.

Cleaning up

Remove any tape and dried excess glue by gently rubbing both sides and all edges of the veneer with a warm damp cloth (1). Lightly scrape off any tape and glue that remains, using a cabinet scraper (2). (The surface must be absolutely clean as no surface finish will adhere to any sticky patches.) Smooth the sides and edges of the veneer with an 80 grit abrasive paper. Then use a finer abrasive paper (3). Apply sanding sealer unless a French polish or oil finish is desired.

1. Rub with a warm damp cloth.

2. Scrape off tape and glue.

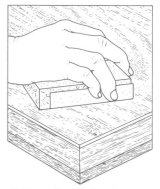

3. Smooth with abrasive paper.

Parquetry

Parquetry is the inlaying of veneers to form a geometrical design. It is a companion craft to marquetry but, whereas in the latter almost all the cutting is done freehand, for parquetry a straightedge guides the knife through the veneer. Most parquetry designs are made from polygon-shaped pieces of veneer — usually diamonds, squares and triangles. Plain figured and straight grained woods are most suitable.

For accurate cutting use a simple jig with a straightedge of steel or wood fixed to it. Cover the jig with vinyl or linoleum to protect the knife blade and, from a single point on the straightedge, mark out 30°, 45°, 60° and 90° angles. A movable fence, used as a guide for the veneer, can be set to the required angle. A fence made of veneer rather than of metal or solid wood is preferable as it can be cut into without damaging the

knife blade. Use a veneer knife with a comfortable handle and with a pointed blade. When cutting veneer, spacers are required to measure and maintain an equal width throughout a strip of veneer. They are made from wooden or metal squares or rectangles and are used in pairs of the same width. They should be positioned against the straightedge at the ends beyond the veneer. Always lay out veneer strips in the order they are cut to avoid inconsistencies in the design and join the pieces with strips of gummed paper tape.

When the design is complete, check that it has right-angled corners. Then add any banding (*see* page 197). Mount the design on a ground, using cauls and resin glue (*see* page 193). Counter veneer if necessary. Lay veneer strips on the edges and then clean up all the surfaces and edges (*see* page 197).

Cutting veneer strips

1. Cut parallel to the grain.

2. Keep strips in order.

Cutting veneer strips

Trim one side of a veneer leaf parallel with the grain. Place it against the fixed straightedge together with spacers of the desired strip width. Set the spacers either side of the veneer leaf. Place a rule against the spacers and with the knife blade at a low angle cut the veneer along the grain (**1**). Twist the blade into the rule slightly. Remove the veneer strip. Continue to cut in this way laying the strips in order as they are cut (**2**). To make a basket-weave pattern, half the veneer strips must be cut across the grain.

Patterns based on squares and rectangles

1. Tape strips together.
2. Trim the ends.

3. Cut the strips across.

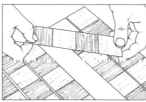
4. Reverse alternate strips.

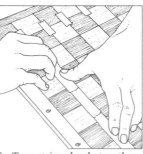
5. Tape strips closely together.

Patterns based on squares and rectangles

Cut four dark and four light strips of veneer of identical width. Use short pieces of gummed paper tape to stick alternate dark and light strips closely together (**1**). Place the veneer fence along the 90° line on the jig. Turn over the taped strips and lie one side flush with the veneer fence. Trim one end of the assembled strips (**2**). Place the trimmed strip ends and the spacers that were used to cut the strips against the jig straightedge. Hold the rule against the spacers and cut the veneers through with a knife. Cut the rest of the veneers in the same way, keeping each strip in the order it is cut (**3**). Discard any waste veneer. To make a chessboard pattern turn around every alternate strip (**4**). Tape each end of the strips closely into position (**5**). Then turn over the assembled strips and tape along them. Remove the positioning tape from the underside. Endless variations can be obtained using this technique. To make triangles halve squares diagonally. Cut rectangles by using different spacer widths.

Pattern variations based on squares and rectangles

Patterns based on diamonds

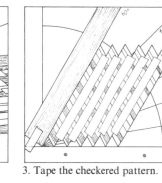

1. Lie the strips together.

2. Cut the strips through.

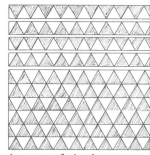

3. Tape the checkered pattern.

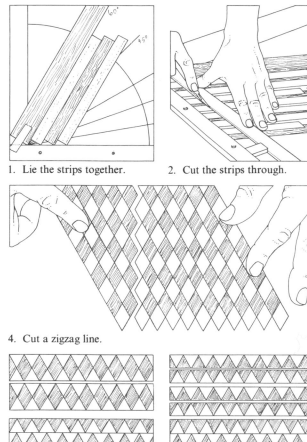

4. Cut a zigzag line.

5. Tape to the opposite side.

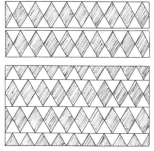

A zigzag pattern.

A chevron pattern.

A pattern of triangles.

Cubes

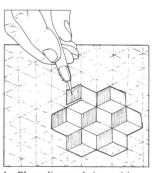

1. Place diamonds in position.

2. Cut around the outline.

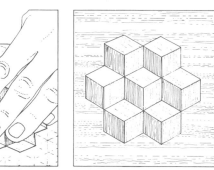

3. Set in the design.

Stars

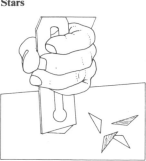

Splitting diamonds.

Star motifs.

Patterns based on diamonds

Cut strips of equal width as for any pattern based on squares. Tape the veneer fence along the 60° line on the jig. Lie alternate light and dark strips tight to the veneer fence so that the corner of each strip touches the straightedge (1). Tape the strips closely together. Trim the corners using the rule and spacers, so that they can be positioned tightly against the straightedge. Using spacers of the same width as were used originally to cut the strips, lie the rule against the spacers and cut the veneers through with several long strokes of the knife (2). To make a checkered pattern move each succeeding strip along by one diamond. Tape the strips closely together (3). To make the overall pattern square, cut a zigzag line down the sides of a row of diamonds (4). Place the straight side against the opposite straight side (5). Tape together. To make a zigzag pattern cut through the centers of all the diamonds of one color and move each alternate strip along half a diamond. To make a chevron pattern, repeat the same procedure described for the zigzag on both sets of diamonds. To make a pattern of triangles turn over every other strip.

Cubes

A cube is made from three 60° angled diamonds of different shades. Cut the diamonds and assemble the design on paper. Pin self-adhesive plastic, sticky side up, on grid paper and lift the diamonds onto it with the point of a knife (1). When the design is complete cut around the outline and through the plastic (2). Hold the design firmly in position on the background veneer and cut around the outline. Set the design into the parquetry (3).

Stars

Star motifs are built up with diamonds. Construct and set them into parquetry in the same way as cubes. To calculate the appropriate angle for a star, divide 360° by the required number of points. A six-pointed star, for example, is made from six 60° angled diamonds. Some star motifs are made from split diamonds cut with a plane blade held vertically across the diamond and tapped with a hammer.

Marquetry

Marquetry is the laying of veneers, cut freehand, to form a design. Using this technique, the endless figure variations found in wood can be exploited.

Originally a fret saw was used to cut awkward veneer shapes, but this is no longer necessary now that veneers as thin as $0.7\,\text{mm}/\frac{1}{32}$ in are available. These can be cut with a craft knife, provided the blade has a fine point. The "window" method is the most satisfactory technique for modern marquetry as the worker can see the exact grain effect of any veneer piece before it is cut.

With the "window" method the design is traced on the veneer selected as the background. A design feature is then cut out from the background veneer and a new piece of veneer of different color or figure is placed under the opening, or "window", in the background veneer. The new veneer is cut to shape, using the window edges as a template, and is then glued into the window.

Depending on the subject, plain or figured veneer can be used for the background. Where the background veneer represents the sky, any darker part should be at the top of the sky. In general, cut the background features first and gradually work through to the foreground, cutting out one feature at a time.

Cut veneers on a piece of hardboard or plywood covered with linoleum or vinyl tiles to prevent the knife blade from dulling. Curved shapes should be cut with the knife almost upright in small stabs made with the knife point. Begin at the nearest end and start each stroke slightly farther away each time. Cut straighter shapes with the knife held at a low angle and with the blade at a 90° angle to the cut. If possible simplify the shape and strengthen the edges of the pieces by overcutting into the borders and into background veneer that is later to be cut away. The picture must be absolutely square so that when the banding is laid the corners can be mitered accurately.

When practising marquetry, use PVA glue, but where this is insufficient to hold a feature in place, stick a little gummed paper tape over the join on the face.

Shading

1. Heat up the sand.   2. Spoon onto the veneer.

3. Remove the hot sand.

Shading
Veneer can be shaded darker by scorching it with hot sand. Place a metal container filled with silver sand over a moderately high flame or hot plate burner (1). Test the sand temperature with a piece of waste veneer. When hot enough, either dip the veneer into the sand or spoon the hot sand onto the area to be tinted (2). Remove after a few seconds (3). The sand will remove moisture, leaving the veneer brittle, so dampen it a little with a moist cloth.

Cutting the design

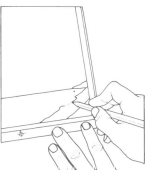

2. Pencil in any gaps.

1. Transfer the background details and registration mark.

3. Overcut the line if possible.

Cutting the design
Draw the design on tracing paper; make a registration mark at the bottom. Choose a suitable veneer for the background, ensuring that it extends over the design borders on every side. Reinforce the end grain on the piece of veneer with some gummed paper tape to prevent splitting. Select which side of the veneer to position at the top. Tape the top of the tracing paper to the top of the background veneer, using gummed paper. Place a sheet of black carbon paper between the veneer and the tracing paper. (Use white carbon for dark veneers.) Trace the background design features, the registration mark and the borders onto the background veneer, using a worn-out ballpoint pen or stylus (1). Fold back the tracing paper and remove the carbon paper. Pencil in any gaps in the traced lines where details will later be positioned (2). Cut around one of the traced features. Overcut into borders or areas that are later to be replaced (3).

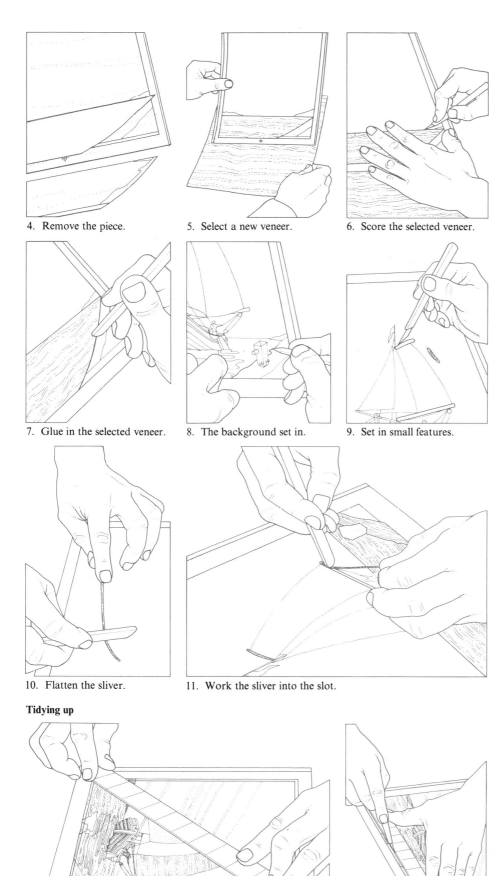

4. Remove the piece.

5. Select a new veneer.

6. Score the selected veneer.

7. Glue in the selected veneer.

8. The background set in.

9. Set in small features.

10. Flatten the sliver.

11. Work the sliver into the slot.

Tidying up

1. Check the diagonals.

2. Cut along the border lines.

Remove the piece (**4**). Select a suitably figured veneer and move it around in the opening, or "window", left in the background veneer (**5**). When the best grain effect shows through the window, hold the selected veneer in place with paper clips or gummed paper tape and score it around the window edges in small stabs, always cutting back into the previous score mark (**6**). Remove the piece of selected veneer from under the window. Then cut along the scored line. With a pointed piece of waste veneer apply PVA glue to the window edges. Fit the selected veneer piece into the window and rub the edges into place with the knife handle (**7**). Replace the tracing paper and carbon paper on the background veneer, aligning the registration marks, and retrace any lines that have been cut away. Cut and set in each feature of the background design, following the process described above (**8**). Trace on the foreground details of the design. Cut and set in the selected veneer pieces for the foreground features. Sometimes it is better to set in small features before adjacent large features; the overcutting will strengthen them (**9**). To cut out very thin lines in the design, first cut on one side of the line and then on the other. Remove the tiny sliver from the background veneer to make a slot. With a rule and knife cut a thin long sliver from a straight-grained veneer and stroke it with the knife handle to flatten it (**10**). Insert one end of the selected sliver into the slot. Put a dab of glue on it to hold it in place. Work in the rest of the sliver and then rub a little glue into the join with the back of the knife handle (**11**).

Tidying up
When all the veneer pieces have been fitted correctly re-mark the lines of the border. The picture must be absolutely square before the banding is laid, so check that the sides of the design are at right angles to each other by ensuring the diagonals are the same length and that the sides are the same length (**1**). Cut along the border lines, using the knife and straightedge (**2**). Add any banding (*see* page 197). Mount the design on a ground, using cauls and resin glue (*see* page 193). Counter veneer if necessary. Veneer the edges and then clean up the surfaces and edges (*see* page 197).

Surface finishes 1

The purpose of applying a surface finish to wood is to protect it from oil, grease, liquids and other general pollution by sealing the pores. The surface finish will also enhance the natural beauty of the wood, if a transparent rather than an opaque finish is applied. To prevent the wood from shrinking and thus the finished surface breaking down, any centrally heated room should contain a humidifier or an abundance of plants to maintain adequate levels of moisture in the air.

Choosing the best materials to apply to a particular piece of work can cause confusion especially as the materials must be compatible. The eventual use of the piece is the prime consideration. Will it be subject to hard wear and tear or have liquids spilled on it? Does it need to be heat resistant? (The qualities possessed by the important surface coatings are described on pages 208–11.) Will the piece look more attractive if it is colored? If so, will a stain be satisfactory or would an opaque finish be more suitable? Is an open-grained surface preferable — or a close-grained woodfilled one?

Before applying any surface finish, it is essential that the wood is thoroughly prepared and its surface is absolutely clean. Any residual grease and dirt will repel the materials and thus give an uneven finish. Plane all protruding joints; punch nails and panel pins below the surface and remove any pencil marks, oil and grease stains, paint and bruises. Fill in any holes later.

Once the wood has been prepared, consult the chart (*right*) for the order in which the appropriate surface finishes should be applied.

Preparing the wood

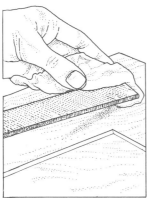

1. Press with a warm iron.

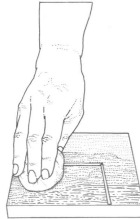

2. Thoroughly clean the wood.

Preparing the wood
Surface finishes tend to accentuate rather than conceal bruises and they are repelled by dirt so special care should be taken with the preparation of the piece of work. Any bruises in the wood must be removed before any finish is applied. Dampen the damaged area with a wet cloth. Heat an old iron file or an electric soldering iron so that the warmth from the iron can be felt 150 mm/6 in from the face. Place a wet cloth over the bruise and press with the iron, moving it about gently (**1**). The resultant steam causes the wood fibers to expand and rise, thus removing the bruise. Smooth the area with sandpaper. To clean the wood, rub the surface vigorously with a cloth and clean water (**2**). The water will raise the grain. When the wood has dried completely, smooth it with a grade 6/0 or 7/0 garnet paper and dust off. The wood is then ready for its first surface finish.

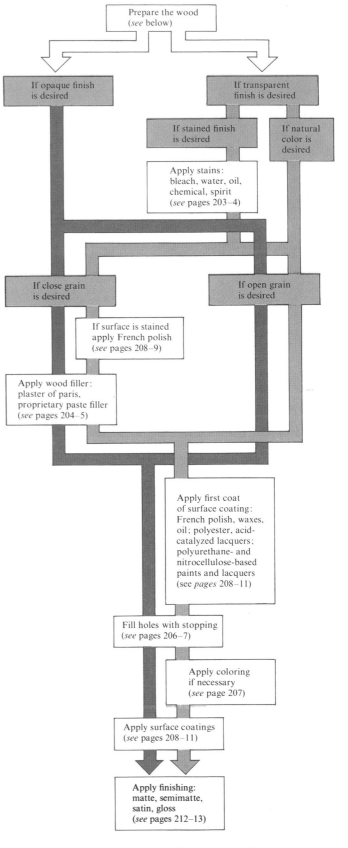

Prepare the wood
(*see below*)

If opaque finish is desired

If transparent finish is desired

If stained finish is desired

If natural color is desired

Apply stains:
bleach, water, oil,
chemical, spirit
(*see* pages 203–4)

If close grain is desired

If open grain is desired

If surface is stained
apply French polish
(*see* pages 208–9)

Apply wood filler:
plaster of paris,
proprietary paste filler
(*see* pages 204–5)

Apply first coat
of surface coating:
French polish, waxes,
oil; polyester, acid-
catalyzed lacquers;
polyurethane- and
nitrocellulose-based
paints and lacquers
(*see pages* 208–11)

Fill holes with stopping
(*see* pages 206–7)

Apply coloring
if necessary
(*see* page 207)

Apply surface coatings
(*see* pages 208–11)

Apply finishing:
matte, semimatte,
satin, gloss
(*see* pages 212–13)

To ensure the appropriate materials are applied in their correct order once the wood has been prepared, follow the arrows in the chart. Starting from the top of the page, follow the shaded boxes, which indicate the type of finish selected, and complete the actions mentioned in the unshaded boxes, in the order in which they appear.

On a transparent surface coating, a stain will often emphasize the natural figure of the wood as well as color it. A stain is unnecessary where an opaque surface coating is to be applied.

There are four different groups of stains: water, chemical, oil and spirit. Only stains in the same group are intermixable. Each group of stains may be diluted by its base solvent.

The art of staining is based on one simple principle: use plenty of the stain at the outset. It is advisable to allow a couple of hours for the wood to dry after staining. Any polishing carried out on wet wood will, at a later stage, cause a reaction that will spoil the final finish. The grain may whiten, pimples may form, a cloudy film may appear or the polish may flake off.

Water stains

These are known as direct dyes. They are made by mixing dry pigment powders with water. The pigment powders are intermixable and so a wide range of shades can be achieved. The depth of the color is determined by the amount of powder used: the more powder, the darker the stain. Pigment powders are comparatively cheap to buy and large quantities of stain can be made up from fairly small amounts of the powder. Some powders may take a while to dissolve, so leave the mixed stain for an hour before using so the final color is certain.

Chemical stains

Some chemicals such as blue copperas and ammonia can be mixed with water and applied to wood as a stain. They react with the tannic acid that is present in varying amounts in the wood and thus change the wood color. The problem is that the woodworker cannot see what color these stains will make until they are applied to the wood. Some chemicals such as bichromate of potash and copper sulphate have a little color content of their own while others, such as ammonia and caustic soda, are colorless.

Oil stains

These are available in a variety of colors, which can be mixed. They are simple to use but are relatively expensive. Oil stains do not penetrate the wood fibers and thus will not raise the grain. Their strength therefore tends to be greater than other stains.

Spirit stains

Spirit stains are made by dissolving dry pigment powder in methylated spirit and then adding French polish in a ratio of four parts methylated spirit to one part French polish. The French polish will act as a binder; if it is not used the methylated spirit will evaporate, leaving behind the dry powder. The rapid evaporation rate of the solvent in spirit stains does tend to make them difficult to use on large surfaces. They also have a tendency to fade.

Bleaches

There are three main bleaches that can be used on wood: sodium hypochlorite, oxalic acid and super bleach.

Sodium hypochlorite is one of the constituents of household bleach and it can be used for removing dark marks, but it will not substantially alter the color of wood. Oxalic acid, which is a poisonous white crystalline powder saturated in water or methylated spirit, has similar uses. It should be neutralized with acetic acid and washed off with water.

Super bleach is extremely powerful. Always wear gloves and avoid splashing it. It is available as a two-part bleach. The first solution is an alkaline, which is applied liberally and allowed to soak into the wood. This may temporarily darken the wood. The second solution is concentrated hydrogen peroxide. When it is applied to the wood after the first solution, the bleaching action takes place. If the overall reduction in color is not sufficient, repeat the whole process.

Whichever bleach is used, wash the wood with clean water afterwards and allow it to dry completely.

The effect of stains on various woods									
	Mahogany	Vandyke	Black	Blue copperas	Bichromate of potash	Ammonia	Sodium hypochlorite	Oxalic acid	Super bleach
Beech	reddish	brown	gray	grayish	light tan	brown	slightly lightens	slightly lightens	almost white
Mahogany	red	brown	subdues red; produces grayish tone	eliminates red; produces browny-gray	deep rich brown	deep brown with grayish tone	removes dark marks	slightly lightens	lightens considerably
Oak	red	deep brown	gray	gray-blue	greenish-brown	slightly greenish deep brown	removes dark marks	slightly lightens	almost white
Pine	reddish	brown	gray	grayish	pale yellow	greenish-brown	removes dark marks	slightly lightens	almost white
Teak	reddish	brown	gray	grayish	yellowish-brown	greenish-brown	removes dark marks	slightly lightens	lightens considerably
Walnut	reddish	brown	gray	grayish	pale yellow	greenish-brown	slightly lightens	slightly lightens	lightens considerably
Manufactured boards	reddish	brown	gray	grayish	deep brown	greenish-brown	removes dark marks	removes dark marks	almost white

Surface finishes 2

Staining with a water stain

1. Test the color of the stain.

2. Tip in the powder.

3. Test the solution.

4. Apply with an absorbent cloth.

5. Brush uneven surfaces.

Staining with a water stain

To test the color of a stain rub some water and a little powdered stain into the palm of your hand (**1**). Then tip some powder from a creased piece of abrasive paper into hot water (**2**). Make a strong solution. Stir and bring to the boil. Test the shade on the back of the hand (**3**). Dilute if necessary. Saturate end grain of wood in clean water. Soak a soft absorbent cloth in the stain solution and work the stain into the pores with a circular motion (**4**). Use a brush on uneven surfaces, ensuring stain does not splash (**5**). Apply the solution to the surface quickly so the whole area is still moist when the operation is complete. While stain is still wet wipe the surface dry in a circular motion with a clean dry cloth. Then wipe dry along the grain. Dry all corners and recesses carefully. Leave to dry for a couple of hours. When completely dry, seal with French polish.

Wood fillers

Wood fillers are used to fill up the pores of wood to obtain a full-grained finish and also for a high-gloss finish; they are rarely used on carved wood. They are based on a variety of chalks to which coloring, oils and solvents are added. Wood fillers are cheaper, quicker and more effective than French polish or any other surface coating that could be used as a filler.

Both plaster of paris and proprietary oil-based fillers are available. Plaster of paris is only used when the work is to be French polished. Combined stain and wood fillers are available, but they are generally unsatisfactory. Some surface coatings require specific wood fillers, so check with the suppliers. Always use a filler that is slightly darker than the wood surface.

Wood fillers can be applied to stained and unstained surfaces. Stained surfaces must be sealed by spreading a thin coat of French polish on the wood before it is covered with wood filler. If left unsealed, the stain might possibly be lifted later on.

The art of successful wood filling is to apply the filler quickly and evenly and to cover the entire surface before the filler has begun to dry. Clean all corners and recesses with a quirk stick and allow the wood filler to dry for at least 4 hours before sandpapering, dusting off and applying the surface coating.

The effect of oil-based woodfillers on various woods							
	Light oak	**Medium oak**	**Dark oak**	**Light mahogany**	**Medium mahogany**	**Dark mahogany**	**Black**
Beech	yellowish with light grain	light tan with slightly dark grain	dark brown with dark grain	pale pink with light grain	pinkish with slightly dark grain	pale red-brown with dark grain	dirty gray with dark grain
Mahogany	yellowish with white grain	yellowish tinge with slightly white grain	darkish brown with slightly dark grain	pinkish with slightly light grain	darkish with slightly dark grain	red-brown with dark grain	grayish-brown with very dark grain
Oak	yellowish with slightly white grain	light brown with slightly dark grain	darkish with dark grain	pinkish with slightly light grain	reddish with slightly dark grain	red-brown with dark grain	gray with very dark grain
Pine	yellowish with light grain	slight tan with slightly dark grain	brownish with dark grain	pale pink with light grain	pinkish with slightly dark grain	brownish with dark grain	grayish with very dark grain
Teak	yellowish with light grain	brown with slightly dark grain	brown with dark grain	pale pink with light grain	pinkish with slightly dark grain	reddish brown with dark grain	grayish with very dark grain
Walnut	yellowish with light grain	darkish with slightly light grain	darkish brown with slightly dark grain	pinkish with light grain	pinkish with slightly dark grain	brownish with dark grain	grayish with very dark grain
Manufactured boards	yellowish with light grain	brownish with slightly dark grain	brown with dark grain	pinkish with light grain	light red-brown with slightly dark grain	red-brown with dark grain	gray-brown with very dark grain

Making a quirk stick

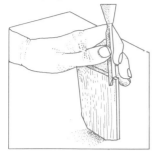

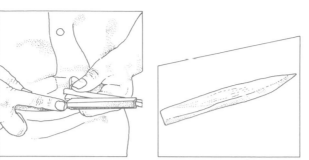

1. Split off a sliver of wood.

2. Shape the wood.

3. The finished product.

Wood filling with a proprietary paste filler

1. Fill the pad with paste.

2. Rub into the surface.

3. Clean out any moldings.

Wood filling with plaster of paris

1. Mix the powders and plaster.

2. Test the color.

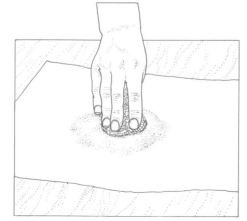

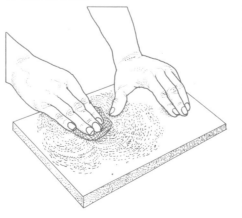

3. Press burlap into the mixture.

4. Rub mixture into the grain.

Making a quirk stick
Split off a piece of hardwood with a chisel and mallet (**1**). Work the quirk stick to shape with a chisel (**2**). The stick may be pointed or rounded, large or small, depending on the shape of the recesses it is intended to clean (**3**).

Wood filling with a proprietary paste filler
Scrape the paste out of its container with a loose pad of burlap (**1**). If too thick, dilute the paste with turps. Rub the paste into the surface of the wood, using a circular motion (**2**). Work the paste well into any corners and moldings. Then, keeping the pad loose, rub the paste across the grain. While the filler is still wet, wipe away the excess with burlap and then a soft, absorbent cloth. With an appropriately shaped quirk stick clean corners, edges and recesses where any wood filler remains (**3**). Leave to dry.

Wood filling with plaster of paris
Tip out some superfine white plaster onto a piece of newspaper and add the required pigment powders. Mix together the powders and plaster by tipping them about in the newspaper (**1**). To test the color, put some water on the back of your hand, dip a finger into the plaster mixture and rub it on the wet hand (**2**). The color will appear much chalkier than desired, but this whiteness will be removed later. Then soak a piece of burlap in cold water. Squeeze it out gently and press it into the plaster mixture (**3**). Rub the filler into the grain, using a circular motion (**4**). When the entire surface has been filled, wipe it clean across the grain with clean burlap held loosely. Clean corners, edges and recesses with a quirk stick. Leave to dry for about 4 hours. The water from the plaster will evaporate and leave a white residue. Remove this with linseed oil on a soft cloth, applied to the surface with a circular motion. Smooth the still wet wood with 7/0 garnet finishing paper, rubbing lightly along the grain. Wipe across the grain with clean dry burlap, pushing the excess filler into the grain pores. Rub down surface with a dry cloth.

Surface finishes 3

The process of filling holes and cracks is known as stopping. For an opaque finish, stopping is done after any wood filler has been applied (if a full-grained finish is desired) but before any surface coating is applied. For any transparent finish, stopping and coloring are done after the wood filler process and after one layer of surface coating has been applied. If wood filler is not used the surface for a transparent finish must still be coated once before stopping and coloring.

There are several materials that are suitable for filling holes. Some, such as beaumontage and shellac, are known as "hard stopping" and are melted for application while others, such as beeswax and Japan wax, are applied cold and are better suited to filling small holes. Shellac is immensely sticky and it is much cleaner to use beaumontage. Any of these stopping materials can be colored by melting them down and mixing in dry pigment powders.

Coloring entails touching up a surface to one overall color. It is best to make up spirit stains full strength and then mix them together. Dilute in a new solution of four parts methylated spirit and one part French polish. Bismarck brown and spirit black are the main spirit stains that, combined in varying quantities, will give warmer or colder shades respectively. To lighten dark spots in wood combine lemon chrome, orange chrome and titanium white. The color should blend into the background and not show up on the wood as an obvious blob of color.

Two thin coats of spirit stain are better than one thick one so bear this in mind when mixing up a color. If more than three applications are required, seal the stain in with a layer of the selected surface coating before continuing with the coloring. Otherwise the stain could be dragged off by later applications.

Fads and rubbers

The professional way to apply French polish is with a rubber or a fad (*see* page 209). Rubbers are also used to apply a pull-over solution (*see* page 211). A rubber is a piece of absorbent wadding covered in soft absorbent material. The wadding is carefully folded so it is pear shaped with a point at one end. Pound wadding is best as it remains soft and springy for a long time whereas absorbent cotton and upholsterers' wadding tend to harden very quickly.

After much use, the wadding in the rubber will become slightly hardened. When this occurs, the cloth covering the wadding is removed. The hardened wadding is then known as a fad and is used by itself to apply polish. Fads and rubbers should be stored in an airtight container such as a lidded jar.

Filling large holes

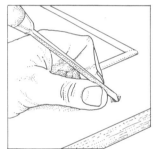

1. Undercut the sides.

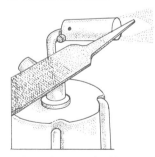

2. Heat the tang of a file.

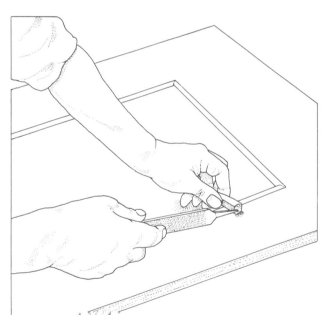

3. Melt the beaumontage stick.

Filling large holes
Proprietary wax filler sticks, made in a variety of colors, can be used; alternatively make a beaumontage stick by melting beeswax and carnauba wax in 10:1 proportions in a double boiler. Add pigment powder. Cool down the mixture and bind with a little French polish. Allow to solidify. Enlarge the hole to be filled by undercutting the sides so the stopping remains lodged in place (**1**). Heat the tang of a file so that when held 150 mm/6 in away from the face the warmth can be felt (**2**). Wipe off the soot. Hold the beaumontage stick over the hole and melt it into the hole with the hot tang (**3**). Press it in well. With a chisel, held bevel downwards, gently scrape away the excess beaumontage (**4**). Smooth the surface by sanding lightly using a sanding block (**5**). Then sand further by hand to finish (**6**).

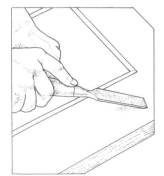

4. Chisel away the excess.

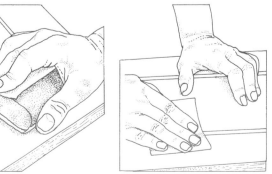

5. Smooth with a sanding block.   6. Smooth by hand.

Filling small holes

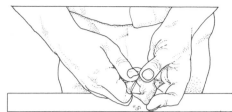

1. Remove shavings of colored wax.

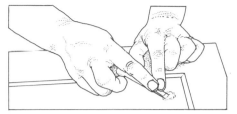

2. Press the shavings into the hole.

Coloring

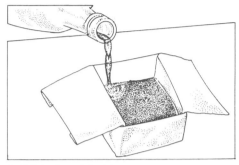

1. Dilute the spirit stain.

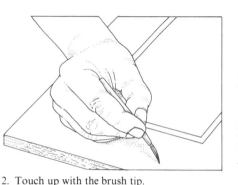

2. Touch up with the brush tip.

Making a rubber

1. Fold into three equal parts.

2. Fold into three again.

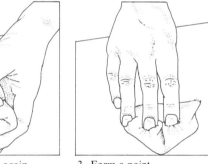

3. Form a point.

4. Fold the sides inwards.

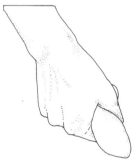

5. Squeeze the wadding.

6. Place the cloth over.

7. Fold one side of the cloth.

8. Secure with the thumb.

9. Twist the cloth firmly.

Filling small holes

Dip beeswax into the required pigment powders. Remove shavings of the colored wax, using a knife (1). Press the shavings well into the hole with a chisel (2). Remove the excess wax with the chisel held vertically. Smooth the surface, first with the back and then with the front of a piece of sandpaper.

Coloring

Make a small "boat" with old sandpaper. Make up spirit stains full strength. In the boat mix the stains to obtain the required color. Dilute with four parts methylated spirit and one part French polish (1). Dip the tip of a fine brush into the new solution and test the color on the back of your hand. Adjust color as necessary. Then touch up any unevenly colored areas with the brush tip (2). Leave to dry. Repeat if necessary.

Making a rubber

Rubbers can be made in various sizes. Small ones are most suitable for awkward corners and carvings while larger ones are convenient for big flat surfaces. The soft absorbent cloth should always be a little larger than the wadding. To make an average-sized rubber use a piece of pound wadding about 225 mm/9 in square and fold it into three equal parts (1). Turn the wadding around 90° and fold the wadding again into three equal parts (2). Fold the top corners into the center middle to form a point (3). Then fold the sides into the center bottom and overlap them (4). Squeeze the wadding into a pear shape with a point at one end (5). Place the square piece of soft absorbent cloth over the wadding (6). Holding the wadding tightly, tuck one side of the cloth under the point of the wadding (7). Secure with the thumb. Then pull the other side of the cloth tightly under the point of the wadding. Again secure with the thumb (8). Wrap the cloth around the rest of the wadding. Twist up the rest of the cloth at the back firmly and hold in the palm of the hand to prevent it trailing across the work (9).

Surface finishes 4

Surface coatings can be used both to protect a surface and to enhance the beauty of the grain. Most protect against dirt and make cleaning easier. All to some extent protect against moisture. Some are heat resistant; others protect against insect or fungal attack. They all have good adhesive qualities provided the surface to which they are applied is absolutely clean. All work should be done in a well-lit room but not in direct sunlight. The room should be clean, and have a constant temperature of about 17°C/65°F.

The important surface coatings are French polish, wax and semiwax, oil, nitrocellulose, polyurethane, acid-catalyzed lacquers and polyesters.

French polish
French polish, though a weak surface coating in that it has little heat and moisture resistance, still remains popular among craftsmen. It is a sticky resinous mixture and so difficult to apply, but when applied correctly it gives one of the best gloss or matte finishes. The aim is to produce a high gloss by applying numerous thin coats of polish; this can then be dulled down if desired, for a matte finish.

The basic ingredients of French polish are shellac, which is a substance exuded by the lac insect, and methylated spirit, which is used as a solvent. The main colors of polish are pale, button and garnet. French polish is also available tinted with dyes of black and red; black-dyed French polish is used to ebonize a surface over black stain. Sycamore, mahogany and birch are some of the best woods for ebonizing.

French polish should be worked in slowly and deliberately with a fad or rubber, gradually building up a film of polish, which is left to dry before the next coat is applied. Always lift the fad or rubber promptly from the surface sides and ends to prevent polish from dripping over the edges. Fadding and bodying up and down the grain is done to prevent the polish from streaking. The slower the polishing process, the more time the previous coat will have had to release its solvents and become dry. It is easier to work on a number of surfaces in rotation so each surface has an opportunity to dry off before the next application of polish. Each coat will partially dissolve the coat beneath so the layers will amalgamate. The quantity of polish worked into a surface determines the degree of gloss; for a full even gloss, apply many thin coats.

French polish

Beech
Use a pale polish on light or natural beech. Button polish may be used after staining.

Mahogany
Use button or garnet polish.

Oak
Use a pale polish on light oaks; button polish on stained oak.

Pine
Pale polish is normally used. Button polish is used for a stained piece.

Teak
Pale or button polish, depending on color desired.

Walnut
Pale polish is normally used. Button polish may be used for a sheen.

Manufactured boards
Pale, button or garnet polish depending on color desired.

Wax and semiwax polish
Wax gives an attractive shine and is easy to maintain, but it has only slight moisture resistance, little or no heat resistance and is easily marked. It also holds dust, which, unless the grain is sealed, will become deeply embedded in the wood. Most proprietary waxes are based on soft paraffin wax and are unsuitable for application to unsealed wood. If using wax as a surface coating, heat 0.45 kg/1 lb of shredded beeswax with 280 ml/$\frac{1}{2}$ pt turpentine and some carnauba wax. It is possible to color beeswax with oil-soluble colors or dry pigment powders. Bleached beeswax is also obtainable. Allow the mixture to cool. Apply the wax paste to the surface with a soft brush or cloth working along the grain. Leave surface to dry. Then burnish with a soft dry cloth. On uneven surfaces, apply the wax with a short-bristled brush made of bear hair or mixed hair and burnish with a dry soft-haired brush. Once the surface has been sealed with the beeswax mixture further coats of proprietary wax may be used to produce a finish with depth and warmth.

To apply a semiwax polish, first seal the wood with two coats of French polish, using a cloth, brush or fad. When dry, smooth surface with 7/0 garnet paper, wipe clean and apply a further coat of French polish. When dry add wax polish to 000 or 0000 steel wool and rub down. Burnish with a soft dry cloth.

Oil
The advantages of oil polish are that it does not crack or blister, or show heat or water marks. Its application, however, takes a great deal of time as oil is not only slow drying but requires numerous coats. Layer upon layer painstakingly applied will give effective protection to wood exposed to the open air and to liquids. It is therefore most suited to garden furniture, table tops, bar tops, counters and liquor cabinets.

The traditional oil to use as a polish is linseed oil; this will produce a high gloss that will be moderately heat and water resistant if applied sparingly over a period of months. Each application must be hardened by oxidization and at least 12 coats are necessary. The addition of a little turpentine and terebene to the oil makes it easier to work with and will speed up the drying process. Using either raw or boiled linseed oil, add 5 per cent turpentine and 5 per cent terebene and simmer the mixture in a double boiler for 15 minutes. Vigorously rub the mixture into the wood with a clean cloth. Allow at least one day for the oil to dry; then smooth surface with fine sandpaper and dust clean. Numerous coats should be applied in this way, leaving at least one week between sessions. Remove any dust that has settled on the surface between sessions and use a clean cloth for every session.

Several proprietary oils containing oxidizing agents are available, but, compared with linseed oil, these give an inferior quality finish.

Teak oil, olive oil and Danish oil are also suitable as polishes for wood, especially teak, oak, mahogany, walnut and cedar, as they enhance the natural grain and coloring. Teak oil is extremely combustible so always soak an oil-filled cloth in water overnight. Then wash it out thoroughly before disposing of it.

French polishing

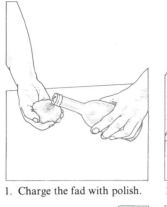

1. Charge the fad with polish.

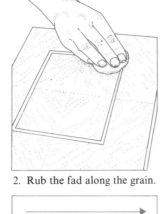

2. Rub the fad along the grain.

3. Smooth the surface.

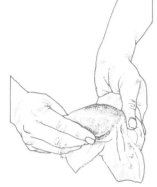

4. Fill the rubber with polish.

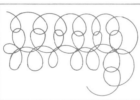

5. Work along the grain.

6. Work in figure-eight pattern.

7. Work in circles.

8. Work along the grain.

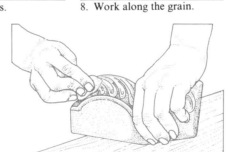

Using the rubber's nose in corners.

Using the rubber's nose in crevices.

Methods of holding work

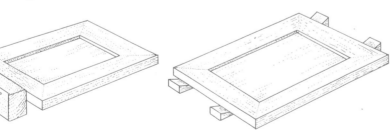

Waste wood nailed to a small door.

Heavy door laid on wedges.

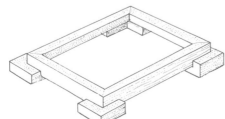

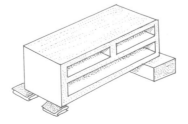

Frame held with bench-sticks.

Carcass propped up.

French polishing

Charge the fad or rubber well with polish. To charge the rubber remove the cloth. Saturate the wadding with polish (1). Replace the cloth and squeeze until the polish oozes through. Work the fad or rubber along the grain with a minimum of downward pressure, overlapping each stroke (2). Too much downward pressure will force out the polish too rapidly, resulting in a sticky surface. Leave to dry for 10 minutes. Rub 7/0 abrasive paper together to make it finer and smooth the surface (3). Dust with a soft cloth. Fill any holes and adjust the color. Then body the surface with a rubber. Charge with polish (4). Work the rubber in strict sequence: first along the grain, overlapping each stroke until the entire surface is covered (5); then in a figure-eight movement (6); then in a circular movement (7); and finally work the rubber along the grain again (8). Leave the surface to dry off. Apply more coats, repeating this process until a high gloss has been achieved. Towards the end of the bodying process, lubricate the hardening rubber with a spot of linseed oil. Remove any excess oil with the dried out rubber, using considerable pressure along the grain. Test the surface by breathing on it: if a mist forms, continue to rub until the oil is gone. With experience, methylated spirit can be used on the dried out rubber, to give a perfectly smooth surface. Leave the surface to harden. Then smooth with very fine abrasive paper.

Methods of holding work

Hold a small door by nailing a piece of waste wood at right angles to the unpolished bottom edge of the door. Lay a large or heavy door on four wooden wedges. The door weight will keep it in position. Lay a table upside down to polish the legs and rails. Then stand it upright to polish the top. Polish a chair from the legs upwards. With a frame, first polish the inner and outer edges, then place the frame flat in homemade bench-sticks nailed to the workbench, and polish the top. Prop up a carcass with a block of waste wood at one end and place folded abrasive paper at the other so the grit is in contact with the carcass and the floor to prevent slipping. Hold a drawer from the inside.

Surface finishes 5

Manufacturers use the words "lacquer" and "varnish" synonymously, so for clarity the term lacquer is used throughout this book. There are four basic types: nitrocellulose, polyurethane, acid-catalyzed lacquers and polyester. Lacquers and paints give a harder and more resistant finish than French polish, wax or oil. Generally they are applied with a brush or a spray gun. The hard finish is produced when the solvents have evaporated on the surface. Always read and follow the manufacturer's instructions and check that the method of lacquer or paint application does not contravene any legal stipulations.

Some lacquers are toxic and inflammable so ensure that the workshop is well ventilated, without being dusty. The humidity should not exceed 70 percent as dampness will cause blooming and chilling. Evaporation time varies according to the lacquer and method of application, but if a lacquer is too slow drying or too thick the surface will remain soft. Lacquers can be cleaned off brushes and spray guns with lacquer thinners. However, brushes may be destroyed by lacquer to such a degree that cleaning proves a waste of time; therefore always use a cheap mixed hair brush. When using a spray gun pay particular attention to cleaning the air cap, horn holes, fluid tip, fluid needle and the inside of the gun body. Use cocktail sticks and pipe cleaners to reach into inaccessible parts. Whether for cleaning or thinning buy a thinner of the same brand as the lacquer.

The best paints are nitrocellulose based; polyurethane paint, however, can also be used. All but non-drip thixotropic paints should be stirred to suspend the pigments. To give a surface the smoothest finish it is best to apply paint with a spray gun. A primer coat should be applied before going on to give a further two undercoats and a top coat. It is advisable to use the same manufacturer's top and base coat throughout the work. Clean oil-based paints off brushes and spray guns with white spirit; use warm water and detergent on water-based paints. Paints can be bought in a matte, semi-matte, satin or gloss finish.

Before painting or lacquering, wood that has not already been stained should be treated with water to raise the grain; it should then be dried and smoothed with an abrasive. Whatever the surface it must be dry, dust free and greaseless before application. The work should be left for at least an hour between coats.

Nitrocellulose

Nitrocellulose may be applied with a brush but is more usually applied with a spray gun. Lacquers made for brushing have slow evaporating solvents to allow for the longer application time. Nitrocellulose is cheap, quick drying and produces a hard heat- and moisture-proof surface. It has an unlimited shelf life. Up to 10 percent polyurethane solution may be added to give a still harder finish. Layers are applied "wet on wet" so that they amalgamate well. Nitrocellulose lacquering is followed by an application of a pull-over solution; this is a diluted nitrocellulose solution, which is applied with a rubber 24 hours after lacquering and before finishing, to give an even surface. Always use a pull-over solution of the same brand as the lacquer.

Polyurethane

Polyurethane may be applied with a brush or spray gun. It is expensive but produces an excellent hard surface with good heat-, moisture- and wear-resistant properties, though it is rarely burn resistant. It can be bought in different forms, sometimes in a twin pack consisting of liquid plastic and a catalyst containing the solvent, which promotes the drying process. The solution may have a shelf life of as little as 8 hours. Oil stained wood should be tested for discoloration. Polyurethane is available both as a transparent and opaque lacquer and its color can be adjusted by mixing the lacquer with spirit stains.

Acid-catalyzed lacquers

Acid-catalyzed lacquers give a very durable heat- and moisture-proof surface. They should be applied with a spray gun. The layers set by polymerization. Most acid-catalyzed lacquers come in a one- or two-part form and once the solution is made up it has only a 1-day shelf life. A surface covered with this lacquer is not easy to repair.

Polyester

Polyester is the hardest and most durable of all the lacquers and has good heat- and moisture-resistant properties. Because it dries fast it can only be applied with a spray gun; one coat is sufficient. Polyester can be burnished to give a high-gloss finish. As with acid-catalyzed lacquers, it is difficult to repair the surface, which usually will have to be stripped and relacquered.

Brush application

1. Make a pool.

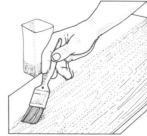

2. Brush along the grain.

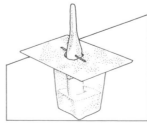

3. Suspend the brush.

Brush application

If lacquer comes in a two-part form, pour both solutions into a container in the recommended proportions. Mix with a wooden stick. Leave for half an hour. Fill the brush with solution. Make a pool on the near side of the work, halfway between the ends (1). Do not allow the solution to drip. Exercising maximum pressure on the brush at the center, brush back and forth along the grain. Reduce the pressure as the brush moves from the center and lift the brush tip promptly at the ends (2). Each stroke must overlap. When the area is covered, brush over the entire surface lightly to straighten the strokes and remove any surplus lacquer. Leave the surface to dry (about 1 hour). Keep the brush supple by suspending it over the solution (3). (After applying polyester or nitrocellulose wash the surface with soapy water.) Smooth with No. 320 wet-or-dry silicon-carbide paper. Dust the surface. Fill holes and adjust the color if necessary. Apply a second coat. Leave to dry; do not smooth. Apply a third coat. Smooth with No. 320 paper; dust. Smooth with No. 400 paper. Dust.

Spray gun application

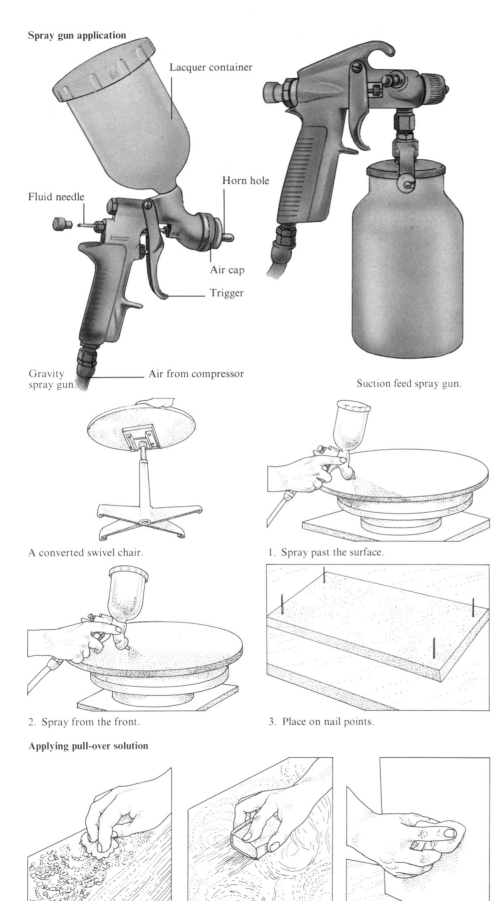

Lacquer container

Horn hole

Fluid needle

Air cap

Trigger

Gravity spray gun.

Air from compressor

Suction feed spray gun.

A converted swivel chair.

1. Spray past the surface.

2. Spray from the front.

3. Place on nail points.

Applying pull-over solution

1. Rub with a soapy cloth.

2. Smooth with abrasive paper.

3. Work with a rubber.

Spray gun application

Spray guns distribute lacquer with a viscous flow that can be carefully controlled to produce a fine spray. To ensure good working order always clean the gun thoroughly. A gravity spray gun has a lacquer container positioned above the gun; the lacquer drains down into the fluid tip. A suction feed spray gun has a lacquer container positioned below the gun; the lacquer is sucked up into the fluid tip. A trigger action controls the air and lacquer flow. The air is forced through two horn holes in the air cap, usually by an air compressor. Guns and air compressors must be compatible. The horn holes can be set horizontally for side-to-side spraying or vertically for up and down spraying. It is best to put the work on a turntable such as a converted swivel chair. Fill the lacquer container with solution. Test the viscosity by spraying some waste wood. Spray at 45 , 150 mm/6 in to 200 mm/8 in away from the surface. Add thinner if necessary. Adjust the horn holes to the appropriate position for the work. Spray the edges of the work first. Always spray in a straight line along the grain, overlapping the previous stroke by 50 percent. Start spraying before the surface and continue past it (1). Then spray the top, working from front to middle (2). Turn the work round and spray the other half. Place the lacquered surface on nail points (3). Then spray the underside. Rack the work carefully. Leaving an hour between coats, apply two or three coats.

Applying pull-over solution

Rub the dry surface in a circular motion with a soapy cloth (1). Smooth with No. 320 fine grade wet-or-dry silicon-carbide paper (2). Rub the edges with wet abrasive paper held flat. Dust the surface. Smooth with No. 400 very fine paper and dust again. Dip a rubber into some pull-over solution. Squeeze until just moist. Work the rubber along the grain. Wipe any excess from the edges. Then work the rubber in a circular motion and finally, with a slightly moister rubber, along the grain (3).

Surface finishes 6

All surface coatings can be treated to give a matte, semimatte, satin or gloss finish. For a matte, semimatte or satin finish a dulling process is used, which diminishes the original shine. For a gloss finish a greater buildup of surface coating than for any other finish is required to produce a surface with a deeper sheen and the brightest possible shine. The process used for a gloss finish is known as burnishing and it may be done by hand or by machine. Dulling brushes have long bristles that are made from bear hair or mixed hair. Whether dulling or burnishing, the more the surface coating is rubbed the greater the effect, so work cautiously. Both processes can be repeated, but to reverse dulling a reviver has to be used on the surface.

After their final layer has been applied, surface coatings should be left for at least 24 hours before they are ready for any finishing treatment. If treated earlier than this, streaks may appear on the surface. Between applying a surface coating and finishing, considerable sinkage may occur, especially where the grain has not been filled with wood filler. If this takes place, smooth the whole surface lightly with a fine abrasive paper, lubricated if necessary. Then wipe clean before starting the final finish.

Matte finish

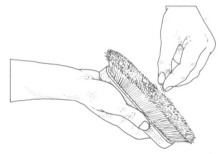

1. Rub the wool along the grain.　2. Rub with a circular motion.　Use brush on uneven surfaces.

Matte finish
Rub the surface along the grain with flattened grade 000 or 0000 steel wool (1). Dust off. Hold the work to the light to check that all the shine has been removed. Pick up wax on a soft absorbent cloth. Rub it onto the surface with a circular motion (2). Leave overnight. Then burnish the surface with a clean soft cloth. Use a brush impregnated with wax on uneven surfaces.

Semimatte finish

1. Put some powder on a brush.　2. Rub along the grain.

Semimatte finish
Sprinkle some pumice powder onto a dulling brush (1). Work the brush up and down the grain with considerable pressure (2). Wipe off frequently. When the polish has been sufficiently dulled, wipe the surface clean with a soft cloth.

Satin finish

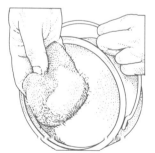

1. Load steel wool with wax.　2. Rub along the grain.　3. Buff up with a soft cloth.

Satin finish
A greater buildup of surface coating is required for a satin finish than for a matte or semimatte finish. Fold some 000 or 0000 steel wool into a pad and load with beeswax (1). Rub along the grain with moderate pressure (2). Too much pressure will dull the satin finish. Turn the pad over and remove the residual wax with the clean side. Then buff up the surface with a clean soft cloth to the required sheen (3).

Gloss finish

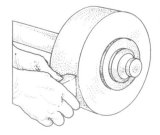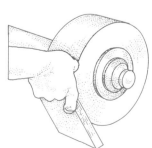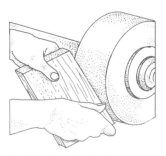

1. Rub the block onto the mop.　2. Hold the surface to the mop.　3. Hold the edge to the mop.

Gloss finish
Rub a block of wax and abrasive powder back and forth against the rotating linen mop of a burnishing machine (1). Hold the work lightly but firmly against the rotating mop in the position shown (2). Move the work slowly from side to side. Place one hand under the work when burnishing the edges (3). Regularly replenish the mop with wax and abrasive powder. Polish with the sheepskin mop.

Burnishing and polishing machines

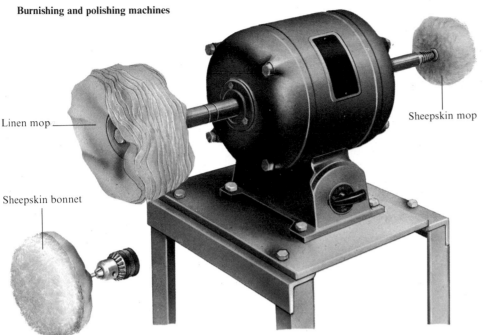

Linen mop

Sheepskin mop

Sheepskin bonnet

A small burnishing machine.

Burnishing and polishing machines
A small burnishing machine can be used to make a gloss finish. The linen mop at one end of the machine is for the initial burnishing and the sheepskin mop at the other is for polishing. Alternatively the surface can be burnished by hand and polished with an electric drill attachment, consisting of a rubber or polyurethane disk over which a sheepskin bonnet is fitted. To burnish by hand use a mild rubbing compound such as those available from motor supply shops and a soft cloth made from natural fibers. Put the compound onto the cloth and rub the surface along the grain. When the compound dries the burnishing action begins. Continue to rub until the finish is bright. Wipe with a clean cloth and polish with the drill attachment. Remove any smears with a reviver.

Stripping

A surface coating may be recognizable by the solvent that will soften it. Application of a reviver, made from equal parts linseed oil, household vinegar and methylated spirit, will also aid recognition as it will conceal scratches on a French polished surface whereas on a lacquered surface the scratches will remain white.

To assess accurately whether the work needs to be stripped, first remove the surface dirt, grease and wax with a reviver. Shake the reviver before applying it to the surface with a soft absorbent cloth; then rub the surface with a clean cloth.

A thorough cleaning of the surface may be all that is required to restore the original patina. If the surface needs a further application of surface coating, clean it with turpentine or a weak solution of soda crystals before rubbing it down with fine steel wool. Then apply the relevant surface coating. If, however, the surface is badly cracked or patchy it will need to be stripped. Should the piece be a valuable antique that has only minimal surface damage, it is usually better to leave it as it is. A surface should not be partially stripped as its coloring will be uneven when it is resurfaced.

Almost all surface coatings can be removed with a proprietary stripper. Protect the floor and workbench with newspaper when using the stripper and wear goggles as well as protective clothing and gloves. Stripper is toxic and inflammable so read all instructions and neutralize any splashes with water.

Before beginning to strip, assemble the stripper, a scraper, a 38 mm/1½ in flat brush, 000 steel wool, some burlap or canvas, a small stiff bristle brush, a tin to hold the stripper, some upholstery wadding and some methylated spirit. A cabinet scraper may also prove useful. Steel wool is sometimes substituted for a scraper if the coating is soft, for example French polish. Burlap will also remove the coatings and is useful for wiping the scraper. The brush will take off the coatings from awkward places.

Any part of the surface may be stripped first, but attempt only a small area at a time. Scrape when the surface finish is soft. Test frequently and strip rapidly as soon as easy removal is possible. Scrape towards or away from the worker. If a surface dries too fast another layer must be applied. The number of stripper layers will vary according to the type and thickness of the surface coating. When the surface has dried, all stripper must be neutralized with methylated spirit or cold water to stop the stripper continuing to react. Leave the work to dry before resurfacing.

Stripping

1. Scrape the broken coating.

2. Strip recesses thoroughly.

3. Rub with steel wool.

Stripping
Brush stripper liberally onto an area. Leave until nearly dry. Scrape the broken coating along the grain (1). Then rub steel wool along the grain. Repeat this process as necessary. Strip recesses thoroughly (2). Neutralize the whole surface with methylated-spirit-soaked upholstery wadding and rub well into the grain with a circular motion. Remove the film with steel wool (3). Then smooth.

Restoring

When embarking on a job of restoration first of all examine the piece of furniture carefully. Look for evidence of insect damage, chipped moldings and beading, worn bearing surfaces and warping. Rock frame structures to see if any joints are loose and check veneers for splits and blisters. Then make a list of all the defects and decide on the order in which they should be tackled. The scope of the work that can be undertaken will depend on the skill and confidence of the worker. Wherever possible surface finishes should be retained, and as a general rule repair is preferable to renewal as it is frequently impossible to imitate the patterns and craftsmanship of the past. The age and value of the piece must also be taken into account. It may be that the piece of furniture has fallen into such a state of disrepair that the extent of work required to restore it would not prove worth while.

To prevent pieces falling into bad condition always attend to a repair immediately, otherwise the one defect may cause strain on another area and lead to a series of further problems. Where it is necessary to replace old wood, match the new wood closely by paying careful attention to species, grain, figure and even, sometimes, the rays so that the repair will be unobtrusive. If at all possible do a repair without dismantling, but where it does prove necessary to dismantle remember to protect the surface finishes with wedge-shaped blocks or battens to prevent bruising. When the joints are knocked apart remove all old glue with a chisel or a shoulder plane.

Removing nails

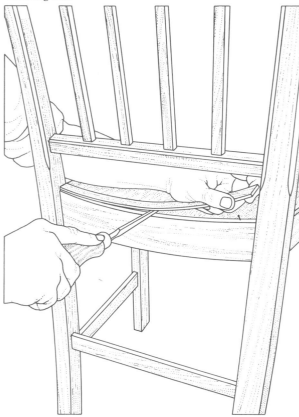

1. Pry up any capping pieces.

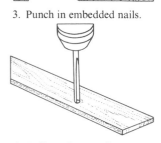

2. Remove pins and loose nails.

3. Punch in embedded nails.

4. Drill out large nails.

Removing nails

Before dismantling a piece of furniture it is important to remove all nails and screws. These may be well concealed, especially in tenons. Remove any capping pieces by prying them up with a chisel (1). Then remove any small nails by gently twisting and levering them out, using side cutters (2). Lift out any loose nails. Nail-heads that are flush or below the surface of the frame should be hammered through using a nail-punch, because levering them out could cause damage to the piece of furniture (3). Where hammering through is not practical, use a brace and plug cutter to penetrate the wood around the nail (4). Then remove the nail with a pair of side cutters.

Dealing with pellets and screws

1. Chisel out any pellets.

2. Heat stubborn screw-heads.

3. Glue loose joints.

Dealing with pellets and screws

Place the piece of furniture on a workbench or floor and protect the points of contact with some cork or a cloth. With a mallet and chisel, gently pry out any pellets (1). Then remove the countersunk screws. Loosen stubborn screws by heating their heads with a soldering iron to make them expand (2). Expansion breaks any seal of rust so that when the screws cool and contract they can be loosened. Prize open loose joints once the screws have been removed. Force glue in, using a hypodermic syringe (3). Replace the screws.

Dismantling

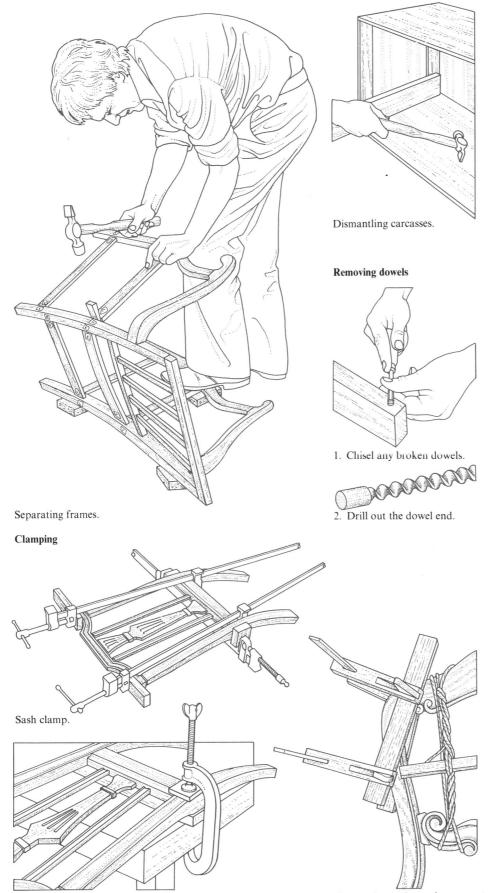

Dismantling

Letter each piece of wood at each end so that matching joints can be identified. Before separating anything, decide on the order of dismantling and check that nothing else is holding the parts together. Always work parallel joints together to prevent straining. Using a hammer or rubber mallet, start to knock frame joints apart using the side of a wedge-shaped block. Place one end of the wedge between the opening formed and gently knock the joints apart. Dismantle carcasses with a hammer and batten, hitting the center of the batten so that the impact is distributed evenly.

Dismantling carcasses.

Separating frames.

Removing dowels

Removing dowels

Chisel any broken dowels until they are flush with the surrounding surface (1). Using the same size drill bit as the dowel diameter, drill into the bottom of the dowel. Once the drill bit has firmly embedded in the dowel, pull it out from the surrounding wood (2).

1. Chisel any broken dowels.

2. Drill out the dowel end.

Clamping

Sash clamp.

C-clamp.

Cam and tourniquet clamps.

Clamping

Awkwardly shaped pieces of furniture were originally clamped with the aid of projections that were later sawn off. Clamping while restoring is far more complicated. The even distribution of pressure in the required places can become a skilled juggling trick and specially shaped clamping blocks may be needed. Always ensure that components are clamped true and at the angle at which they were originally designed. All surfaces must be protected with wooden blocks or cork. Sash clamps are often used with wooden bearers or specially cut saddles to spread the load evenly. C-clamps are generally for holding parts together away from edges. Cam clamps are especially useful to a restorer as an extra edge clamp can be added for three-sided clamping pressure. Tourniquet clamps are for rounded corners or extremely awkward shapes.

Natural defects

Wood beetle infestation

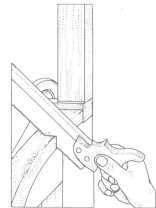

2. Cut into the infected area.

3. Remove the infected wood.

1. Clamp the wood firmly.

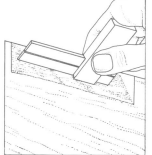

4. Check for flatness.

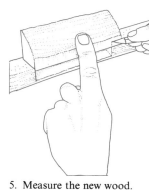

5. Measure the new wood.

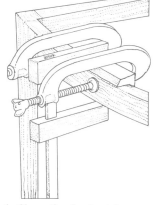

6. Clamp the glued patch.

7. Plane the patch flush.

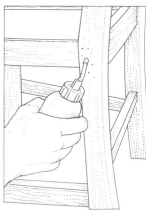

8. Treat with an insecticide.

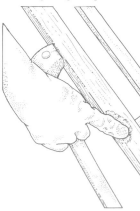

9. Fill the holes with wax.

Wood beetle infestation

In temperate zones, the pin-hole borer, the furniture beetle, the longhorn and the death-watch beetle are among the insects that burrow into timber, particularly sapwood. Their telltale marks are the flight holes appearing on the surface of the wood. Where the attack is severe, the infected area will need to be replaced with good wood, retaining the original front surface if possible. Clamp the piece of furniture firmly to the bench (1). Ensure the infected area is accessible. Using a dovetail saw, make a sloping cut at the edge of the heavily infected area (2). Do not mark the front surface. Make a similar cut at the far end of the infected area. With a chisel, carefully remove the infected wood between the two sawcuts (3). Whether working on a curved or straight piece of wood, always chisel the surface flat. Check the cutaway area for flatness with a try square or straight-edge, both along and across (4). Take a piece of wood of similar grain direction and figure to the main piece and measure it against the cutaway area (5). Making it slightly wider and higher than the cutaway area, cut the patch to shape. Plane its sides. Apply PVA glue to the patch and tap it into the cutaway area. Clamp all around for 30 minutes to dry (6). Take off the clamps and plane both sides so the patch lies flush with the main piece (7). Then smooth with an abrasive paper. Color the patch so it blends in with the adjacent wood. If the infestation has only just begun to appear, or when the above treatment for severe infection has been completed, the entire piece of furniture should be treated with an insecticide. Use a proprietary liquid, in a bottle with a nozzle, and squirt the liquid into every visible flight hole (8). Sometimes liquid appears from another flight hole some distance from the hole that is being treated. This is evidence of a multitude of galleries running just below the surface. Block the flight holes after treatment; wear gloves while working some colored wax into the holes (9). When the wax has dried, buff the area with a soft cloth. Whatever time of year the initial treatment was done, always respray the flight holes the following spring, when the surviving insect eggs may be hatching out.

Joint defects 1

Loose joints

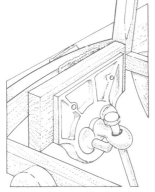

1. Grip the wood firmly.

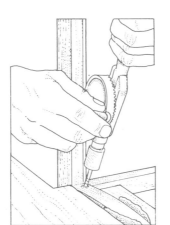

2. Bore a hole.

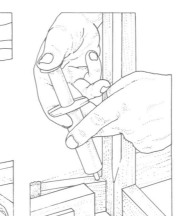

3. Insert animal glue.

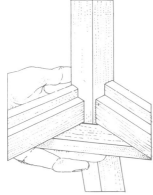

4. Select a piece of wood.

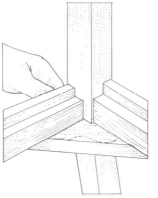

5. Mark a curve at the front.

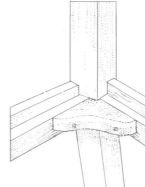

6. Screw in position.

Loose joints

Loose joints can frequently be remedied by reinforcement gluing and the addition of a corner block. Protecting the surface finish with some cork, grip the piece of furniture in the bench vise (1). With a hand drill, bore an inconspicuous hole in the loose joint cavity (2). Insert animal glue into the hole, using a hypodermic syringe (3). Do not substitute a resin glue as it would dry too fast. Clamp the reglued joint so that the pressure is distributed evenly and leave for 8 hours. For the corner block, select a piece of wood with its grain running between adjacent sides of the frame (4). The edges may need to be beveled to fit the frame closely. Mark a curved front on the corner block to facilitate the entry of screws (5). Cut out the block, using a bow saw, and drill clearance holes for the screws. Glue and screw the block into position, countersinking the screws (6).

Over-wide mortises

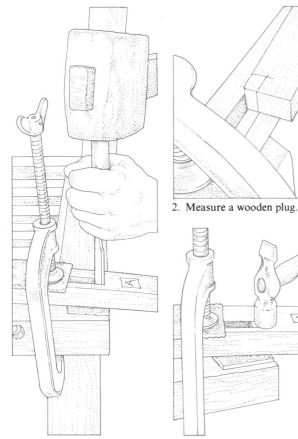

1. Clean out the mortise.

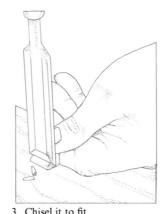

2. Measure a wooden plug.

3. Chisel it to fit.

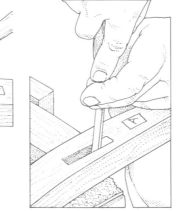

4. Hammer in the glued plug.

5. Lever the plug to one end.

Over-wide mortises

Where a dismantled joint reveals a loose tenon in an over-wide mortise, clamp the component to the bench and clean out the mortise with a chisel (1). Then choose a piece of wood with matching grain direction to the main piece and mark out the length, width and depth of a wooden plug that will fit into the mortise and hold the tenon closely (2). Saw out the plug and chisel it to fit (3). Glue the plug and hammer it into the mortise (4). Lever the plug tightly against one end of the mortise, using a tool such as a screwdriver (5). When the glue has dried, plane the plug flush with the surface. Smooth with an abrasive paper. Stain and polish where necessary.

Joint defects 2

The general principle of how to splice on new wood applies both to repairing broken legs and rails and to cutting a new mortise or new tenon. The restorer must maintain as much as possible of an original piece of furniture and every effort should be made to preserve, rather than replace, components. Splicing is a means of putting this principle into practice. Broken components and joints should be attended to as soon as possible when the breaks are fresh, leaving no time for grease and dirt to accumulate, for pieces to be mislaid or for surfaces to break further. New wood should be tapered to 15° to 20° to provide as large a gluing area as possible and to make the joint strong and inconspicuous. It should also have the same grain pattern as the main piece.

Having dismantled the failed piece, check to see if it has been set square or at an angle in the wood. Cut the new piece in such a way as to leave it a little above the surface when glued in position. When dry, plane the piece flush with the adjoining surface, taking care to remove as little of this surface as possible. Always smooth with an abrasive paper to ensure the surface is absolutely flat before staining and polishing the new piece to match the original wood.

Where dismantling proves difficult or causes undue strain on other parts of the structure, false tenons can be used to repair broken mortise and tenon joints. This is particularly common with chair back rails and stretchers. Before cutting new ones, it is important to check whether the original tenons were set square or at an angle and whether the mortise is over-wide. There are several kinds of false tenon, such as stepped ones, and these reflect the variety of mortise and tenon joints, so always check which type is required.

New mortises

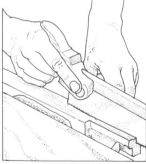

1. Check the wood carefully.

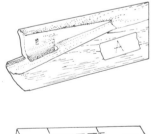

2. Chisel away old tenon.

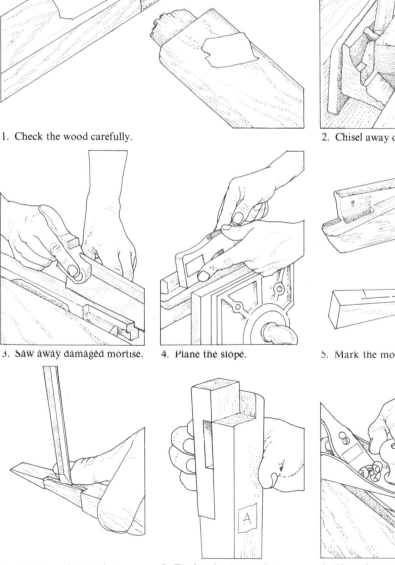

3. Saw away damaged mortise.　4. Plane the slope.　5. Mark the mortise.

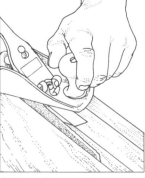

6. Chisel out the mortise.　7. Fit the glued new piece.　8. Plane the new piece.

New mortises
Where a mortise has been severely damaged the entire area may need to be replaced (1). Grip the component in the bench vise with cork for protection. Saw away any tenon protruding from the mortise. Remove the rest of the tenon from the mortise with a chisel and mallet, looking out for nails (2). Clean away the old glue with a chisel. Consider carefully how best to cut into the good wood to allow for maximum strength when the new piece is fitted. Saw away the damaged area, making a shallow slope onto which the new piece will be spliced (3). With a shoulder plane, smooth the slope (4). Check it for flatness with a try square or straightedge. Select some new wood of matching grain pattern to the old wood. Cut the wood to fit the cutaway area, but extend it above the surface. Plane the sloping side smooth. Mark the position of the mortise (5). Saw and chisel out the marked area (6). Glue the sloping side and fit it in position against the component and clamp firmly (7). When the glue has dried, take off the clamps. Using a smoothing plane, level to just above the surface of the adjacent wood (8). Smooth the new piece flush, using an abrasive paper. Color the repair carefully to match the original wood.

New tenons

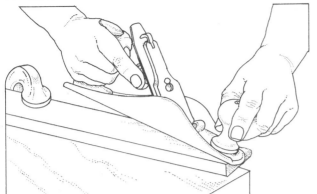

1. Plane the damaged wood.

2. Clamp the new glued piece.

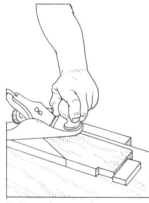

3. Plane the new piece.

4. Gauge the new tenon.

5. Saw the new tenon.

False tenons

1. Glue in the false tenon.

2. Slot the rail over glued false tenons.

Clamped joints

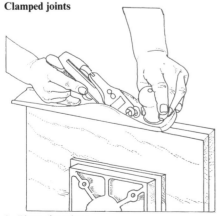

1. Plane the edges.

2. Clamp the glued edges.

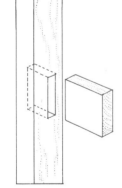
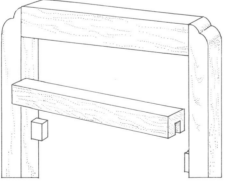
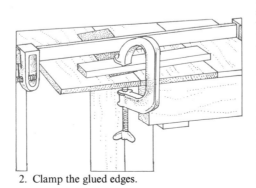

New tenons

Saw or plane away the damaged area around the broken tenon, making a shallow slope onto which a new piece of wood can be spliced (**1**). Choose a piece of wood with a grain that matches the old piece and cut it almost to size. Grip it in the vise and plane one edge to fit the sloping side of the component. Glue the component and new piece together and clamp (**2**). When the glue has dried, remove the clamp. Plane the new piece flush; take care to remove as little of the finish on the adjoining surface as possible (**3**). Smooth with an abrasive paper. Set a mortise gauge to the dimensions of the mortise and then use it to gauge the size of the new tenon (**4**). Grip the component, protected with cork, firmly in the bench vise and saw the marked tenon (**5**).

False tenons

Clean out the old mortise with a chisel and mallet. To make a false tenon, cut a rectangular piece of wood to twice the depth and the exact width and thickness of the mortise. Glue the wood into the mortise (**1**). Mark the measurements of the protruding false tenon onto the rail end and underside. Saw from the bottom, at an angle along the marked lines; then chisel out the waste. Vertically chisel away the rest of the waste, making the slot square. Repeat the process at the other end. Glue the false tenons and slot the rail over them (**2**).

Clamped joints

Remove the relevant parts and place them in the bench vise. Plane the edges straight and square (**1**). Take out of the vise. Glue the edges and lay them flat on a board covered with newspaper to prevent them sticking to the board. Place some paper and a batten over the joint and clamp it down. (For wide boards, clamp a second batten on the opposite side.) Clamp the two edges together with one or more sash clamps (**2**). Leave the joint to dry. Remove the clamps and smooth away any excess glue.

Surface defects 1

Worn surfaces

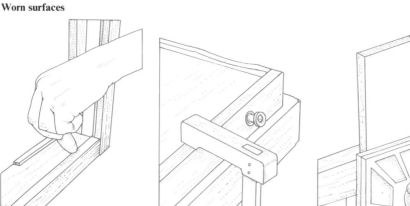

Rub candlewax along the runner.

1. Clamp the drawer.

2. Secure in the vise.

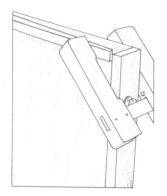

3. Clamp the glued new piece.

4. Plane to shape and size.

Split surfaces

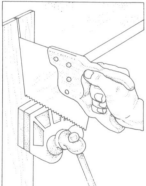

1. Saw in the split wood.

2. Plane a thin strip of wood.

3. Hammer in the glued strip.

Worn surfaces

Where a drawer is sticking, it may be sufficient to rub a little candlewax along the runner or drawer side to enable the drawer to move smoothly. Where the bottom edge of a drawer side has been worn away, it will need to be built up with a new piece of wood. Clamp the drawer upside down, leaving the worn surface free to be worked on (1). Plane the worn surface flat with a shallow sloping side. For the new piece, choose wood similar to that of the drawer side and of the same grain direction. Secure it in the vise and saw off a tapered piece that will closely fit the sloping edge of the drawer side (2). As the inside edge will become inaccessible after gluing, plane this edge at this stage. Apply PVA glue to the planed worn surface. Place the new piece with its inside edge in position and clamp it to the drawer (3). When the glue is absolutely dry, remove the clamps and plane the new piece to the required shape and size (4).

Split surfaces

Splits running along the grain often occur in the bottoms of drawers. Remove the drawer bottom and grip it in the bench vise. Using a ripsaw, cut down to the base of the split (1). Select a thin strip of wood of similar type and grain direction to the drawer bottom and plane it against a bench stop until it fits into the sawcut (2). Glue the strip and insert it into the sawcut. It may be necessary to hammer it into position (3). When the glue is absolutely dry, plane the strip flush on all sides and stain if necessary.

Bowed surfaces

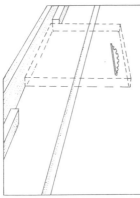

1. Nail stops to the fence.

2. Lower the wood.

3. Cut even, parallel grooves.

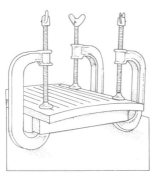

4. Clamp open the grooves.

5. Square off the groove ends.

6. Hammer in the glued strips.

Bowed stiles and rails

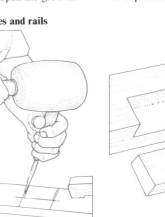

1. Chisel out the marked area.

2. Cut the patch to fit.

3. Clamp the glued patch.

Warped doors

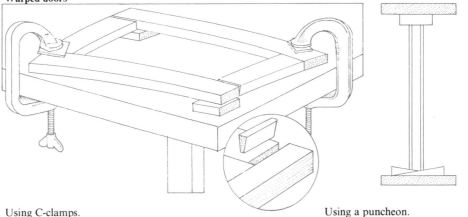

Using C-clamps.

Using a puncheon.

Bowed surfaces

Set the blade of a table saw to three-quarters the thickness of the bowed wood. Position the guide fence so that the saw blade will cut 19 mm/¾ in in from the side and nail wooden stops to prevent the wood being cut nearer than 13 mm/½ in from both ends (1). Switch on the saw. Lower the wood onto the blade, holding it against the fence and nearest wooden stop (2). As the saw cuts, push the wood gently towards the far wooden stop. Having reached the end, switch off the saw. When the blade stops rotating, lift off the wood. Make similar parallel grooves 19 mm/¾ in apart. Stop 19 mm/¾ in from the far side (3). For a convex bow, clamp the wood over gently curved bearers in the opposite direction to the bow (4). (For a concave bow clamp the grooved surface flat.) Cut strips of wood to the thickness of the clamped grooves. Chisel the groove ends square to take the strips (5). Glue the strips and tap into the grooves (6). When the glue has dried, remove the clamps. Plane the strips flush. Smooth with an abrasive paper.

Bowed stiles and rails

With the concave side uppermost, clamp the bowed wood flat. Where the bow is greatest, mark out a tapered area with undercut sides. It should be no longer than 75 mm/3 in and no deeper than half the thickness of the main piece. Cut away and level the marked area (1). Select a piece of wood similar to the rail or stile and of the same grain pattern. Bevel it to fit tightly into the cutaway area (2). Glue the patch and push it well in. Grip it in the vise (3). When dry, remove; plane the patch flush and color to match.

Warped doors

Lay the door flat with wooden blocks beneath the straight or downward-bent corners. Clamp upward-bent corners with C-clamps. Lay large doors on the floor. Put wooden blocks under the straight and downward-bent corners. Place puncheons between the upward-bent corners and wooden planks laid against the ceiling. From opposite sides knock the thin ends of two triangular wedges between the puncheon and the door. Force glued wedges into the open joints. Release the pressure and plane the wedges flush.

Surface defects 2

Damaged moldings

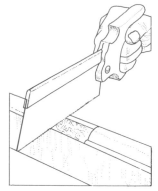

1. Make two beveled kerfs.

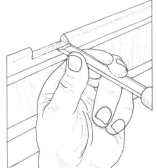

2. Chisel the damaged area.

3. Check the flatness.

4. Check the patch for fit.

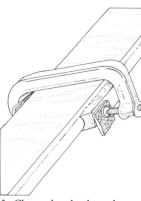

5. Clamp the glued patch.

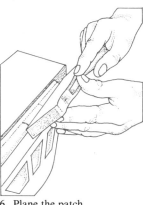

6. Plane the patch.

Veneer blisters

1. Slit open any blisters.

2. Insert glue in both sides.

3. Clamp until dry.

Making new pellets

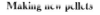

1. Mark the required shape.

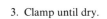

2. Pare the pellet to fit.

3. Chisel the pellet flush.

Damaged moldings

Where an area of molding is missing or damaged it can usually be replaced by a new piece. Make a beveled kerf on either side of the damaged area so that the top opening in the molding is wider than the bottom and so that the front is narrower than the back (**1**). Chisel out the area between the two kerfs, working across the grain (**2**). Test for flatness with a try square or straightedge (**3**). Select a piece of wood with a grain that matches the molding. Making it slightly larger than the cutaway area, plane it to shape to make a perfect fit in the cutaway area (**4**). Glue the patch, insert it firmly into the cutaway area and clamp (**5**). When the glue has dried, remove the clamp and plane the patch to shape with a shoulder plane for convex curves and a molding plane for concave ones (**6**). Smooth with an abrasive. Then color the patch to blend in with the surrounding molding.

Veneer blisters

Brush over the surface with the palm of the hand and tap suspected air pockets and blisters with a finger. Make a slit along the grain through the middle of the blister (**1**). Gently lift the veneer with the knife point and insert animal glue under each side (**2**). Push the veneer down so the excess glue is squeezed out. Clamp until dry, protecting the veneer with cork (**3**).

Making new pellets

Where pellets have been damaged, new ones must be made from suitable wood with a matching grain to the main piece. Where possible, use the old pellet as a template to mark the new one (**1**). Pare the wood, slightly tapering it. Keep it attached to the main piece for as long as possible (**2**). Put some glue in the hole and gently tap in the tapered pellet until it fits tightly. When dry, chisel the pellet flush with the adjacent surface (**3**). Smooth with an abrasive paper. Stain and polish are required.

Patching holes

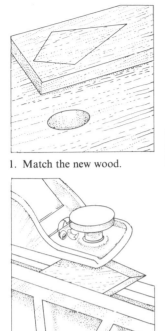

1. Match the new wood.

2. Plane the patch.

3. Chisel the hole to shape.

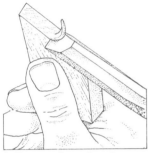

4. Remove the under edge.

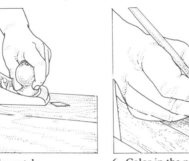

5. Plane the patch.

6. Color in the patch.

Patching holes

Where removed fittings have left holes, the surface must be patched. The wood should be carefully matched: even the ray direction is important (1). Cut out an irregular diamond-shaped patch. An asymmetrical shape will blend in well with the adjacent surface and so make the patch less conspicuous. Slightly taper the edges of the patch with a block plane (2). Lay the patch over the hole on the surface. Align the grain and rays as perfectly as possible; then draw around the patch with a very sharp pencil. With a mallet and chisel, cut inside the marked lines; gradually move outwards until the chisel is working on the marked lines (3). Take special care when cutting into the acute angles. Chisel the edges with a slight taper. On the patch, remove the sharp under edges with a chisel (4). Glue the patch with PVA. Spreading the pressure with a wooden block, hammer the patch into the hole. Plane the patch almost flush with the adjacent surface (5). Smooth the patch flush with the surface using an abrasive paper. Color and polish as necessary (6).

Treatment of surfaces

Reviving surfaces
To remove grease and dirt, apply a thin coat of proprietary soft furniture wax. Rub off with a soft cloth to leave a thin film. An alternative method is to clean the surface with a soft cloth dampened with soap and warm water. Allow it to dry. If the surface remains dull after this treatment, apply a mixture of equal parts linseed oil, household vinegar and methylated spirit. Rub the surface gently. Where deep scratches occur, omit the oil. Gently wipe off the mixture, using a clean cloth. Then apply a coat of good-quality wax polish and rub well.

Stains caused by water
When wood is dry, rub area well with genuine turpentine. Restain where necessary.

Stains caused by alcohol
Where a surface is polished, treat as for water. Where varnished, smooth lightly with an abrasive and rebuild surface finish with varnish.

Stains caused by grease and oil
Sprinkle talcum powder over stains. Cover with tissue paper or blotting paper and warm with an iron to draw grease and oil from wood. Repeat until marks disappear. On non-veneered surfaces apply trichlorethylene to the stain. Never use this chemical on veneered surfaces and manufactured boards, as it attacks glue.

Ink stains
Bleach with oxalic acid. Afterwards, neutralize acid with a weak soda-crystals solution.

Heat marks
Rub camphorated oil into marks with soft cloth. Wipe the surface clean.

Cigarette burns
Smooth away or scrape to below the burned surface. Stain as necessary and rebuild the surface finish. Paint out deeply ingrained burn marks with spirit stains.

Light scratches
Smooth with very fine sandpaper dipped in linseed oil.

Deep scratches
Fill with shavings of suitably colored beeswax. Press the shavings into the scratch with a knife. When dry, remove excess wax and smooth with fine abrasive paper.

Dents
Cut through finish with fine knife, making several parallel cuts along the grain. Fill depression with water and allow to soak in. Apply cloth and hot iron to swell the wood. Allow to dry, and refinish with the appropriate surface coatings.

Materials

An extensive study of the most sympathetic of all materials

Growth and structure

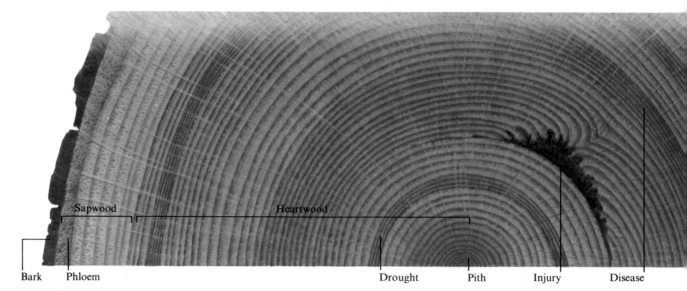

Sapwood   Heartwood

Bark   Phloem   Drought   Pith   Injury   Disease

The **bark** is the outermost layer of the tree. Its function is to protect the vital inner tissue from extremes of temperature and some diseases. It covers the inner bark or **phloem** which conducts synthesized food (the product of photosynthesis) from the leaves to the rest of the tree. It also allows vital gaseous exchanges.

The **cambium** is the growing layer of the tree. It is a thin delicate layer of tissue responsible for the production of new wood both outwards to the phloem and inwards to the sapwood, adding girth to the trunk and branches throughout the growing season; in temperate areas this is spring and summer.

The main function of the vessels of the **sapwood** is to conduct water-soluble mineral salts from the roots to the leaves. The sapwood is usually distinguishable from the heartwood — the main part of the tree — by its paler color. It can measure up to 50 mm/2 in wide, and in tropical species it can be up to 200 mm/8 in.

The **heartwood** lies within the sapwood. It is composed of dead cells that give rigid support and store food. Most timber comes from this part of the tree. At the center of the heartwood lies the **pith** or medulla, which is usually invisible but can sometimes be seen with difficulty in certain species.

The tree is unique in the plant kingdom for two reasons: its size and longevity. There are some 44,000 known species of tree varying in height from 1 m/3¼ ft — the dwarf pine of Alaska — to over 120 m/394 ft — the California redwood. The record girth measurement is 53 m/174 ft and belongs to a Mexican cypress known as the Giant Tree of Tula, which is thought to be three trees that have grown together. Some species such as the lowland oak can produce a root system almost twice as extensive as their branch system or crown. The variation in life span is even more dramatic. The English silver birch is considered overmature at 80 years of age, whereas a bristlecone pine felled in 1964 in Nevada had lived for 4,900 years. Other relatively long-lived species such as the oak and California redwood can live for 1,000 years or more.

Like all other plants, trees are dependent on the process of photosynthesis, which is necessary for the production of new cells and, subsequently, growth. In a tree the growing points are called primary meristems and these are responsible for increasing the height. They are situated at the tip of the trunk and each branch and twig. To increase the girth of the tree the secondary meristems, or cambium, produce new phloem and sapwood during the growing season. The new wood is laid down in sheathlike formations called growth rings. Under ideal conditions a tree growing in fertile soil will produce wider growth rings than a tree growing in infertile soil. The study of a tree's growth-ring formation, known as dendrochronology, can provide an accurate record of these environmental conditions as well as the biological and traumatic incidents that occur throughout a tree's life.

At the center of the tree is the heartwood. This acts as a rigid support structure and also contains many substances such as oils, resins, gums, dyes and latex. Many of these are extracted from the stumpwood after the tree is felled, but some, such as latex, are tapped from the living tree. In spite of the introduction of synthetic materials many industries such as paint and pharmaceutical manufacture are still reliant on these natural substances.

Some substances in the wood, such as gums and resins, affect the wood color. The numerous dark-colored veins in African walnut, for example, actually enhance the wood and give it a characteristic figure. Less beneficial, from the woodworker's point of view, are mineral deposits, which may cause hard and soft patches in the wood, making it difficult to work.

Although all trees grow and produce wood in a similar way, some trees belong to a more primitive plant type than others. Trees are therefore divided into two main classifications. The first and more primitive are the gymnosperms, or conifers; these are cone-bearing needle-leafed trees known as softwoods. The second are the angiosperms; these are subdivided into two classes, the monocotyledons, which include the palms and grasses and are, therefore, of little interest to the woodworker, and the dicotyledons. It is the dicotyledons, or broad-leafed deciduous trees, that include the most varied species of trees known as hardwoods. The terms hardwood and softwood are botanical terms and do not reflect the physical properties of the tree. The softwoods pitch pine and yew, for example, are much harder than the hardwoods balsam, willow, lime, obeche and gaboon.

Cambium

The three main reference planes by which wood can be recognized are illustrated. A transverse section is on the end grain, a radial section is on a quarter-sawed surface and the tangential section is on a plain-sawed surface.

The wide range of wood textures reflects the many varied patterns of growth among trees. As a general guide, when the wood is even textured there is little contrast between the wood produced at the beginning of the growing season (earlywood) and that produced later on (latewood). A high contrast results in a wood of uneven texture. Wood is also described as fine or coarse textured, the difference here being defined by the size of the cells (be they vessels in hardwoods or tracheids in softwoods) and the ray widths. Wood with large cells and wide rays is coarse textured and that with small cells and narrow rays, fine textured. The transverse sections show a fine even-textured wood (*right*) and a coarse uneven-textured wood (*far right*).

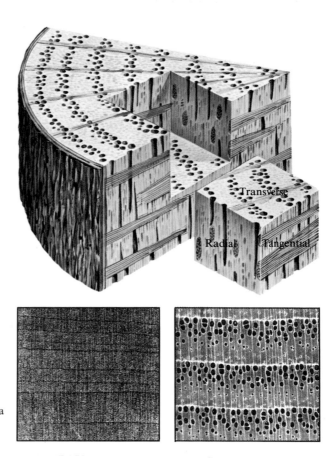

The **growth rings** are the annual increments of wood laid down by the cambium. Faults in the pattern are caused by **disease** and **drought**, which inhibit the growth of a tree, resulting in the production of narrow growth rings, and traumatic injury such as fire, which can leave a scar buried in the tree.

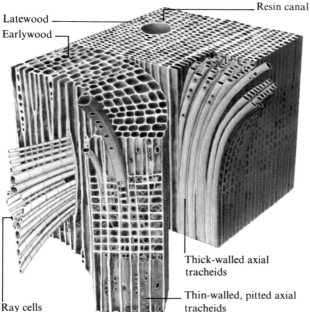

Latewood
Earlywood
Resin canal
Thick-walled axial tracheids
Thin-walled, pitted axial tracheids
Ray cells

The cellular structure of softwoods is relatively simple. There is one main type of cell known as the axial tracheid, which is long, thin and hollow. The larger tracheids have pitted thin walls and occur in the earlywood. The smaller ones have thicker walls and are found in the latewood, and they give the timber its strength. The rays appear as long thin horizontal lines

radiating outwards from the pith to the phloem of the tree. In most softwoods they are very fine and barely visible. Food storage occurs mainly in the parenchyma cells, which can also produce resin. Some species, such as true pine and larch, have special resin canals that run horizontally and vertically throughout the structure and are lined with parenchyma cells.

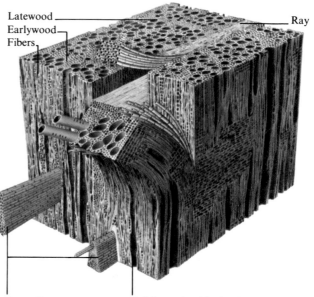

Latewood
Earlywood
Fibers
Ray
Ray cells
Axial vessels with pitted walls

The cellular structure of hardwoods is more complex than that of softwoods, the main difference being the presence of axial vessels, which aid conduction of water and soluble substances throughout the tree. The vessels are usually of relatively even size and distribution in the growth ring; this is known as diffuse porous wood (*above*). On rare occasions vessels formed in the

earlywood are larger than those of the latewood; the wood is then termed ring porous. The support cells, which give hardwood its strength, are known as fibers and can be thin or thick walled. As in softwoods the rays run from the pith to the phloem and help distribute the soluble food throughout the the tree. Food storage occurs in the parenchyma cells.

227

Forestry and distribution

Wood is generally superior in its fabricating and aesthetic qualities to many of its competitors. The two main types of wood of interest to the woodworker are softwoods and hardwoods. The hardwoods are the more popular and versatile as well as being the more plentiful of the two woods, making up 60 per cent of the world's forests.

There are over 44,000 known species of timber, of which 12,000 are in use as commercial timber. Many extractive materials, such as resins, gums and latex, also provide the raw material for many separate but equally important industries that, for example, manufacture chemicals and medicines. As new areas of tropical and subtropical forests become accessible, many more species will become available.

At one time two-thirds of the world's surface was covered by natural forest. Man's intervention has reduced this to only one-third of the surface area. This shrinkage of natural forest, coupled with the growing needs of industry, has resulted in an increase in the cultivation of man-made forests. Success in this venture depends on the forester's knowledge of topography, to ensure the correct climatic conditions, and his skill at silviculture, to ensure that the right species are planted in the most beneficial soils.

To ensure as high a yield as possible, the forester must plant the trees using a system of rotation and in planned cropping areas, known as settings, to allow for natural and artificial re-seeding. Seeds may be planted by hand, machine or by air from a helicopter.

Forests cover one-third of the world's land surface. They can be divided into three main zones (*right*). Softwoods tend to be used as building materials and for packaging and paper pulp, while the hardwoods are preferred for specialized uses such as cabinetmaking. The total amount of wood supplied to the world market in 1979 was nearly 2,000 million cubic metres/6,562 million cubic feet, and demand is increasing, despite the introduction of many wood substitutes such as plastic laminates. Through the cultivation of man-made forests, man is trying to meet both present and future demands for timber, within the limits of the natural cycle of the timber crop.

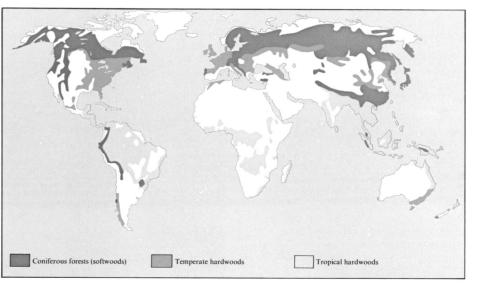

Coniferous forests (softwoods) Temperate hardwoods Tropical hardwoods

To ensure a sustained yield of trees, a forest is divided into settings (*left*). A selected setting is cleared and then replanted with seeds or young trees. Once the cleared setting reestablishes itself, an adjacent setting may be cut and replanted. One of the major hazards to the forest is fire, both natural and man-made. One method of combating fires is from a small plane or helicopter, which "bombs" the blaze with retardant chemicals (*below left*).

The harvesting of larger trees from the higher slopes of mountains has, in the past, proved impossible because the areas have been inaccessible to transport. With the advent of bulldozers, however, it has been possible to build roads so trucks could bring in machinery such as the tower yarders (*right*). These are self-powered, telescopic steel towers, which may extend to 33.5 m/110 ft. They are equipped with enough cable to haul in logs from ranges of up to 800 m/880 yd.

The harvesting of trees requires different techniques, depending on the location and species of the tree. Selecting trees to harvest from man-made conifer forests that contain only a few species is relatively easy, but in the natural tropical forests of Africa, south-eastern Asia and South America, where hardwood species tend to occur only in isolated groups, the process of selecting and marking is much more difficult and expensive, requiring more manpower and considerably more time.

Part of modern forestry management involves protecting trees from insects, diseases, fungi and fire. Nine out of ten forest fires are caused by man and should therefore be preventable. Even though watchtowers are manned continually throughout the summer, extensive damage is caused annually in the huge forest tracts of northern America, Africa and Australia. Although fire can devastate forests completely, together insects and disease destroy more timber than fire. Diseased trees can be detected by infrared photography from a satellite, and spread of disease can be counteracted by spraying the trees with chemicals.

The achievements of modern forestry management have only partially dispelled fears that the world's rapidly increasing population would necessitate large forest depletion for both food production and firewood. Also, there is still debate as to whether the world's forestry resources will be sufficiently large to support all the world's forseeable requirements and still provide for its ecological needs.

Felled trunks (*above*) are trimmed and then crosscut (bucked) into logs, using chain saws. Also available are machines, such as the multiprocessor, that can fell, de-limb and buck the trunks in one operation — and infinitely faster than a team of woodsmen.

Enormous 20-wheeled trucks (*above*), weighing up to 40 tonnes/44 tons, can be loaded with a dozen or more logs that together may add a further 80 tonnes/88 tons or more to the load. The cost of transporting logs, including the building and maintaining of the access roads, is one of the greatest expenses incurred by timber companies.

In many parts of south-eastern Asia and west Africa water remains the cheapest form of transporting the logs (*above*). In order to cross still lakes and wide estuaries, the logs are fastened together and held by large poles (booms) to form a raft. They are then towed by a small tug to their destinations, which may well be the conversion mills.

Hardwoods 1

Acacia melanoxylon

Acer pseudoplatanus

Afzelia spp.

Aucoumea klaineana

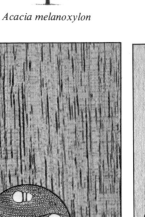

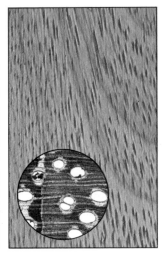

Blackwood
Indigenous to mountain areas
of eastern and south-eastern
Australia

Sycamore
Indigenous to Europe

Doussié
Indigenous to Cameroon, east
Africa, Mozambique, Nigeria

Gaboon
Indigenous to equatorial
Africa

Buxus sempervirens

Cardwellia sublimis

Carya spp.

Castanea sativa

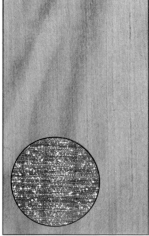

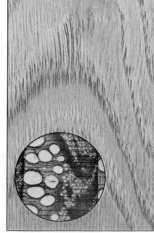

European boxwood
Indigenous to south and north-
western Europe

Silky oak
Indigenous to brush forests of
north-eastern Australia

Hickory
Indigenous to eastern and
southern North America,
southern Canada

European chestnut
Indigenous to Europe, Middle
East, north Africa

Castanospermum
australe

Ceratopetalum
apetalum

Chlorophora excelsa

Dalbergia latifolia

Black bean
Indigenous to north-eastern
Australia

Coachwood
Indigenous to brush areas of
the Australian east coastal
districts

Iroko
Indigenous to west Africa

Indian rosewood
Indigenous to southern India

Diospyros spp.

Dryobalanops spp.

Endiandra palmerstonii

Entandrophragma
cylindricum

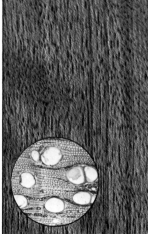

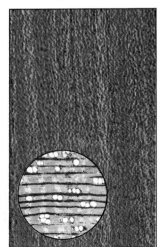

Ebony
Indigenous to the Andaman
Islands, India, Sri Lanka and
Africa

Kapur
Indigenous to Malaysia

Queensland walnut
Indigenous to coastal areas of
north-eastern Australia

Sapele
Indigenous to Cameroon, east
Africa, Gold Coast, west
Africa

231

Hardwoods 2

Entandrophragma utile

Eucalyptus spp.

Fagus sylvatica

Fraxinus excelsior

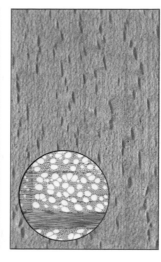

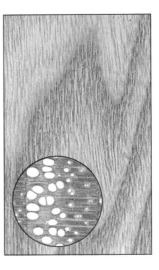

Utile
Indigenous to west and central
Africa

Tasmanian oak
Indigenous to south-eastern
Australia

European beech
Indigenous to Europe

European ash
Indigenous to Europe

Gonystylus bancanus

*Gossweilerodendron
balsamiferum*

Juglans regia

Khaya ivorensis

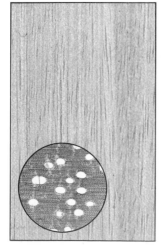

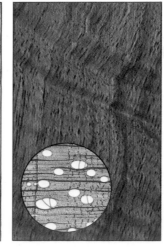

Ramin
Indigenous to freshwater
swamps in Malaya, Sarawak

Agba
Indigenous to Angola, Nigeria

European walnut
Indigenous to Europe

African mahogany
Indigenous to Ghana, Ivory
Coast, Nigeria

Lovoa trichilioides

Microberlinia brazzavillensis

Pericopsis elata

Platanus acerifolia

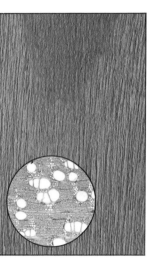

African walnut
Indigenous to the rain forests of Cameroon, Ghana, Ivory Coast, Nigeria

Zebrano
Indigenous to Cameroon

Afrormosia
Indigenous to Ghana, Ivory Coast, Zaire

European plane
Indigenous to Europe

Prunus avium

Pterocarpus macrocarpus

Quercus robur

Quercus rubra

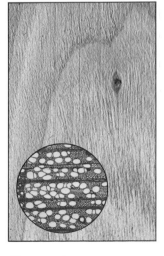

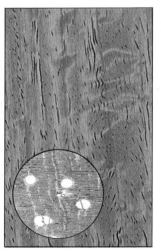

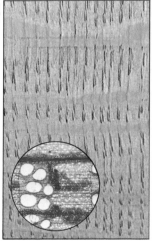

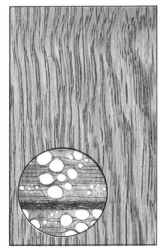

European cherry
Indigenous throughout Europe, western Asia, the USSR and the mountains of north Africa

Burma padauk
Indigenous to northern Burma

European oak
Indigenous to Europe and north Africa

American red oak
Indigenous to North America

Hardwoods 3

Shorea pauciflora

Swietenia macrophylla

Tectona grandis

Terminalia ivorensis

Dark red meranti
Indigenous to Malaysia

Honduras mahogany
Indigenous to Brazil, British Honduras, Costa Rica, Mexico, Nicaragua, Peru

Teak
Indigenous to India, south-eastern Asia

Idigbo
Indigenous to deciduous and rain forests of Cameroon, Ivory Coast, Nigeria

Tieghemella heckelii

Tilia × vulgaris

Triplochiton scleroxylon

Ulmus procera

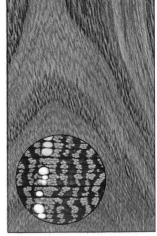

Makore
Indigenous to the rain forests of Ghana, Ivory Coast, Nigeria, Sierra Leone

Lime
Indigenous to Europe

Obeche
Indigenous to Cameroon, Ivory Coast, Nigeria

European elm
Indigenous to Europe

Softwoods

Araucaria angustifolia

Cedrus spp.

Larix decidua

Pinus sylvestris

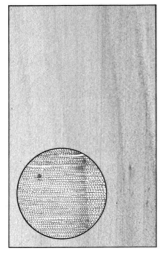

Parana pine
Indigenous to Argentina and Brazil

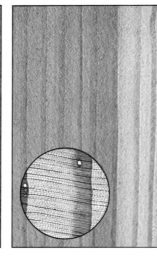

Cedar
Indigenous to the Himalayas, the Middle East and North America

European larch
Indigenous to the mountains of Europe

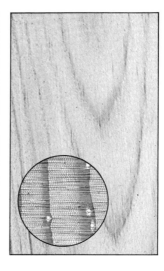

European redwood
Indigenous to Scandinavia, Scotland, the USSR and the mountains of Europe

Pseudotsuga menziesii

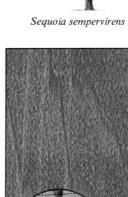

Sequoia sempervirens

Taxus baccata

Thuja plicata

Douglas fir
Indigenous to western North America

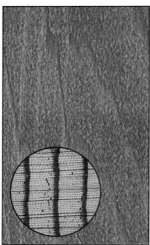

California redwood
Indigenous to western North America

Yew
Indigenous to Asia, Europe and north Africa

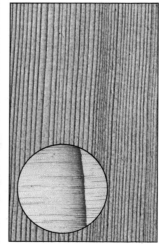

Western red cedar
Indigenous to western North America

Conversion and seasoning

The process of converting trees into marketable-sized boards takes place at a sawmill. The first stage, primary conversion, is when the log is cut, by a band saw or a circular saw. This is done in one of two ways: either plain sawing or quarter sawing.

Plain sawing, also referred to as "through-and-through" or slash sawing, is the simplest and most common way. The logs are gripped and sliced perpendicularly. It is a fast, economical process yielding the maximum number of usable boards. Quarter sawing requires greater skill on the part of the sawyer. The logs are first cut into quarters and then sawed into boards, using radial or approximately radial cuts. Wood cut in this way is more stable in use than plain-sawed boards and is therefore more suitable for structural work. The disadvantage with this method is cost as a higher percentage of wood is wasted when making truly radial cuts.

After primary conversion, the boards are usually fed into the chipper canter — a machine developed to remove the waney edges — although some timber is left with the waney edges intact. The boards are usually dried or seasoned before resawing (secondary conversion) to specific dimensions. Before the boards are fit to use, they must then be dried or seasoned to levels suitable for their required end use (see table opposite). Seasoning also minimizes subsequent shrinking and provides resistance to insect and fungal attack. Strength properties are improved and the wood is easier to work and more able to take a surface finish.

Water in the living tree is contained as free liquid in the cell cavities and bound liquid in the cell walls. Once the timber has been converted into boards, it will start to lose this moisture by evaporation until a level known as fiber saturation point (about 30 percent moisture content for most wood) is reached, in which the cell cavities are comparatively dry but the cell walls are still saturated. Subsequent loss of moisture will be at a slower rate until the boards are in balance with the relative humidity of the surrounding atmosphere. At this stage the wood is said to be at its equilibrium moisture content.

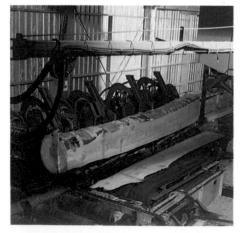

One method of debarking (*above*) is to rotate the logs in a cylindrical drum, bombarding them with jets of water. Another method is to place the logs between spiked rollers that tear off the bark.

After debarking, the logs in this modern sawmill (*right*) pass under an electronic scanner that assesses the taper of the log and computes the best lengths for crosscutting to ensure maximum timber yield.

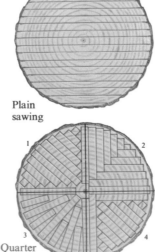

Plain sawing

Quarter sawing

A vertical band saw (*above*) cuts the log to yield plain-sawed boards, which are then removed by a conveyor belt to be stacked. Quarter-sawed timber (*right*) is first cut into quarters and then converted by one of the four methods shown.

Converted boards piled in layers for air drying are said to be "in stick." Quarter-sawed boards may be sticked in neat stacks (*right*) and plain-sawed boards in "boules" — the order in which they were cut from the log (*above*).

If the humidity of the surrounding atmosphere changes so will the wood's equilibrium moisture content. The moisture content of boards stored in the open but protected from direct sun and rain will vary from approximately 17 percent in dry weather to 23 percent in wet weather, in temperate countries.

The two most usual methods of seasoning wood are air drying, which uses the ambient heat from the sun, and kiln drying, which uses a combination of hot air and steam. For moisture levels below 17 percent, kiln drying is necessary.

For natural seasoning (air drying), the unseasoned boards must be stacked exactly over each other on a firm, level base of concrete or creosoted timber with "stickers" of the same or neutral wood, such as poplar or fir, between each board. Spacing is crucial: the boards should be wide enough apart to allow adequate circulation of air but not so far apart that the boards will sag and warp; nor should they be so close that mold and fungal attack are encouraged. The top of the stack should be weighted with waste wood and the whole stack covered with a lean-to roof to protect the boards from rain and direct heat from the sun. To dry boards to their normal equilibrium moisture content of 17 to 23 percent takes approximately one year for every 25 mm/1 in of wood thickness for hardwoods — less for softwoods.

Kiln drying has to be carefully monitored. To prevent too rapid drying, and therefore degradation in the timber, steam is injected into the kiln to temper the rate of moisture evaporation from the boards. The rate of kiln drying is controlled by many factors, such as the quality of the timber and the method of conversion. This type of drying is very much dependent on the skill of the operator. The advantage of kilning is that the drying time is reduced from years to weeks.

After seasoning, the converted boards are usually planed. Other finishing processes available for specific purposes are mortising, spindle molding, tenoning, drilling, routing and tongue and grooving. These latter processes, however, are normally done by the retailer or the woodworker rather than the timber merchant.

Grain direction
There are various ways of determining the grain direction on a board: by feeling or planing the surface for rough and smooth direction; and by examining the edges of the board to see in which direction the grain is slanting.

Across the grain

Against the grain

With the grain

Along the grain

Moisture content of timber for various purposes		
Air drying sufficient %	27%	Appreciable shrinkage starts at about this point
	25%	Suitable moisture content for creosoting
	20%	Decay safety line (wet rots)
	18%	Exterior joinery
Kiln drying necessary %	16%	Garden furniture
	14%	Woodwork for use in places only slightly or occasionally heated
	12%	Interior woodwork with regular intermittent heating
	11%	Woodwork in continuously heated buildings
	9%	Woodwork used in close proximity to sources of heat, e.g., radiator covers, mantelpieces

It is advisable to know the average moisture content required for wood in various environments (*left*), as it is important not to vary these by more than 2 percent if degradation is to be avoided. The moisture content of a piece of wood is first found by weighing an unseasoned sliver, drying it in an oven and reweighing it. It is then calculated as: wet weight less dry weight over dry weight, multiplied by 100. Wood is not a homogenous material and the rate of shrinkage is unequal along the three main axes (*below*). Circumferential shrinkage along the length of the growth rings is roughly twice as much as that between the rings. Thus the shrinkage of plain-sawed boards across the width is almost exactly twice that of quarter-sawed boards.

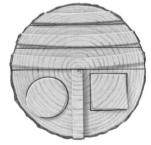

Veneers and manufactured boards

Veneering is the term given to the method of laying thin slices of wood (veneers) on a groundwork of plain but structurally sound wood. There are two main uses for veneer: the first is as a decorative surface in traditional and fine cabinetmaking; the second is as a facing or base in the construction of manufactured boards, such as multi-ply, blockboard, battenboard and laminboard. Manufactured boards can be bought ready surfaced with veneer or unsurfaced, in which case a decorative veneer can be applied separately.

Selecting trees that will yield well-figured wood is extremely difficult. Apart from external signs such as burls and stumps, there is little to indicate in the standing tree the extent of the figure contained within, until it is actually sawed. There are various methods of sawing that are suitable for revealing the figures in a particular tree. The main methods are flat, quarter and half-round slicing. Rotary-cut veneers are used mainly when making three-ply and multi-ply.

Manufactured boards are usually made of beech or birch. They are becoming increasingly popular as a substitute for solid wood. They can be categorized into plywoods (for example, multi-ply and blockboard), particle boards (for example, chipboard) and fiberboards.

One of the most utilized and particularly versatile plywoods is obtainable in three-ply or multi-ply up to 19 layers; each layer is glued with the grain at right angles to the previous one. Alternating the plies in this way improves strength and stability by counteracting any tendency to distort through shrinkage or swelling. It also eliminates the possibility of splitting along the length or the width, as the board has no natural line of cleavage. This is a valuable quality when nailing, screwing or fastening on the outer veneer or face, but take care not to work near the edge of the wood or the surface will delaminate. It can also be cut and bent more easily than solid wood.

Three other types of plywoods are blockboard, battenboard and laminboard. They differ from three- and multi-ply in having a solid core made of strips of wood combined together to form a slab, which is then sandwiched between outer veneers. Blockboard has strips of wood up to 25 mm/1 in wide, but when faced with veneers there may be a tendency for the line of the strips to show through the facing; for this reason it is not normally used for high-quality work. A cheaper

Manufactured boards

modification of this type of board is battenboard, in which the core strips vary from 50 mm/2 in to 75 mm/3 in in width. Laminboard is of similar construction, but the strips making up the core are much narrower, between 1.5 mm/$\frac{1}{16}$ in and 6 mm/$\frac{1}{4}$ in thick. As it is usually made of hardwood and is generally used for higher quality work, it tends to be more expensive than any other type of manufactured board.

It is important to assess the end use when selecting plywoods. Different types of adhesives are used to bond plywoods, depending on whether the materials are to be used inside or outdoors. Standard plywoods are not bonded with a waterproof adhesive, but other grades of plywoods are available that will not only resist water but extremes of weather too.

Particle board is not as strong as plywoods, but it is still of sufficient quality for a variety of nonstructural uses such as in panels, ceilings and cladding for partitions. It can be made of flax shives (flaxboard) or sugarcane (bagasse board), but the best quality board is manufactured from compressed wood particles (chipboard). All these boards are made from waste wood obtained from forest thinnings and shavings from other wood manufacturing processes, bonded together with a synthetic resin glue under heat and pressure. Various methods of processing provide three basic types of chipboard, including a special type of furniture board that has a gradation from coarse to fine chips towards the surface, giving a smooth close-textured finish. This can then be faced with quality veneers. The other types of particle boards have either a single layer of chips all of a similar size, giving the board uniform density and strength, or a core of coarser chips between two outer layers of finer chips (sandwich board). As with plywoods, movement is quite small compared to solid wood. Particle board, however, is very absorbent so all its surfaces must be sealed if it is to be used in damp areas. It is available faced with a variety of veneers or plastic coatings.

A useful manufactured board for low-grade nonstructural work is fiberboard. There are two basic types: insulation board and hardboard. Insulation board is usually made from bagasse, flax shives or straw and has little application in furniture making. Hardboard is manufactured almost exclusively from wet wood fibers bonded together under extreme pressure to produce a dense finished board.

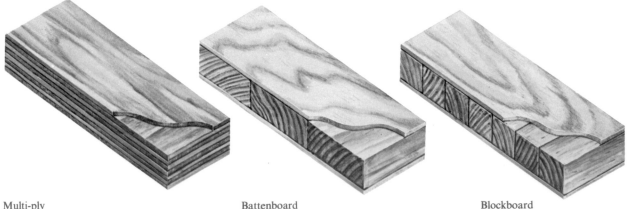

Multi-ply   Battenboard   Blockboard

Methods of obtaining veneers

Butt veneer
A highly decorative veneer found at the swollen base of trees such as walnut (*above*).

This veneer is obtained by halving the log longitudinally and converting it by using the half-round slicing method.

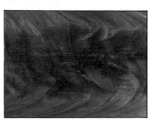

Curl veneer
This attractive veneer is found at the crown and the base of the branches. Walnut and

mahogany (*above*) are highly valued for this type of figure. Other variations of curl are feather and swirl.

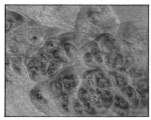

Burl veneer
Irregular outgrowths called burls are found on many trees and often yield a highly

ornamental figure. They are caused by injury to the tree and are commonly cut from oak, ash and walnut (*above*).

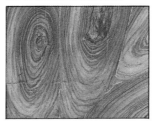

Oyster veneer
The striking pattern of this veneer is achieved by cutting transversely at an angle of 45°

from the branches of trees such as kingwood (*above*) and laburnum. It is used mainly on reproduction furniture.

Flat-sliced veneer
This is the most common method of cutting, in which the half log is sliced longitudinally,

producing landscape or cathedral figures. The rio rosewood sample (*above*) shows marbled heartwood.

Quarter-cut veneer
A quartered log is sliced at right angles to the tree's growth rings. This is the best

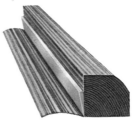

method to obtain the striped effects and ray figures that are a feature of some woods. The sample (*above*) is sapele.

Half-round sliced veneer
With this method, which is also known as semi-rotary, the log is mounted eccentrically on

a rotary lathe. This produces a pattern with a broad-grain figure. The sample illustrated (*above*) is yew.

Rotary-cut veneer
When the log is mounted centrally on a lathe and then rotated against the blade, a

continuous sheet of veneer is produced. This method is often used to obtain the "bird's-eye" effect in maple (*above*).

Laminboard   Chipboard   Flaxboard

Selecting and buying timber

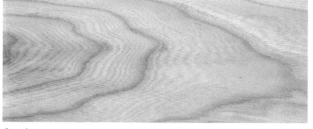

Landscape

Ray, or flame

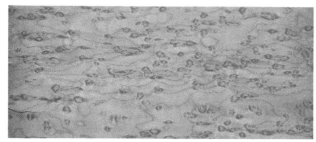

"Bird's-eye" maple

Blister

Quilt

Stripe

Ribbon

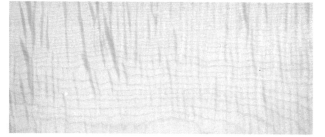

Fiddleback

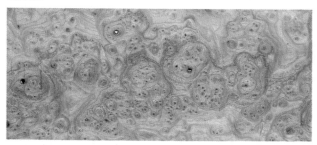

Burl

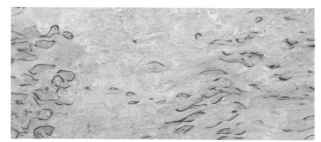

Masur birch

Mineral streaks

Gum veins

Landscape

Growth rings are one of two structural features that can produce an attractive figure, depending on the method of conversion. The elongated growth ring pattern known as landscape or cathedral figure is seen only on plain-sawed surfaces. Quarter-sawed surfaces will show the rings as parallel lines in the wood.

"Bird's-eye" maple

This figure is seen on the tangential surfaces of only a few species, the best known being Canadian rock maple. The pattern is caused by fungi, which attack the cambium. This retards the complete formation of the annual rings, leaving conical depressions that, on conversion, appear as circles resembling pin knots.

Quilt

This form of blister figure is caused by irregularities in the tree's growth rings. The pattern is most effectively seen on plain-sawed surfaces, where the incident light is reflected from the irregular grain at different angles, giving a three-dimensional appearance to wood such as African mahogany (*illustrated*).

Ribbon

A growth characteristic of many tropical hardwoods is interlocked grain. This produces a striped figure on quarter-sawed surfaces. If the wood is moved in relation to the light source the color of the stripes will appear to alternate. This optical effect is not due to color variation in the wood.

Burl

Burls are the result of injury to trees such as common elm (*illustrated*). The highly irregular grain, which contains many small knots, also contributes to the pattern of the figure. Burl is extremely difficult to season and convert and for this reason the wood is best utilized as a veneer rather than as solid timber.

Mineral streaks

These are localized discolorations of timber, in the form of streaks or patches usually darker than the natural color, which do not impair the strength of the piece. The color of the streaks can vary from the dark red common in parana pine (*illustrated*) to yellow or white, as seen in red meranti and keruing.

Ray

The second structural feature that plays an important role in the production of figure is rays. As they occur radially throughout the tree, they are best seen (if they are visible at all) on radial or quarter-sawed surfaces. Decorative figures of this type in woods such as oak and maple (*illustrated*) are used mainly as veneers.

Blister

This type of figure, in common with the "bird's-eye" figure in maple, is caused by injury to the cambium, with the result that large depressions are formed in the growth rings. The figure is seen in softwoods such as pitch pine and hardwoods such as kevazingo (*illustrated*) and American whitewood.

Stripe

This figure is most commonly seen in ebonies such as macassar (*illustrated*). The stripes are due entirely to color variation within the wood, caused by alternation of the pale sapwood with the darker colored heartwood. Where the color is less evenly distributed the pattern is known as marblewood.

Fiddleback

This figure is seen on the quarter-sawed surfaces of woods such as sycamore (*illustrated*). This particular wood is often used in the manufacture of violins, hence the name of the figure. As with the ribbon figure the apparent color changes occur because of the way in which the light is reflected from the grain.

Masur birch

This distinctive figure is found in the butt logs of birch from Finland and Norway. It is caused by the larvae of a gnat-like fly of the genus *Agromyza*, which tunnels into the wood. The tree heals the damage by producing resin to fill the bore channels. Masur birch is used mainly in the production of treen.

Gum veins

These are caused by traumatic incidents such as wounding or forest fires. They are normally considered to be a defect, but in Italian ancona walnut they are a natural characteristic that actually enhances the appearance of the wood. The gum veins can show as streaks as in paldao (*illustrated*) or as flecks.

When selecting wood from the 12,000 species commercially available, the worker will usually be considering two distinct uses — wood for decoration or wood for structural purposes. The appearance, or figure, of wood is determined by inherent coloring substances and the way in which the different tissues or grain grow. These characteristics are best seen in veneers that are bought purely as decoration, which are laid on a groundwork of structurally sound timber or manufactured boards. The direction of the grain is also a good indication of the structural properties of timber, that is, its bending and compressive strength (*see* pages 244–51). Any marked deviation of the grain from the vertical axis of the tree will affect the wood's strength. This may not matter for some uses.

There are five naturally occurring types of grain: straight, irregular, wavy, spiral and double spiral or interlocked. Straight-grained timbers usually have the greatest overall strength but may be least figured. Such wood, therefore, is best used for structural purposes. Diagonal grain can occur naturally, but it also frequently results from incorrect conversion. Grain that has been deflected from a straight course by, for example, a knot is described as irregular. Such timber still has many uses for the woodworker. Wood with a wavy grain has an attractive appearance that, used in joinery, outweighs any loss of strength. On quarter-sawed surfaces this distinctive figure sometimes shows as fiddleback. Spiral grain is the term applied to grain that follows either a left- or right-handed diagonal course up the tree. Such timber is not suitable for cabinetwork. Double spiral or interlocked grain inclines obliquely first one way, then the other. It is difficult to work and there is some loss in strength.

A combination of fiddleback and ribbon is sometimes found in interlocked grain, producing a mottled figure called roe. Interlocked grain and wavy grain combined can produce many beautiful figures.

Types of grain

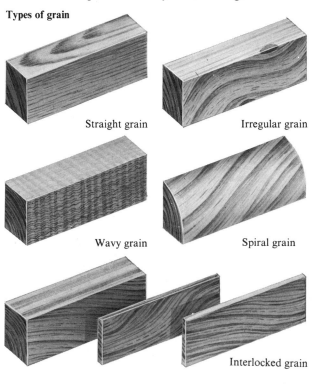

Straight grain

Irregular grain

Wavy grain

Spiral grain

Interlocked grain

Defects

Defects in timber can be divided into two basic categories: man-made defects, caused by incorrect seasoning or conversion, and natural defects. The latter can be divided into biodegradation, which is the collective term for attack by living organisms, such as woodborers, fungi and rots, and inherent defects, such as knots and pitch pockets, which occur as the tree is growing. Only experience will provide the knowledge to assess good-quality wood. A selection of the defects that may be encountered in furniture or when buying timber and storing it is outlined below.

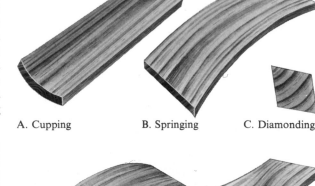

A. Cupping   B. Springing   C. Diamonding

Man-made defects

The most common seasoning defects are caused by the wood warping as a result of too rapid drying. They include cupping (A), where the board hollows across the width; bowing (E), where it curves like a bow along its length; springing (B), where the board lies flat but bends along its edges; "in winding" (F), where the board is twisted along its length; and diamonding (C), where the squared material is distorted. Rapid drying can also cause case-hardening, where the outside of the wood dries more quickly than the inside, setting up stresses within. This may be remedied if the wood is steamed again and redried. If this defect is not remedied quickly the wood will set, forming interior splits known as honeycombing. The condition is unalterable and renders previously sound wood useless. If green wood is dried too quickly the wood cells flatten or collapse, resulting in a concave surface across the width of the board. Badly converted wood can result in diagonal grain, that is, grain that is inclined obliquely to the longitudinal axis of the tree. If the grain is very eccentric, it can impair impact and bending strength. Roller check (D) and machine burn (H), depending on the severity of the fault, can be remedied by planing. Boards that have retained the natural trunk shape of the tree are known as waney edged (G).

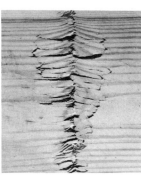

D. Roller check   E. Bowing   F. "In winding"

G. Waney edge

H. Machine burn

Natural defects

Attack by wood-borers, such as the furniture beetle, is a common form of biodegradation. This pest, also known as woodworm, can attack and completely riddle all wood, seasoned or unseasoned. Marine borers (K, R), such as the shipworm and gribble, occur in temperate and tropical countries, devastating marine timbers. Termites (Q) are the most voracious wood predators in tropical and sub-tropical countries. The wood

J. Brown rot

K. Temperate water shipworm

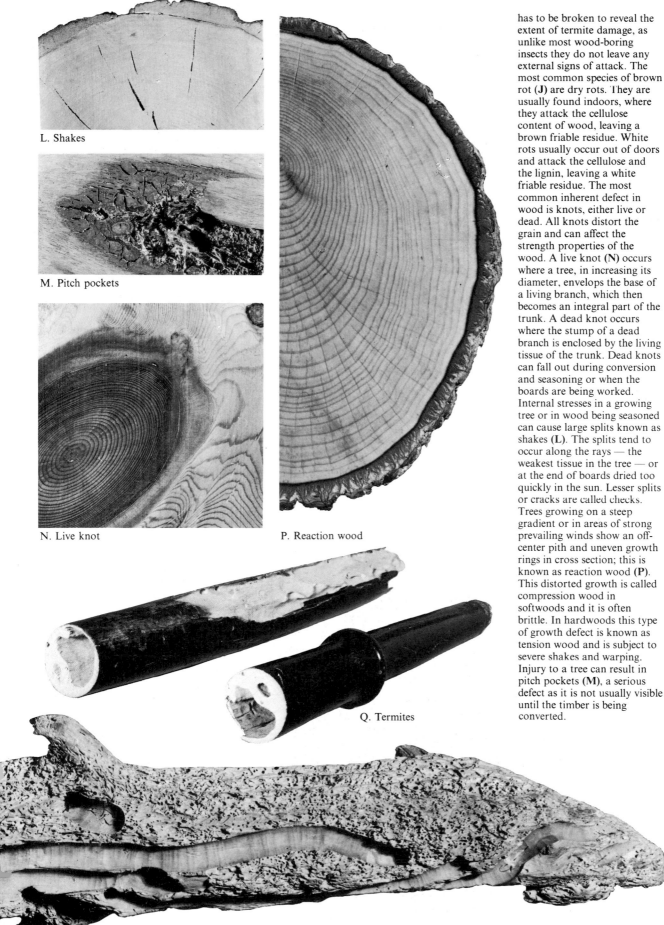

L. Shakes

M. Pitch pockets

N. Live knot

P. Reaction wood

Q. Termites

has to be broken to reveal the extent of termite damage, as unlike most wood-boring insects they do not leave any external signs of attack. The most common species of brown rot (**J**) are dry rots. They are usually found indoors, where they attack the cellulose content of wood, leaving a brown friable residue. White rots usually occur out of doors and attack the cellulose and the lignin, leaving a white friable residue. The most common inherent defect in wood is knots, either live or dead. All knots distort the grain and can affect the strength properties of the wood. A live knot (**N**) occurs where a tree, in increasing its diameter, envelops the base of a living branch, which then becomes an integral part of the trunk. A dead knot occurs where the stump of a dead branch is enclosed by the living tissue of the trunk. Dead knots can fall out during conversion and seasoning or when the boards are being worked. Internal stresses in a growing tree or in wood being seasoned can cause large splits known as shakes (**L**). The splits tend to occur along the rays — the weakest tissue in the tree — or at the end of boards dried too quickly in the sun. Lesser splits or cracks are called checks. Trees growing on a steep gradient or in areas of strong prevailing winds show an off-center pith and uneven growth rings in cross section; this is known as reaction wood (**P**). This distorted growth is called compression wood in softwoods and it is often brittle. In hardwoods this type of growth defect is known as tension wood and is subject to severe shakes and warping. Injury to a tree can result in pitch pockets (**M**), a serious defect as it is not usually visible until the timber is being converted.

R. Tropical water shipworm, also showing smaller holes made by the gribble

Selected world timbers 1

Wood	Natural characteristics	Technical qualities	Natural durability	Reaction to preservative	Seasoning qualities
Acacia melanoxylon **Blackwood** (page 230)	Medium even texture; straight to interlocked grain; fiddleback ray figure; pronounced growth rings; a natural luster	Strong; good impact properties; pliable	Fair	Very resistant	Good; stable when dry
Acer pseudoplatanus **Sycamore** (page 230)	Smooth texture; the grain may be close and wavy but is more usually straight; fiddleback ray figure; shows a satiny sheen	Quite strong; good bending strength	Poor	Resistant	Stable when dried slowly
Afzelia spp. **Doussié** (page 230)	Medium texture; straight grain with some irregularity; yellow deposits may cause surface flecks	Moderately good bending and compressive strength; resistant to splitting; very stiff	Good	Very resistant	Dries slowly; prone to slight distortion and checking
Araucaria angustifolia **Parana pine** (page 235)	Fine even texture; straight grain; fairly knot free	Varies from soft to hard	Poor	Fairly resistant	Good; but dry carefully to avoid distortion and splitting
Aucoumea klaineana **Gaboon** (page 230)	Medium coarse texture; often irregular grain; few figures; nondescript appearance	A weak wood in all properties	Poor	Poor	Fair; slight tendency to distort
Buxus sempervirens **European boxwood** (page 230)	A very fine even texture; straight and irregular grain	A hard timber; excellent compressive strength; fair bending properties	Fair	Resistant	Dry slowly; a tendency to surface checks; splits badly if dried in the round
Cardwellia sublimis **Silky oak** (page 230)	Coarse but even texture; fairly straight grain; good silver grain ray figure on quarter-sawed surfaces; occasional dark lines	Strong and stiff; fair bending strength	Fair	Poor	Fair, if dried slowly; some checking and distortion
Carya spp. **Hickory** (page 230)	Coarse texture; straight grain but can be wavy; well-defined growth rings	Good steam-bending properties; very strong; hardness varies according to growth rate	Poor	Fairly resistant	Dry very slowly due to high rate of shrinkage
Castanea sativa **European chestnut** (page 230)	Coarse texture; some straight but often spiral grain; growth rings prominent; tannin content	Reasonable bending strength and shear; stiff but not as good as oak; splits easily	Good	Very resistant	Dry very slowly; prone to collapse, internal cracking and distortion
Castanospermum australe **Black bean** (page 231)	Medium texture, somewhat greasy; straight to slight interlocked grain	Stiff and strong but inclined to be brittle; hard	Fair to good	Poor	Dry carefully; prone to distortion, collapse, checking and internal cracking
Cedrus spp. **Cedar** (page 235)	Fine even texture; conspicuous growth rings; aromatic when freshly sawed	Medium strength; brittle	Very good; resistant to fungi	Poor	Fair; prone to distortion
Ceratopetalum apetalum **Australian coachwood** (page 231)	Fine texture; fairly straight grain; decorative growth ring figure	Good bending strength	Poor	Poor	Good; stable when dry
Chlorophora excelsa **Iroko** (page 231)	Fairly coarse texture; interlocked grain gives typical ribbon figure on quarter-sawed surfaces; white flecks caused by calcareous deposits often present; good acid and fire resistance; sometimes likened to teak	Good bending strength; hard; resistant to shock loading	Excellent if the grain is straight	Fairly resistant	Quite good; minimal checking and distortion

Air dried weight	Working qualities	Fixing qualities	Finishing qualities	Relevant uses	Woods with similar qualities	General comments
656 kg/m³	Reasonable, but some hard patches; tends to crumble on the end grain	Good	Excellent	Cabinetmaking; furniture; fine interior joinery; woodwind instruments; gun-stocks	Black bean	An excellent multi-purpose wood
640 kg/m³	Fair	Care needed with nails and screws; good with glue	Excellent	Fine interior joinery and veneers; stringed instrument backing; fine furniture; turning; an excellent flooring wood	Rock maple (weight 800 kg/m³) has a distinctive "bird's-eye" figure	Susceptible to stains, when being kilned; stack vertically for air drying; never put in stick formation
816 kg/m³	Fairly hard to work; keep cutting angle low	Poor with nails and glue; fair with screws, pre-boring required	Fair, but may need filling	Fine interior joinery; staircases; flooring; sills; doors, etc.	None	Not suitable for domestic use as it stains yellow in damp conditions
544 kg/m³	Good	Good	Good	Fine interior joinery; framework; drawer sides; moldings	Western yellow pine	Prone to movement, so ensure stable humidity conditions; it is also used as a general purpose plywood
432 kg/m³	Abrasive to saws and tools due to silica content	Poor with nails and screws; good with glue	Fair	Door skins; drawer sides; partition work	Red meranti	Used to face three-ply, multi-ply, blockboard and laminboard
912 kg/m³	Fairly resistant to saws but cuts cleanly; fairly hard to work; tendency to tear on irregular grain so keep cutting angle low; tendency to machine burn	May split with nails; pre-boring required with screws; fair with glue	Excellent	Tool handles; mallet heads; croquet heads; skittles; rulers; chessmen; checkers; fancy carvings	Species of boxwood from central America, South and North Africa, western Asia, Iran	A good wood for carving and turning
528 kg/m³	Good; some grain pickup on quarter-sawed surfaces caused by wide rays; lower cutting angle by 20° for best results	Good	Fairly good	Furniture; interior joinery; paneling; flooring	American red oak	A good, lightweight, decorative wood, similar to oak in appearance
816 kg/m³	Fair; some grain pickup; inclined to blunt tools and saws	Difficult with nails and screws; good with glue	Fair	High impact strength makes it useful for pick and hammer handles; sports equipment	European ash	Now often replaced by metals and synthetic materials
544 kg/m³	Good; tendency to bind on saws	Good	Excellent but may need filling	Furniture; turning; kitchenware; cleft fencing	Oak	Because of its resemblance to oak is often used as an oak substitute; stains in contact with ferrous metals
704 kg/m³	Fair; prone to hard and soft patches	Fair with nails and screws; good with glue	Good	Good all-round cabinetmaking; fine interior joinery; doors; panels	Blackwood	A decorative wood
560 kg/m³	Good	Care needed with nails and screws; good with glue	Good	Fine interior joinery; nonstructural parts of furniture		Has a decorative appearance; Cedar of Lebanon is the best-known commercial species
640 kg/m³	Good	Good	Excellent	Interior joinery; cabinetmaking; furniture	Ash; hickory	One of Australia's first coach-building woods
640 kg/m³	Good, but calcareous deposits may blunt saw and tool edges	Good	Good, but needs filling	Fine interior joinery; exterior joinery; boat building; counter and bench tops; garden seats; parquet flooring even where underfloor heating systems are installed	Afrormosia; teak	Not so attractive a wood for furniture as teak but has a good combination of properties and is comparatively inexpensive

Selected world timbers 2

Wood	Natural characteristics	Technical qualities	Natural durability	Reaction to preservative	Seasoning qualities
Dalbergia latifolia **Indian rosewood** (page 231)	Coarse to very fine even texture; generally straight grain but some slight interlock; irregular black markings give an unusual appearance	Very strong and hard; good bending, compression and impact strength; stiff	Excellent	Poor	A tendency to surface checking; dry slowly
Diospyros spp. **Ebony** (page 231)	Very fine texture; straight grain; very hard; subspecies, e.g., macassar, have a pronounced figure	African spp. have good strength; the Asian spp. are variable	Good	Poor	African spp. season well if dried carefully, but Asian spp. have a tendency to fine surface checking
Dryobalanops spp. **Kapur** (page 231)	Coarse even texture; reasonably straight grain; whitish resin ducts show as lines mainly on plain-sawed surfaces; a distinct camphor smell when freshly sawed	Reasonably hard; good bending properties; stiff	Excellent	Very resistant	Dries well; has a slight tendency to check
Endiandra palmerstonii **Queensland walnut** (page 231)	Medium even texture; some interlocked grain producing checkered markings on quarter-sawed wood; lustrous surface; distasteful odor when freshly sawed	Stronger than European walnut; hard; stiff; good bending strength	Poor	Poor	Care needed; a tendency to surface and internal checking and distortion; surface collapse possible
Entandrophragma cylindricum **Sapele** (page 231)	Fine texture; interlocked grain; striped roe figure on quarter-sawed surfaces; growth rings pronounced on plain-sawed surfaces; cedarlike scent when freshly sawed	Good strength properties; equal to oak in bending, stiffness and shock resistance; hard; good splitting resistance	Fair	Poor	Dry carefully; seasoning needed to avoid distortion
Entandrophragma utile **Utile** (page 231)	Rather coarse texture; interlocked grain shows a ribbon figure on quarter-sawed surfaces	Reasonably hard; fair shock resistance; poor bending strength	Fair	Very resistant	A tendency to check and distort; steady drying recommended
Eucalyptus spp. **Tasmanian oak** (page 232)	Coarse texture; straight grain; similar in appearance to ash or oak, but minus the characteristic silver grain of oak	About the same hardness as European oak; good bending properties; good stiffness; tough; excellent resistance to shock load	Fair	Resistant	Careful drying needed; a distinct tendency to surface and internal cracking, collapse and distortion
Fagus sylvatica **European beech** (page 232)	Fine even texture; straight grained; a "seed pip" ray figure shows on tangential surfaces; a small flecked figure on quarter-sawed surfaces	Excellent bending and impact strength; stiff; shear; resistant to splitting	Poor	Good	Fairly good; dry carefully; inclined to check and distort; high shrinkage factor
Fraxinus excelsior **European ash** (page 232)	Coarse to medium fine texture; straight grain; conspicuous growth rings	Good strength; tough; excellent bending qualities; resistant to splitting	Poor	Fairly resistant	Fairly good but inclined to distortion and end-grain splitting at high temperatures
Gonystylus bancanus **Ramin** (page 232)	Moderate to fine texture; straight to slightly interlocked grain; can cause skin irritation; prone to staining	Fairly good; nearly comparable to home-grown beech; good compressive strength but prone to splitting; not a good bending timber and rather weak in shear	Poor	Good	Prone to end-grain splitting but shows little distortion; care needed to prevent staining
Gossweilerodendron balsamiferum **Agba** (page 232)	Fine texture; generally straight grain; resinous; similar appearance to gaboon or light-colored mahogany	Quite good, comparable to American mahogany; subject to brittle heart; moderately good bending strength	Quite good	Resistant	Quite good, with little tendency to split or distort, but gum exudation can be a problem

Air dried weight	Working qualities	Fixing qualities	Finishing qualities	Relevant uses	Woods with similar qualities	General comments
848 kg/m³	Quite hard to work; calcareous deposits dull blades; turns well	Poor with nails; good with screws; fair with glue	Excellent	Furniture, solid and veneer; interior joinery; shallow carving; knife and brush handles, etc; musical instruments, e.g., guitars	Cocobolo kingwood; South American rosewood	Becoming more popular; a decorative wood
African spp. 1,008 kg/m³ Asian spp. 1,168 kg/m³	Difficult; inclined to be brittle; blunting effect on tools; use a low-angled plane blade	Very difficult with nails and screws; fair with glue	Excellent	Small work, e.g., door handles, knobs, cutlery handles, tool handles, chessmen, organ stops, decorative inlays, finger boards of stringed instruments; a good turning wood; domestic flooring	African blackwood	A good multi-purpose wood for small intricate work
704/768 kg/m³	Fairly easy to work; saws may be blunted; some grain pickup when planing	Fair	Fair	Flooring; interior joinery	Keruing	Stains when wet if in contact with ferrous metals
672 kg/m³	A severe dulling effect on saws and tool cutting edges; keep well sharpened; use a low-angled plane blade	Poor with nails; good with screws and glue	Excellent	Fine interior joinery; cabinetmaking; furniture; decorative paneling	European walnut; paldao	Not botanically related to walnut, but similar in appearance, texture and weight
624 kg/m³	Good; some grain pickup when planing quarter-sawed surfaces; keep cutting angle low	Good	Excellent	Furniture, solid and veneer; fine interior joinery; flooring	African mahogany; Honduras mahogany; utile	A well-utilized, decorative and sound timber
656 kg/m³	Works well; some tendency to grain pickup during planing due to interlocked grain	Good	Good, but needs filling	Furniture, solid and veneer; cabinetmaking; interior joinery	African mahogany; sapele	An attractive mahogany-type wood; for constructional and decorative work
Variable around 656 kg/m³	Moderate, similar to ash	Slight tendency to split with nails; good with screws and glue	Good	Furniture, solid and veneer; interior joinery; flooring; spade handles; ladder rungs	Ash; European oak	A good general-purpose timber
720 kg/m³	Variable; some wood tends to bind and burn when sawing or drilling; usually reasonable to work	Fair with nails and screws; good with glue	Good	Furniture, especially chairs; fine interior joinery; toys; small models; small turning	Chilean beech; silver beech; Southland beech from Australasia	A good multi-purpose timber in plentiful supply
688 kg/m³	Fair; tendency to wastage when planing flat-sawed wood due to distortion	Fair with nails and screws; good with glue	Excellent; may need filling	Furniture; fine interior joinery; tennis, squash, badminton frames; gymnasium equipment; drum frames; turning	Hickory; Japanese ash, North American ash	An excellent and attractive general-purpose wood
656 kg/m³	Tends to tear on quarter-sawed surfaces; keep cutting angle low; wear goggles and gloves when sawing unseasoned wood	Tends to split with nails; fair with screws and glue	Good	Interior joinery; turning; moldings; handles; toys	None	A generally good utility wood; can be quite attractive
512 kg/m³	Good; gum can sometimes bind saw and tool blades; interlocked grain can tear, so keep cutting angle low	Fair with nails and glue; good with screws	Fair; gum can affect finish	Interior joinery; furniture linings and backings	African mahogany (except for color)	A good multi-purpose timber; do not use in damp areas, e.g., kitchens, bathrooms; can produce objectionable odors

Selected world timbers 3

Wood	Natural characteristics	Technical qualities	Natural durability	Reaction to preservative	Seasoning qualities
Juglans regia **European walnut** (page 232)	Usually fine texture; a general naturally wavy grain; clear of knots except on burl veneers, which have pin knots	Hard, tough; good splitting and impact resistance; good bending properties and shock loading strength	Fair	Resistant	Dries slowly but well; tendency to checking and internal cracks with thick wood
Khaya ivorensis **African mahogany** (page 232)	Medium texture; shows a ribbon figure on quarter-sawed surfaces; subject to black gum line figure	Medium strength, but some heartwood is prone to brittleness; boards subject to hairline cracks due to compression failure	Fair; subject to pinhole and bark borer attack	Resistant	Good; stable when dry
Larix decidua **European larch** (page 235)	Coarse to medium texture; resinous; high contrast between early- and latewood; small knots; faintly aromatic	Strong; hard; not pliable	Fair	Fairly resistant	Good; stable when dry
Lovoa trichilioides **African walnut** (page 233)	Fairly coarse texture; interlocked grain produces fair ribbon figure on quarter-sawed surfaces; fairly clear of knots	Fairly hard and stiff; fair compressive strength; moderate bending and shock loading qualities; fairly resistant to splitting	Good	Very resistant	Good; dries fairly quickly without much checking or distortion; reasonably stable
Microberlinia brazzavillensis **Zebrano** (page 233)	Normally coarse texture; interlocked grain; striped figure on quarter-sawed surfaces; color variations give an attractive figure	Strong; shock resistant	Good	Resistant	Prone to distortion, so slow drying necessary
Pericopsis elata **Afrormosia** (page 233)	Fine texture; interlocked grain; good acid and alkali resistance	Strong and hard	Excellent	Very resistant	Dries well, if slowly, with little degradation
Pinus sylvestris **European redwood** (page 235)	Coarse texture; resinous; clear growth rings; knotty	Resilient; flexible; good bending and compressive strength	Poor; prone to insect and fungi attack	Good	Good; dries quickly
Platanus acerifolia **European plane** (page 233)	Smooth even texture; usually straight grain; on quarter-sawed surfaces the dark brown ray figure shows as flecks — it is then known as lacewood	Reasonably good properties in bending; hard; resistant to splitting, although the knots are inclined to split	Very poor	Fairly resistant	Seasons well, but sometimes there is slight distortion
Prunus avium **European cherry** (page 233)	Fine even texture; straight grain; gum marks sometimes visible on surface	A tough wood; good bending and impact strength; stiff; resistant to splitting; knots are usually small	Fair, but do not use externally unless treated	Poor	Tendency to distort, so care needed; reasonably stable when dry
Pseudotsuga menziesii **Douglas fir** (page 235)	Coarse texture; resinous; conspicuous growth rings	Medium strength	Poor; fairly resistant to decay	Poor	Good; stable when dry
Pterocarpus macrocarpus **Burma padauk** (page 233)	Fine smooth texture; slightly interlocked grain; narrow ribbon figure on quarter-sawed surfaces; few knots	Very strong and hard	Good	Poor	Seasons slowly but well, with little surface checking or distortion; very stable when dry
Quercus robur **European oak** (page 233)	Coarse texture; straight grain; well-defined growth rings; relatively knot-free timber; quality varies according to locality; wide rays very noticeably on quarter-sawn surfaces giving a silver grain figure	Very strong, used as a yardstick for other hardwoods for this reason; bends well, but care must be taken to avoid rupture on inner face of curved wood	Fair; subject to lyctus attack	Very resistant. Do not treat with a salt chemical, which will turn the wood bluey-black	Must be dried slowly; a tendency to distort; some collapse and inner checks

Air dried weight	Working qualities	Fixing qualities	Finishing qualities	Relevant uses	Woods with similar qualities	General comments
640 kg/m³	Good	Good	Excellent	Furniture, solid and veneer; fine interior joinery; good for carving and turning	African walnut; American black walnut; Australian walnut; Indian laurel	A highly desirable wood, often well figured, especially ancona Italian walnut; decorative work can be done using burl veneers
512 kg/m³	Good; some grain pickup due to interlocked grain	Good	Mainly good	Furniture, solid and veneer, especially reproduction work; fine interior joinery; paneling; turning	Other mahoganies	A common African redwood; not a true mahogany but an excellent substitute; used in the manufacture of plywoods
592 kg/m³	Good	Care needed with nails; good with screws and glue	Good	Fencing; boat building	All species of larch	Of limited use, but excellent
544 kg/m³	Fair; some grain pickup when planing quarter-sawed surfaces; keep cutting angle low	Good	Fair	Furniture, solid and veneer; interior joinery; cabinet-making	African mahogany (except for color)	A good all-round decorative wood; in fairly good supply
Variable around 720 kg/m³	Good	Fair with nails and screws; good with glue	Good	Furniture, mainly in veneer form; small items, e.g., tool handles	Berlinia	Used mainly as a veneering wood
688 kg/m³	Fairly good, but inclined to blunt cutting edges; keep cutting angle low; some grain pickup on quarter-sawed surfaces	Poor with nails; good with screws if pre-bored; good with glue	Excellent	Furniture, solid and veneer; fine interior joinery; boat superstructure	Iroko; teak	A first-class wood
528 kg/m³	Good	Good	Good	Interior joinery; furniture carcasses; flooring	Western yellow pine; pitch pine	Resinous knots can be troublesome; either pretreat or drill out and plug. Also known as Scotch pine
624 kg/m³	Saws well; tends to bind; keep cutting angle low on quarter-sawed surfaces	Good	Good	Decorative work, e.g., inlay; marquetry	White beech; planes indigenous to Iran, North America and Turkey	The Scots call their sycamores "plane"; the Americans call their plane trees "sycamore"; in America their species are more prolific and used for furniture and cabinetmaking
608 kg/m³	Very good	Good	Excellent	Furniture, solid and veneer; cabinetmaking; interior joinery	American black cherry; other fruitwoods, e.g., apple, pear, plum	The European wood is used mainly for veneering; American cherry is less heavy
544 kg/m³	Good; with care	Good	Poor	Interior joinery; laminated boards	None	Prone to windblow defects
848 kg/m³	Fair	Not easy with nails; fair with screws if pre-bored; good with glue	Excellent	Furniture; interior joinery; flooring	African padauk; Andaman padauk	Burma padauk is one of the heaviest, strongest and hardest of the South East Asian woods
704 kg/m³	Good; some grain pickup on lowland oak timber	Good	Good, but needs filling	Furniture, solid and veneer; fine interior joinery; flooring, both strip and block	Other oaks, especially American white, Japanese, Turkish	German slavonia considered to be the finest oaks; stains occur when in contact with ferrous metals in damp conditions; one of the most utilized hardwoods

Selected world timbers 4

Wood	Natural characteristics	Technical qualities	Natural durability	Reaction to preservatives	Seasoning qualities
Quercus rubra **American red oak** (page 233)	Coarse but even texture; fairly straight grain; not usually a knotty timber; fairly fire and acid resistant; pronounced growth rings	Variable according to growth conditions; good bending qualities; quite strong	Poor	Poor	Reasonable; dries slowly; prone to checking and distortion
Sequoia sempervirens **California redwood** (page 235)	Texture varies from fine to coarse; straight grain; knot free; prominent growth ring figure; resistant to acid and alkali	Medium strength	Good	Fairly resistant	Good; stable when dry
Shorea pauciflora **Dark red meranti** (page 234)	Fairly coarse texture; grain tends to be slightly interlocked, giving ribbon figure on quarter-sawed surfaces; often has conspicuous resin ducts	Good strength; used as a substitute mahogany; poor bending quality	Good	Poor	Fairly good, but a tendency to slight checking and distortion
Swietenia macrophylla **Honduras mahogany** (page 234)	Reasonably fine, even texture; straight grain, but some interlocked grain shows a ribbon figure on quarter-sawed surfaces; growth rings clearly defined; natural surface luster	Quite strong; good for bending and compression	Fair; subject to pinhole wood borer attack	Fairly resistant	Good; dries well with little tendency to checking and distortion
Taxus baccata **Yew** (page 235)	Smooth fine texture; irregular grain	High strength; very hard	Excellent	Resistant	Good
Tectona grandis **Teak** (page 234)	Coarse greasy texture; prominent growth ring figure especially on plain-sawed surfaces; often smells like old leather; few knots; acid, alkali and fire resistant	Good impact and excellent bending strength; stiff	Excellent; resistant to insect and fungi attack	Very poor	Seasons well but slowly; otherwise, prone to checking and distortion
Terminalia ivorensis **Idigbo** (page 234)	Medium to coarse texture; some interlocked grain, but fairly straight with a few irregularities; distinct growth rings	Moderate strength properties; similar to African mahogany, but a tendency to brittle heart; subject to small knots; contains a yellow pigment that stains when wet	Fair, but prone to insect attack; fairly resistant to fungi and termites	Very poor	Seasons rapidly and well, with little tendency to check or distort; stable when dry
Thuja plicata **Western red cedar** (page 235)	Soft texture; straight grain; prominent growth rings; subject to bruising; aromatic	Low strength	Excellent	Poor	Good; stable when dry, but thicker stock may collapse
Tieghemella heckelii **Makore** (page 234)	Fine texture and generally straight grain; some interlock grain; the roe figure gives a checkered appearance on quarter-sawed surfaces when interlocked grain is present; a lustrous sheen on plain-sawed surfaces	Medium strong; reasonable bending qualities	Excellent	Very poor	Stable when dry; some tendency to split around knots
Tilia × vulgaris **Lime** (page 234)	Fine even texture; good straight grain; rather soft	Good bending and compressive strength, stiffness	Poor	Good	Good; fairly stable when dry, but some tendency to distort
Triplochiton scleroxylon **Obeche** (page 234)	Moderately coarse but even texture; interlocked grain shows ribbon figure on quarter-sawed surfaces; lustrous surface; bad smell when freshly sawed; prone to blue stain	Good bending strength; resistant to shock loads	Poor	Poor	Seasons well, but may split around knots; prone to slight distortion
Ulmus procera **European elm** (page 234)	Coarse texture; often coarse grain	Good bending strength; resistant to splitting	Poor; subject to heart rot and insect attack	Poor	Variable; subject to checking and distortion

Air dried weight	Working qualities	Fixing qualities	Finishing qualities	Relevant uses	Woods with similar qualities	General comments
768 kg/m³	Fair	Good	Good, but needs filling	Furniture, solid and veneer; interior joinery; flooring	American white; European; Japanese and Turkish oak	The other species listed are slightly better quality than the American red oak
416 kg/m³	Good; with care	Good	Good	Exterior joinery; window and door stock	None	A lightweight timber; not suitable for heavy structural work
688 kg/m³	Good; with care	Good with nails and screws; fair with glue	Fair, but may need filling	Joinery (interior and exterior)	Light red meranti; white meranti; yellow meranti	A good multi-purpose wood that can be used as a substitute mahogany. Generally, dark red meranti is a better quality wood and is easier to work than the other merantis.
544 kg/m³	Good; some grain pickup when planing	Tends to split with nails; good with screws and glue	Excellent	Fine interior joinery; paneling; furniture, solid and veneer; cabinetmaking; turning	Honduras and Cuban mahogany are the best-quality woods but other redwoods are similar, e.g., African mahogany	This wood replaced Cuban mahogany on the world market; Cuban mahogany is much darker and denser, but has been in short supply for many years
672 kg/m³	Good, but high wastage	Care needed with nails and screws; good with glue	Good	Veneers; mallet heads; long bows; objets d'art; turning	None	Has a relatively low yield; the hardest known softwood
640 kg/m³	Slow machine speeds needed to prevent surface burning and tool edges blunting; care required with end-grain work	Fairly good with nails and screws; fair with glue	Fair	One of the best woods for furniture, solid and veneer; fine joinery (interior and exterior); sheds; gates; turning	Afrormosia, iroko and, for nonstructural work, western red cedar	The best-known timber for exterior use, especially in marine work
544 kg/m³	Good; some tearing on quarter-sawed surfaces; keep cutting angle low	Fairly good with nails and screws; fair with glue	Fair; if filled, use teak oil	Fine interior joinery; exterior joinery; flooring; boat building	Indian laurel	Stains yellow in damp conditions; do not use for draining boards, kitchen utensils or tables
384 kg/m³	Good	Care needed with nails and screws; good with glue	Good; an oil base is best	External siding; wooden shingles; interior paneling; gates; sheds	None	Not a true cedar, but one of the best and most durable exterior softwoods available for non-structural work.
624 kg/m³	Works well, but tends to blunt saws and tools due to abrasive silica content; fine dust tends to irritate eyes and nose; use goggles and mask	Tends to split with nails; good with screws; fair with glue	Excellent; in a few cases, filler required	Furniture, solid and veneer; fine interior joinery; exterior joinery; flooring	Cuban mahogany; sapele	A good timber for structural and decorative work
544 kg/m³	Good	Good	Good	Brush handles; toys; models; an excellent turning and carving wood	Obeche	A very good multi-purpose wood
384 kg/m³	Good; some tendency to crumble on end grain	Will not hold well in tough usage with nails and screws; fair with glue	Fair	Framing for cabinetmaking; drawer sides; cupboard linings	Meranti	A good wood – but not strong or hard enough for constructional use
560 kg/m³	Fair	Variable with nails and screws; fair with glue	Good, but needs filling	Structural parts of furniture; veneers; turning; chopping blocks	Other elms, e.g., American white, Dutch, rock, wych	Rock elm, in particular, is noted for its strength

Fixtures and Fittings

A detailed compendium of all the universal woodworking accessories

Screws, nails and adhesives

Screws are used for fastening wood to wood and for fitting hardware. They are also used to fix joints, particularly those made with knock-down fittings. Screws are ordered by length and gauge (thickness). The higher the gauge number, the thicker the screw. Screws are usually made from brass, steel or aluminum alloy. They may be coated with chromium plating, zinc, black japan, cadmium or bronze.

When joining wood to wood the threaded portion of the screw should penetrate the bottom piece. When joining two pieces of equal thickness, the screw length should be just less than the combined thickness. When joining very thick pieces, counterbore the top piece. Drill pilot holes when there is a danger of the wood splitting. Lubricate screws with grease or soap so they turn more easily, are protected against corrosion and can be withdrawn later. Always use brass screws in oak and sycamore, as steel is attacked by the acid in the wood and will corrode, staining the wood.

Both nails and screws are used in combination with glue when making permanent joints. Nails are made from steel, brass, aluminum alloy, iron and copper.

Screw parts

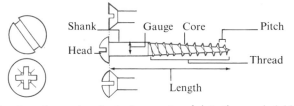

Shank • Gauge • Core • Pitch
Head • Thread • Length

Countersunk, round- and raised-head screws have straight and cross slots. Length is that part sunk into the wood. Add 3 to the gauge number for the drill size in $\frac{1}{64}$th of an inch.

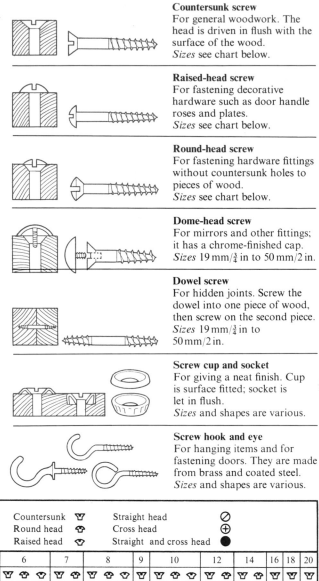

Countersunk screw
For general woodwork. The head is driven in flush with the surface of the wood.
Sizes see chart below.

Raised-head screw
For fastening decorative hardware such as door handle roses and plates.
Sizes see chart below.

Round-head screw
For fastening hardware fittings without countersunk holes to pieces of wood.
Sizes see chart below.

Dome-head screw
For mirrors and other fittings; it has a chrome-finished cap.
Sizes 19 mm/$\frac{3}{4}$ in to 50 mm/2 in.

Dowel screw
For hidden joints. Screw the dowel into one piece of wood, then screw on the second piece.
Sizes 19 mm/$\frac{3}{4}$ in to 50 mm/2 in.

Screw cup and socket
For giving a neat finish. Cup is surface fitted; socket is let in flush.
Sizes and shapes are various.

Screw hook and eye
For hanging items and for fastening doors. They are made from brass and coated steel.
Sizes and shapes are various.

Screw lengths and gauges																	
Countersunk ⌁ Straight head ⊘		Round head ⌁ Cross head ⊕		Raised head ⌁ Straight and cross head ●													

Gauge number	0	1	2	3	4	5	6	7	8	9	10	12	14	16	18	20
Length	⌁⌁	⌁ ⌁⌁	⌁⌁	⌁⌁⌁	⌁⌁⌁	⌁⌁⌁	⌁⌁⌁	⌁⌁	⌁⌁⌁	⌁	⌁⌁⌁	⌁⌁	⌁⌁	⌁⌁	⌁	⌁
6 mm/$\frac{1}{4}$ in	⊘ ⊘	⊘ ⊘ ⊘		⊘ ⊕												
10 mm/$\frac{3}{8}$ in		⊘ ⊘ ⊘	● ●	● ● ⊘	⊕	● ●		⊘ ⊕								
13 mm/$\frac{1}{2}$ in		⊘	● ●	● ● ●	● ● ●	⊘	● ● ●	⊘	● ●		⊘					
16 mm/$\frac{5}{8}$ in			⊘	● ● ⊘	● ● ●	● ● ●	● ● ⊘		● ●		⊘ ●					
19 mm/$\frac{3}{4}$ in			⊘	● ● ●	● ● ●	● ● ⊘	● ● ●	● ● ⊘	● ● ⊘		● ●	⊘ ⊘				
22 mm/$\frac{7}{8}$ in			⊘				● ⊕ ●		● ⊕							
25 mm/1 in			⊘	● ● ⊘ ●	⊘	● ● ●	● ● ⊘	● ● ●	● ⊘	● ● ⊘ ● ●	⊘				
28 mm/1$\frac{1}{8}$ in					●		●								
31 mm/1$\frac{1}{4}$ in			⊘	⊘	● ● ●	● ⊘	● ● ● ●	● ● ⊘	⊘ ● ⊘	⊘					
38 mm/1$\frac{1}{2}$ in			⊘		● ● ●		● ● ● ●	● ● ● ●	⊘ ⊘ ⊘					
44 mm/1$\frac{3}{4}$ in					●	⊘	● ● ⊕	● ● ● ●	⊘ ●	⊘				
50 mm/2 in					● ⊘	⊘	● ⊘ ●	● ● ● ●	● ● ⊘ ⊘ ⊘ ⊘ ⊘				
56 mm/2$\frac{1}{4}$ in					⊘		●	●	●	●			
63 mm/2$\frac{1}{2}$ in					⊘		● ⊘	⊘ ● ⊘	● ⊘ ●	⊘			
69 mm/2$\frac{3}{4}$ in							⊘	⊘	⊘	⊘			
75 mm/3 in					⊘		●	● ⊘	● ● ●	⊘ ⊘		
81 mm/3$\frac{1}{4}$ in								⊘	⊘			
88 mm/3$\frac{1}{2}$ in							⊘	⊘	⊘		
100 mm/4 in							⊘	⊘	⊘ ⊘ ⊘		
113 mm/4$\frac{1}{2}$ in								⊘	⊘	⊘	
125 mm/5 in								⊘	⊘	⊘	
150 mm/6 in									⊘	⊘	⊘	

Adhesives

Type	Uses	Description	Strength	Water resistance	Setting time	Requires clamping	Gap filling
Animal glue	interior woodwork; veneers; leather; paper	soften cake pieces in cold water overnight. Heat these pieces, powder or granules in cold water. Apply hot	good	very poor	12 hours	no	poor
Gummed paper tape	temporary fixing of veneers	obtained in rolls	good	poor	instant	no	no
Impact (contact)	sheet materials; wood; cork; metal, etc.	ready to use in syrup form. Is a flammable vapor. Not suitable for joints or veneer	very good	good	instant	no	poor
Synthetic resin (four types)	most materials except PVC	add epoxy resin to hardener in equal parts. Fast- and slow-setting types available	very good	very good	12 hours fast type; 3 days slow type	no	very good
	interior woodwork; glass; cork; hardboard	PVA (polyvinyl acetate) is a ready-to-use liquid. Does not stain. Unlimited shelf life	very good	poor	2–3 hours	yes	fair
	interior and exterior woodwork	add resorcinol syrup or powder to liquid hardener	very good	very good	4–8 hours	yes	fair
	interior and exterior woodwork	mix urea formaldehyde powder with water and add hardener	very good	good	4–8 hours	yes	very good

Nails

Round wire nail
For work where strength is more important than looks.
Sizes 19 mm/$\frac{3}{4}$ in to 150 mm/6 in.

Annular nail
For fastening manufactured boards, especially three-ply. Its barbed surface aids grip.
Sizes 19 mm/$\frac{3}{4}$ in to 75 mm/3 in.

Oval wire nail
For joinery. Its head will minimize splitting.
Sizes 13 mm/$\frac{1}{2}$ in to 150 mm/6 in.

Twisted square-head nail
For general use. Its shank provides a screw-type grip.
Sizes 19 mm/$\frac{3}{4}$ in to 100 mm/4 in.

Finishing
For general use. The head is driven below the surface.
Sizes 13 mm/$\frac{1}{2}$ in to 150 mm/6 in.

Corrugated fastener
For framing and batten joints.
Sizes 6 mm/$\frac{1}{4}$ in to 22 mm/$\frac{7}{8}$ in deep; 22 mm/$\frac{7}{8}$ in to 31 mm/1$\frac{1}{4}$ in long.

Panel pin
For cabinetwork and moldings. It is also available with a twisted shank.
Sizes 13 mm/$\frac{1}{2}$ in to 50 mm/2 in.

Small brad
For carpentry. Used in place of round-head nails when extra holding power is required.
Sizes 13 mm/$\frac{1}{2}$ in to 19 mm/$\frac{3}{4}$ in.

Veneer pin
For temporarily holding veneers in position on the ground before they are taped.
Sizes 13 mm/$\frac{1}{2}$ in to 50 mm/2 in.

Dowel nail
For fastening end to end and for hidden joints.
Sizes 38 mm/1$\frac{1}{2}$ in to 50 mm/2 in.

Hardboard pin
For hardboard. Its diamond-shaped head grips strongly.
Sizes 19 mm/$\frac{3}{4}$ in to 38 mm/1$\frac{1}{2}$ in.

Needle point
For fastening small moldings. The head is cut off flush. It has a very fine gauge.
Sizes 13 mm/$\frac{1}{2}$ in to 50 mm/2 in.

Masonry pin
For fastening wood to masonry. It is made from hardened steel.
Sizes 19 mm/$\frac{3}{4}$ in to 100 mm/4 in.

Upholstery nail
For upholstery. It is finished in copper, chrome and bronze.
Sizes 3 mm/$\frac{1}{8}$ in to 13 mm/$\frac{1}{2}$ in head.

Knock-down fittings

Knock-down fittings can be used to assemble furniture made from both solid wood and manufactured boards. The fittings, usually of plastic or of metal, are especially suitable for manufactured boards, such as plastic-laminated and wood-veneered chipboard, that cannot be joined in the traditional way. Furniture with these fittings is usually bought ready to be assembled by using a screwdriver or wrench. Such furniture can easily be assembled or dismantled. The best knock-down fittings remain firm and stable when correctly fitted and tightened, and they distribute the load evenly throughout the structure.

Knock-down fittings are of three basic types: those with threaded fittings, in which a bushing lines the hole in one part and another part is attached to it with a dowel or screw; those with interlocking fittings, in which both parts are surface mounted and locked by wedge action; and those with cam-action fittings, either onset or inset, in which a hook that is attached to one part engages in a cam-operated fitting on the other part and is tightened with a screwdriver.

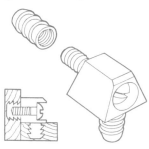

Knock-down dowel fitting
For making a right-angled butt joint.
Fitting Surface.
It is made from hardened plastic. Drill a 10 mm/⅜ in hole in each piece of wood; then drive in the fitting. Tighten the joint with a dome-headed machine screw to give a strong rigid joint.
Size 19 mm/¾ in.

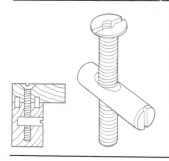

Bolt and cross dowel
For joining frames such as those on tables and chairs.
Fitting Sunk flush.
The steel bolt has an electro-brass finish and a solid brass collar. Use the bolt in conjunction with a cross dowel and two guide dowels. Tighten the joint with an Allen wrench.
Sizes 50 mm/2 in and 75 mm/3 in.

Cross dowel and guide dowel
For use with the bolt (*above*) when constructing frames.
Fitting Recessed.
The steel cross dowel has a tapered entry, allowing the bolt to engage more easily.
Sizes 13 mm/½ in and 19 mm/¾ in.
The brass guide dowel prevents the joint swiveling.
Size 10 mm/⅜ in diameter.

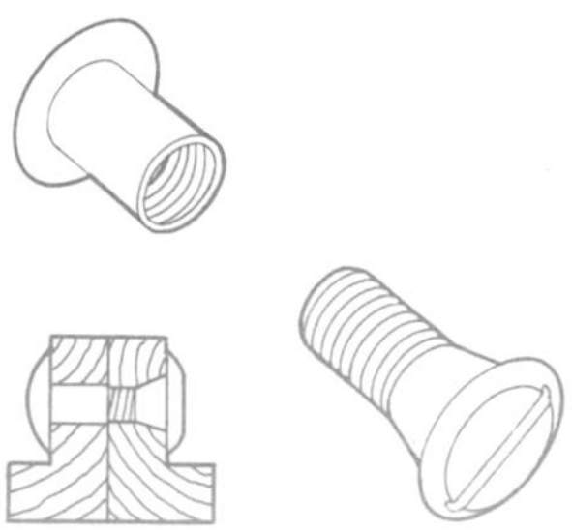

Cabinet connecting screw
For joining shelf units or cabinets together, especially kitchen fitments.
Fitting Recessed.
It is designed for use where the overall thickness of the two boards to be joined is between 31 mm/1¼ in and 38 mm/1½ in. Once the screw is in position cover with a plastic cap.
Size 50 mm/2 in.

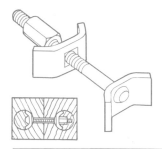

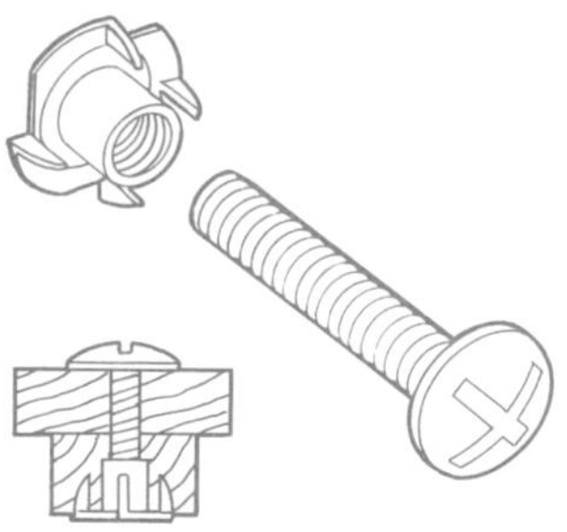

Pronged T nut and bolt
For fastening frames.
Fitting Recessed.
Drill a hole through both pieces of wood and press the nut home through one piece. The prongs will bite into the wood, providing a firm anchor for the nut. Place the bolt through the other hole. Screw it into the nut.
Size 6 mm/¼ in.

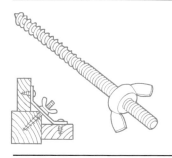

Panel butting connector
For butt joining panels edge to edge.
Fitting Recessed.
Made from steel, this fitting is particularly useful for anything that requires a tight joint such as a counter top. Once the nut is tightened, the panels will not twist and buckle. The connector is better concealed.
Size 63 mm/2½ in.

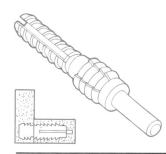

Hanger bolt and wing nut
For use with a corner bracket when constructing furniture.
Fitting Recessed.
The bolt is made of steel and has a 6 mm/¼ in machine screw thread on one end and a standard wood screw thread on the other. The joint is tightened with a wing nut on the machine-screw end.
Size 50 mm/2 in.

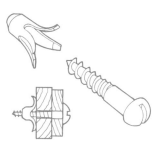

Chipboard dowel rivet
For fastening manufactured boards, especially chipboard, of a greater thickness than 16 mm/⅝ in.
Fitting Recessed.
The barbs on the dowel are in opposite directions to hold the joint together. The fitting is strong and permanent.
Size 38 mm/1½ in when fastened.

Hollow door fitting
For anchoring hooks and other fittings to doors with cavities, such as those faced with hardboard or five-ply.
Fitting Sunk flush.
Pass the anchor through a pre-drilled hole in the door. As the screw is driven in, the anchor arms grip the inside of the cavity.
Size 16 mm/⅝ in long.

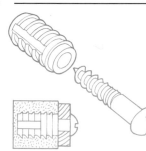

Plastic bushing
For preventing chipboard from crumbling when a screw is driven in.
Fitting Recessed.
As the screw is inserted, the outer threads of the bushing expand and grip the board, ensuring a tight fit. It may be used for end or face fastening.
Sizes 13 mm/½ in and 25 mm/1 in.

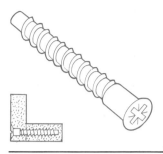

Chipboard screw
For joining frames and panels.
Fitting Recessed.
A pilot hole the thickness of
the core is required to sink the
bolt. Countersinking is
necessary. The thread is
designed to fasten chipboard
without it crumbling. It is
made from steel, with a cross-
head that can be covered.
Size 50 mm/2 in.

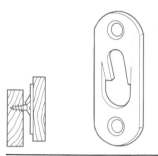

Keyhole plate
For concealed hanging of wall
cabinets or removable frame
components, such as bedheads.
Fitting Surface.
The plate is fastened onto the
cabinet and is hung from a No.
10 gauge wood screw driven
into the surface to which the
cabinet is to be attached.
Size 44 mm × 16 mm/1¾ in ×
⅝ in.

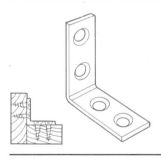

Right-angled bracket
For joining and supporting
frame and furniture panels.
The bracket, which is made of
zinc-plated steel, is used where
appearance is not important,
for example, in greenhouses.
Fitting Surface.
Sizes 19 mm × 19 mm ×
19 mm/¾ in × ¾ in × ¾ in and
38 mm × 38 mm ×
16 mm/1½ in × 1½ in × ⅝ in.

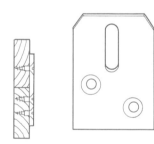

Shrinkage plate
For edge or end fixing of solid
wood. The screws should be
fixed at the appropriate end of
the slot to allow the wood to
shrink across the grain. It can
also be used as a fastener on
furniture made of
manufactured boards.
Fitting Surface.
Size 44 mm × 31 mm/1¾ in ×
1¼ in.

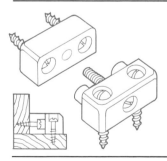

Block joint
For making a right-angled butt
joint; ideal for 13 mm/½ in to
16 mm/⅝ in thick boards.
Fitting Surface.
Screw the plastic blocks to the
panels. Join the two blocks
with a machine screw. It is
usually necessary to have two
fittings per joint.
Size 28 mm × 38 mm ×
16 mm/1⅛ in × 1½ in × ⅝ in.

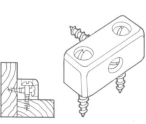

Modesty block
For light construction work or
for strengthening right-angled
butt joints on long panels,
shelves and frames. Made from
hardened plastic, it is ideal
where a neater, less obtrusive
fitting than the block joint is
required.
Fitting Surface.
Size 28 mm × 13 mm ×
13 mm/1⅛ in × ½ in × ½ in.

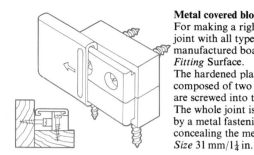

Metal covered block joint
For making a right-angled butt
joint with all types of
manufactured boards.
Fitting Surface.
The hardened plastic block is
composed of two parts, which
are screwed into the panels.
The whole joint is then covered
by a metal fastening plate, thus
concealing the method of fitting.
Size 31 mm/1¼ in.

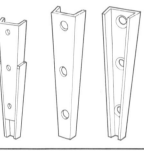

Flushmount fitting
For hanging lightweight
cabinets onto walls and other
cabinets. It is made from zinc-
plated steel.
Fitting Surface.
There are two identical parts,
each with a slightly raised
tongue. One part interlocks
with the other.
Size 3 mm/⅛ in thick when
assembled.

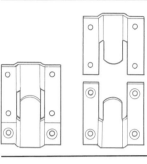

Taper connector
For hanging wall cupboards
and joining heavy framed
components. Often used for
bed frame constructions.
Fitting Surface.
The parts slide together and
lock. It is made from heavy
gauge, cold-rolled steel.
Sizes 100 mm/4 in and
150 mm/6 in long; 6 mm/¼ in
deep.

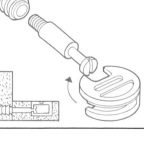

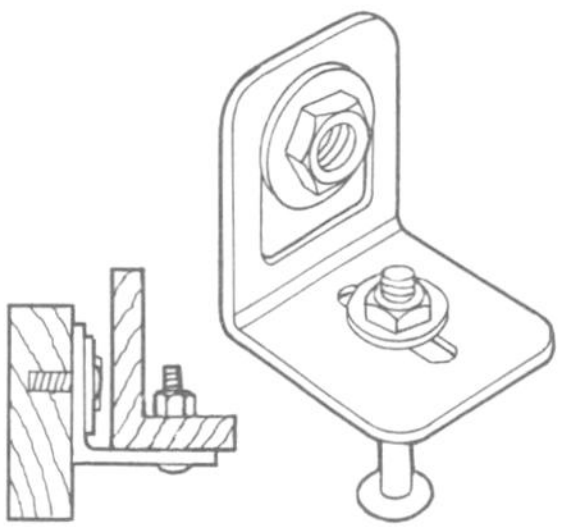

Cabinet hanger
For hanging cabinets.
Fitting Surface.
Once fitted, the cabinet can be
moved up to 13 mm/½ in from
side to side, and up to
13 mm/½ in from back to front.
To obtain vertical adjustment
to within 10 mm/⅜ in, twist the
nut on the nylon cam. It is a
simple, effective device.
Size 44 mm/1¾ in square.

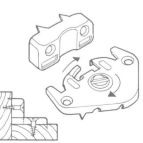

Bolt and cam fitting
For joining panels.
Fitting Bolt is recessed; cam is
sunk flush.
This fitting is composed of a
plastic bushing, a connecting
bolt and a plastic or metal
cam. Once in place, tighten the
cam with a 90° turn of the
screwdriver. Plastic caps are
available to cover the cam.
Size 28 mm/1⅛ in bolt length.

Hooked cam fitting
For joining panels and
attaching cabinets to walls and
shelving.
Fitting Surface.
The cam action is operated by
turning the central slot with a
screwdriver. This draws the
two parts together. It is made
in various metal finishes.
Size 56 mm × 35 mm/2¼ in ×
1⅜ in.

257

Hinges

The traditional butt hinge, set into a hand-cut rectangular recess remains one of the most popular hinges for hanging cabinet and room doors. The development of rapid assembly furniture, however, has resulted in the introduction of many new types that are easier to fit. These are fitted as surface mounted hinges or as recessed hinges, which are set into holes that have been bored or routed out. On inset doors, hinges are positioned 150 mm/6 in from the top and 175 mm/7 in to 225 mm/9 in from the bottom. On framed doors hinges should line up with the rail lines. However, doors are usually hinged the hinge length from the top and bottom. It is the position of the center of the knuckle that determines the throw of the door. Hinges that open 180° or more are often incorporated on kitchen units to save space.

Hinges are available in a range of finishes. Brass hinges tend to be used for fine cabinetwork and are of two types, solid or extruded brass being superior to folded or pressed brass. Steel and nylon hinges tend to be used with cabinets, such as kitchen units.

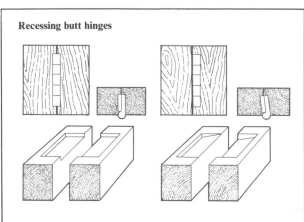

Recessing butt hinges

A simple method to recess a butt hinge is to let both flaps equally into the frame and carcass. Alternatively, the total thickness of the knuckle may be recessed into the door and one flap tapered into the carcass. If the swing of the door is restricted, move the hinge pivot point forward (*see also* page 159).

Common butt hinge
For all doors.
Fitting Surface and recessed.
Sizes Narrow suite, 25 mm × 16 mm/1 in × $\frac{5}{8}$ in to 75 mm × 35 mm/3 in × 1$\frac{3}{8}$ in; broad, 38 mm × 22 mm/1$\frac{1}{2}$ in × $\frac{7}{8}$ in to 100 mm × 60 mm/4 in × 2$\frac{3}{8}$ in; strong, 38 mm × 25 mm/1$\frac{1}{2}$ in × 1 in to 100 mm × 75 mm/4 in × 3 in; extra broad, 25 mm × 25 mm/1 in × 1 in to 75 mm × 63 mm/3 in × 2$\frac{1}{2}$ in.

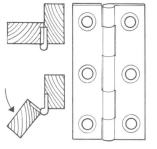

Rising butt
For room doors. As the door is opened, the hinge rises and the door is lifted clear of the carpet. The hinges are either left- or right-handed. The top of the door should have a taper beveled to prevent it from jamming against the frame.
Fitting Recessed.
Sizes 75 mm/3 in to 100 mm/4 in.

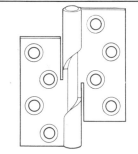

Lift-off hinge
For room and cabinet doors. This hinge, which can be left- or right-handed, is made in two parts so that one half can be removed, yet still be attached to the door. The leaf with the pin must be fixed to the frame, countersunk side facing.
Fitting Recessed.
Sizes 38 mm/1$\frac{1}{2}$ in to 150 mm/6 in.

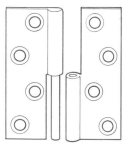

Loose-pin hinge
For cabinet and room doors. The whole of the knuckle protrudes to throw the door clear of the carcass. The ball tips may be unscrewed at one or both ends to remove the pin if the door has to be taken off.
Fitting Surface and recessed.
Sizes 25 mm × 16 mm/1 in × $\frac{5}{8}$ in to 75 hmm × 41 mm/3 in × 1$\frac{5}{8}$ in.

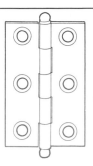

Piano hinge
For supporting long lengths such as chest flaps, piano lids and table flaps. It is available either as a drilled and countersunk flap or as an undrilled flap.
Fitting Recessed.
Sizes 1,800 mm/72 in and 825 mm/33 in long (can be cut to the required length); 25 mm/1 in and 31 mm/1$\frac{1}{4}$ in wide.

Back flap hinge
For table leaves and fall flaps on desks, secretaries and cabinets. Its wide flaps provide strong support.
Fitting Surface and recessed.
Sizes Light pattern, 25 mm × 41 mm/1 in × 1$\frac{5}{8}$ in to 75 mm × 125 mm/3 in × 5 in; heavy pattern, 31 mm × 75 mm/1$\frac{1}{4}$ in × 3 in to 75 mm × 150 mm/3 in to 6 in.

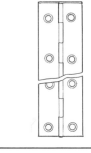

Rule joint hinge
For table extension flaps. The extra long flap is designed to clear the hollow in the cove of the rule joint. The knuckle is on the reverse side to a normal butt hinge to permit a full 90° drop. It has a bronzed finish.
Fitting Recessed.
Sizes 31 mm × 63 mm/1$\frac{1}{4}$ in × 2$\frac{1}{2}$ in to 38 mm × 63 mm/1$\frac{1}{2}$ in × 2$\frac{1}{2}$ in.

Counterflap hinge
For counters and tables with lift-up leaves. The hinge allows the flap to lie back flat along the counter. The most common type has a double pin connected by a smooth link let in flush with the hinge surface.
Fitting Recessed, knuckle down.
Sizes 31 mm × 75 mm/1$\frac{1}{4}$ in × 3 in to 38 mm × 113 mm/1$\frac{1}{2}$ in × 4$\frac{1}{2}$ in.

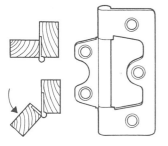

Flush hinge
For all types of door. This thin one-flap hinge has part cut away to act as a second flap. It is a weak hinge but has the advantage that it will support light doors without the need for recessing. It is made from coated steel.
Fitting Surface.
Sizes 38 mm/1½ in to 75 mm/3 in.

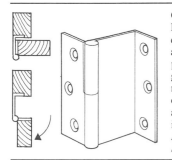

Cranked hinge
For lay-on doors. This good-quality decorative hinge is available mainly in highly polished extruded brass. It gives an opening of 270°. It may be used on either a right- or left-handed door. It is also available double cranked and in a lift-off pattern.
Fitting Surface.
Size 38 mm/1½ in.

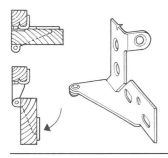

Pivot hinge
For inset or lay-on doors; particularly suitable for kitchen cabinets. This hinge will throw the door clear of the carcass and through an angle of 270°, tucking it out of the way. A small angled cut is needed at the top and bottom of the door for a flush finish.
Fitting Recessed.
Size 16 mm/⅝ in.

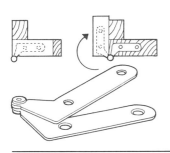

Necked pivot hinge
For doors that need to be thrown clear of a carcass or through an angle of 270°; suitable for kitchen units. The hinges are fitted into the top and bottom of the door, the pivot plates being recessed into both the frame and the door.
Fitting Recessed.
Sizes 50 mm/2 in to 75 mm/3 in in the closed position.

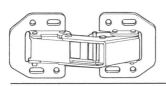

Lay-on concealed hinge
For cabinet and cupboard doors. The hinge, which opens at 90°, is available spring-loaded to ensure a secure hold in either the opened or closed position. It can also be used to hold small lift-up flaps in the horizontal position.
Fitting Surface.
Size 113 mm × 41 mm/4½ in × 1⅝ in.

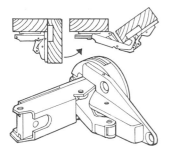

Standard 170° opening hinge
For lay-on doors. This wide-angled hinge makes it possible to open the door without disturbing an adjoining closed door by projecting it outwards. The hinge, which is made from metal and plastic, is also available spring-loaded.
Fitting Surface.
Size 110 mm/4⅜ in along the straight edge.

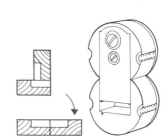

Flush-fitting flap hinge
For fall-flap desks and liquor cabinets. The pivot action allows the open flap to finish flush with the inner surface of the cabinet. This hinge is often used as an alternative to the traditional back flap and counterflap hinges. It is made from metal and plastic.
Fitting Recessed.
Size 35 mm/1⅜ in diameter.

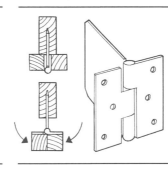

Oni semi-mortise hinge
For hanging two doors from a central division; for kitchen cabinets and storage units. The long flap is held by six wood screws. Double- and single-leaf hinges are available.
Fitting The long flap is mortised; the narrow flaps are recessed.
Size 50 mm/2 in and 75 mm/3 in.

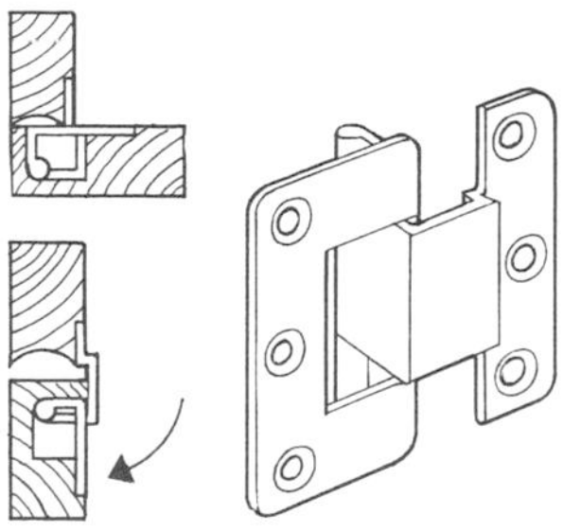

Oni semi-invisible hinge
For all types of door, particularly those made from manufactured boards. The cabinet edge has to be hollowed slightly for the hinge to clear it. It is a strong hinge made from extruded brass.
Fitting Recessed.
Size Available in one size only for a minimum door thickness of 19 mm/¾ in.

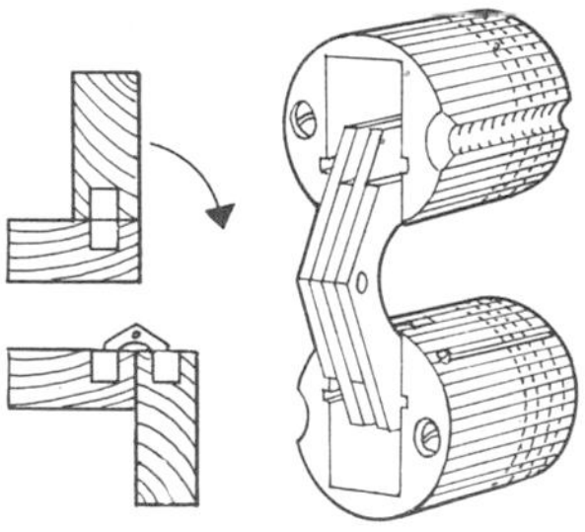

Cylinder hinge
For inset and lay-on doors, fall-flaps, flush tops and cabinet folding doors. It is designed to improve the appearance of furniture, as the hinge is completely concealed when the door is closed. It has an 180° angle of opening.
Fitting Recessed.
Sizes 13 mm/½ in to 16 mm/⅝ in diameter.

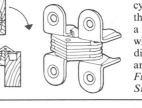

Soss hinge
For all types of light door. It is the original invisible hinge and was the forerunner of the cylinder hinge. It operates through the interconnection of a number of scissor joints, which improves the load distribution of the door. It has an 180° angle of opening.
Fitting Recessed.
Size 44 mm/1¾ in.

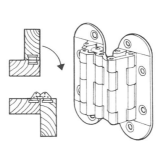

Sepa hinge
For inset and lay-on doors, fall flaps, flush tops. This invisible hinge has a unique seven-pivoted action that allows an inset door to be thrown through an angle of 90° and a lay-on door through an angle of 180°. It is made from extruded brass.
Fitting Recessed.
Size 50 mm/2 in.

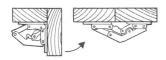

Stays, casters and handles

Stays are used to support the fall flaps of cabinets and writing desks, to restrict the swing of wardrobe doors and to support and hold lift-up flaps and box lids. Modern stays have plastic slides or stay housings, for a quiet action, and mechanisms that control the fall so that the flap is lowered smoothly. Some stays are handed, that is, they can only be used on either the left- or the right-hand side of the carcass. Most stays can be fitted on either side. The fitting of a stay is dependent on the carcass to which it will be attached, but as a rough guide a stay should be positioned at approximately 45° to a carcass or flap that is opened at 90°.

Caster wheels, balls or rollers are fitted to furniture to make it more maneuverable. Caster wheels are made from hard or soft rubber or plastic. The ball and roller types are made from hardened plastic, solid brass and coated steel. Less common are wooden rollers and porcelain wheels.

Glides have a greater bearing surface than casters, but create more friction. They are used to protect floor surfaces against scarring and are commonly made from nickel-plated steel and plastic. Height adjusters, or levelers, similar in design to glides, are fitted onto legged furniture to obtain even floor contact.

Handles, knobs and pulls are available in a wide range of shapes and finishes to suit most types of furniture. The number and placement of handles on furniture is not critical as long as the door or drawer functions smoothly and efficiently. Handle styles should ideally be seen as an integral part of the furniture design. Once these considerations have been met, the final choice is a matter of personal taste.

Stays

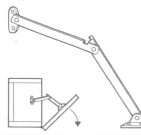

Cranked stay
For fall flaps or doors. The steel arm is hinged at the center, enabling the stay to fold into the cupboard when the flap is closed. It is made right- and left-handed in nickel-plate and electro-brass finishes.
Size 250 mm/10 in long.

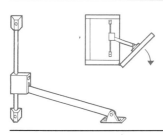

Friction stay
For all fall-flap applications. The movement of the arm is controlled by adjusting the screw, which responds to finger pressure, on the plastic stay housing. It is made left- or right-handed from nickel-plated steel.
Size 250 mm/10 in.

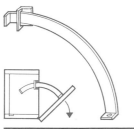

Quadrant stay
For fall flaps on writing desks and cabinets. A slotted bracket is attached to the carcass. A curved arm screwed to the flap slides through the bracket into a groove cut in the bottom of the carcass. It is made from steel and extruded brass.
Size 113 mm/4½ in.

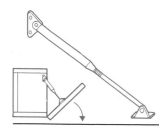

Telescopic stay
For fall flaps, of shallow carcasses in particular. Screwed to the carcass and flap, the supporting arm slides into a hollow metal rod. Once closed, the leaf is held tight; thus extra catches are not needed.
Sizes 125 mm/5 in and 200 mm/8 in.

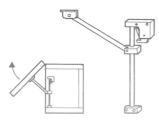

Self-locking flap stay
For lift-up flaps. To open, lift the flap until it locks. To close, lift the flap slightly and then lower it. One stay is normally sufficient for the average-sized flap. It can be adapted for left- or right-handed use by adjusting the plastic housing.
Size 150 mm/6 in arm length.

Cranked lift-up stay
For lift-up flaps on small storage units. The wide, sturdy arm locks in the open position. To close, lift the flap slightly before lowering. Its friction mechanism prevents slamming. It is made left- and right-handed from nickel-plated steel.
Size 125 mm/5 in.

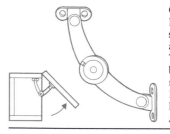

Heavy-duty flap support stay
For attaching foldaway tables, shelves and work tops to a wall. Made from steel, it locks automatically when fully opened. To fold, lift slightly and press the middle of the stay. Stays are sold in pairs.
Size 300 mm × 300 mm/12 in × 12 in.

Up-and-over door mechanism
For wide access to wall units. It comprises two telescopic stays and two straight stays. The mechanism moves the door out of the way — a useful safety feature. No hinges are required.
Size 160 mm/6⅜ in when stays extended but door closed.

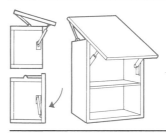

Lid stay
For lids that need to be retained open, for example, on hi-fi cabinets. The controlled braking of this stay ensures smooth, quiet closing of the lid. It is made left- and right-handed from nickel-plated steel with a plastic housing.
Size 169 mm/6¾ in arm.

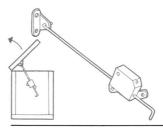

Silent stay
For cupboard and wardrobe doors. The nylon slide gives a quiet action. It can be fitted to suit the chosen opening angle of the door and it is made from steel with an electro-brass finish.
Sizes 216 mm/8⅝ in and 253 mm/10⅛ in.

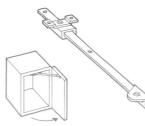

Casters

Methods of fixing casters

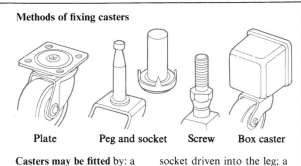

Plate   Peg and socket   Screw   Box caster

Casters may be fitted by: a plate made in various shapes, fastened to the base of furniture; a peg and socket driven into the leg; a screw; or a box caster cupped over the end of a tapered or splayed leg.

Ball caster
For all types of furniture. A heavy-duty ball-bearing caster capable of supporting loads up to 230 kg/500 lb.
Fitting Plate, peg and socket, screw-in.
Size 50 mm/2 in diameter.

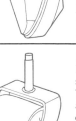

Roller caster
For easy moving of furniture such as divans and bed frames. The roller is made from plastic and is sometimes slightly crowned to facilitate easy swiveling.
Fitting Plate, peg and socket.
Size 60 mm/2$\frac{3}{8}$ in roller width.

Wheel caster
For trolleys and beds. If the cart is to be used to carry heavy loads, wheel casters should be used together with the ball casters.
Fitting Plate, peg and socket, screw-in.
Size 41 mm/1$\frac{5}{8}$ in diameter.

Adjustable swivel glide
For legged furniture. This combination of protective glider and leveler compensates for uneven floors or legs. The steel pin adjusts automatically to the leg angle.
Fitting Steel pin.
Size 6 mm/$\frac{1}{4}$ in pin length.

Handles

Open-pull handle
For doors, drawers and flap pulls. It is available in wood, plastic and metal with numerous finishes and patterns. It can be fitted horizontally or vertically.
Size 88 mm/3$\frac{1}{2}$ in.

Drawer pull handle
For drawers. It was originally designed for use on military chests but it can be equally effective on modern furniture. It is made from brass and steel.
Size 88 mm × 44 mm/3$\frac{1}{2}$ in × 1$\frac{3}{4}$ in.

Swan-necked handle
For high-quality furniture, such as chests of drawers. It is commonly available in solid brass and steel with various finishes and patterns.
Sizes 88 mm/3$\frac{1}{2}$ in to 113 mm/4$\frac{1}{2}$ in.

Extruded aluminum pull handle
For doors and drawers, such as on built-in cupboards. It can be fitted vertically or horizontally.
Sizes It is made in multiples of 75 mm/3 in, up to a maximum of 900 mm/36 in.

Door handles

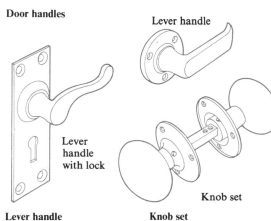

Lever handle

Lever handle with lock

Knob set

Lever handle
For room doors. Most are available spring-loaded for automatic return; the door can then be opened with the elbow or the hand.
Size 100 mm/4 in.

Knob set
For room doors. The most popular type of handle for rim and mortise locks, it is usually sold in pairs and is made in various shapes.
Size 63 mm/2$\frac{1}{2}$ in diameter.

Recessed pull handle
For sliding doors and drawers. On sliding doors, it is designed to give a convenient gripping. It is made from wood, metal and plastic in numerous finishes.
Size 78 mm × 31 mm/3$\frac{1}{8}$ in × 1$\frac{1}{4}$ in.

Knob
For cupboards, drawers and doors. It is available in many shapes — round, oval, cylindrical and polygonal. It is made from wood, metal and plastic in numerous finishes.
Size 50 mm/2 in.

Ring handle
For cupboards and drawers. The handle and its round- or square-shaped surround are made from solid brass and coated steel.
Size 50 mm × 38 mm/2 in × 1$\frac{1}{2}$ in.

Locks, bolts and catches

When fitting locks, bolts and catches for security, always fit the best quality available in each category. Locks made from solid brass are superior in quality to those made from folded or pressed brass or steel. Handed locks are also available. When viewed from the opening side, these are either left- or right-handed, according to the direction in which the bolt shoots when the key is turned.

Catches are required to fasten doors or boxes and to hold them closed. Many are made from plastic or coated steel with plastic rollers for a smoother, quieter action on doors in constant use.

Unless specified below, the escutcheons for locks and striking plates for all the fittings are usually bought separately from the main fitting. They are available in a wide range of sizes and finishes.

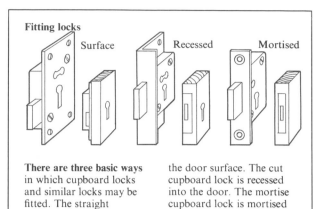

Fitting locks

Surface   Recessed   Mortised

There are three basic ways in which cupboard locks and similar locks may be fitted. The straight cupboard lock is fitted to the door surface. The cut cupboard lock is recessed into the door. The mortise cupboard lock is mortised into the door thickness.

Locks

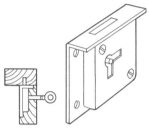

Drawer lock
For flush-fitting drawers.
Fitting Recessed.
Some locks have two keyholes placed at right angles so that the lock can be used upright, in which case they are handed.
Sizes 38 mm × 31 mm/1½ in × 1¼ in to 75 mm × 47 mm/3 in × 1⅞ in.

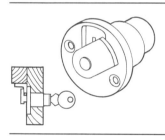

Fall-flap lock
For fall-flap desks and cabinets; it can also be adapted for small cabinet doors.
Fitting Mortised.
The actual lock is made from hardened plastic but has a nickel-plated mechanism. Recessed locks are also made.
Size 19 mm/¾ in.

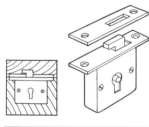

Piano lock
For securing lids of long lengths such as those on pianos.
Fitting Mortised.
The bolt shoots upwards and sideways to lock the lid. The mortise must be lengthened to accommodate the bolt.
Size 81 mm/3¼ in.

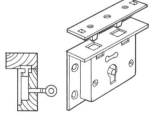

Box lock
For lift-up lids on boxes.
Fitting Recessed.
The striking plate has two projecting tabs that fall with the lid and are engaged sideways by a bolt in the main part of the lock.
Sizes 38 mm/1½ in to 75 mm/3 in.

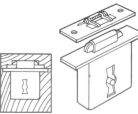

Bird's-beak lock
For rolltop desks originally, but can be used on other furniture such as television cabinets.
Fitting Mortised.
The two bolts move upwards and sideways to lock. It is supplied with an escutcheon.
Size 81 mm/3¼ in.

Spring lock
For drawers and cupboards.
Fitting Recessed.
This spring-loaded lock can close without a key. It has a striking plate and is made left- and right-handed.
Sizes 75 mm × 38 mm/3 in × 1½ in to 75 mm × 47 mm/3 in × 1⅞ in.

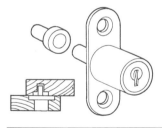

Hook lock
For inset sliding room doors.
Fitting Recessed and mortised.
A back hook lock is also available for lay-on doors. Both are available as left- and right-handed locks.
Sizes 50 mm × 25 mm/2 in × 1 in to 75 mm × 38 mm/3 in × 1½ in.

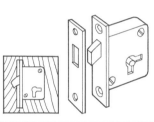

Sliding-door lock
For wooden sliding doors.
Fitting Sunk flush.
A socket set into the rear frame receives the bolt from the lock when the key is turned. This lock is supplied with an escutcheon. It is finished in nickel plate.
Size 19 mm/¾ in long.

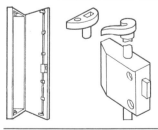

Espagnolette lock
For tall cupboards and wardrobe doors.
Fitting Surface.
Top and bottom bolts are shot or withdrawn in one movement. It is supplied with an escutcheon.
Size 1,200 mm/48 in bar length (can be cut to any length).

Escutcheon
For protecting the wood against scarring by the key.
Fitting Pressing an open-based escutcheon into a hole (*left*); pressing a serrated escutcheon into a hole (*center*); fixing a plate to the surface with escutcheon pins (*right*).
Sizes and finishes are various.

Bolts

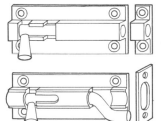

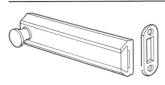

Barrel bolt
For cabinet doors.
Fitting Surface.
Made from extruded brass with a steel shoot, the bolt is available either straight or necked for rabbeted doors.
Sizes 50 mm × 25 mm/2 in × 1 in to 150 mm × 25 mm/6 in × 1 in.

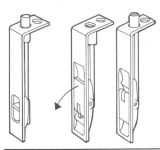

Flush bolt
For cupboard doors.
Fitting Recessed.
The bolt can be operated by a knob or a concealed lever. It is made from brass or chromium-plated steel.
Sizes 150 mm × 19 mm/6 in × ¾ in to 900 mm × 25 mm/36 in × 1 in.

Flat bolt
For light cabinet doors.
Fitting Surface.
Made from solid brass, it is supplied with a catch plate.
Size 50 mm × 16 mm/2 in × ⅝ in.

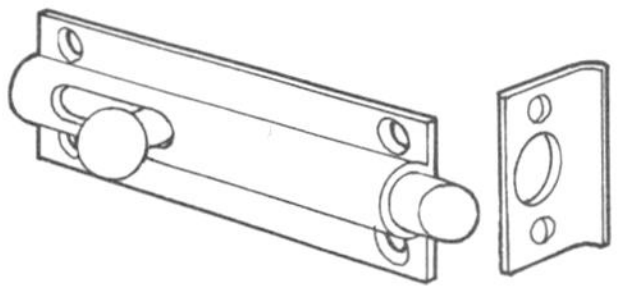

Spring bolt
For doors that require a safety catch.
Fitting Surface.
To open, pull out the knob.
Size 16 mm × 69 mm/⅝ in × 2¾ in.

Catches

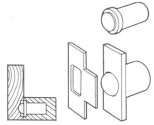

Single ball catch
For cabinet doors.
Fitting Mortised.
It is made from brass and has a spring-loaded steel ball at the top. The edging gives a neat finish. It is supplied with striking plate.
Sizes 3 mm/⅛ in to 16 mm/⅝ in diameter.

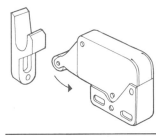

Automatic latch
For opening cabinet doors automatically.
Fitting Surface.
Once fitted, light pressure on the door releases the spring-loaded roller clear of the hook.
It is made from steel and has a plastic roller.
Size 44 mm/1¾ in.

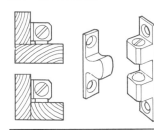

Double ball catch
For high-quality furniture.
Fitting Surface.
The two spring-loaded steel balls are adjusted by turning an integral screw to vary the pull of the door. It is made from solid brass, steel or plastic with a variety of finishes.
Size 50 mm/2 in.

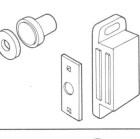

Magnetic catch
For cupboards and doors.
Fitting Surface and recessed.
It is made from hardened plastic and steel. Cylindrical recessed magnetic catches are suitable for lay-on doors.
Sizes 44 mm × 16 mm/1¾ in × ⅝ in to 60 mm × 16 mm/2⅜ in × ⅝ in.

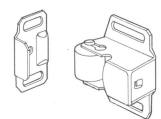

Single roller catch
For kitchen units.
Fitting Surface.
This spring-loaded catch and its striking plate are made from plastic or steel with a rubber roller. Both parts have elongated screw-holes for adjustment when fitting.
Size 35 mm/1⅜ in.

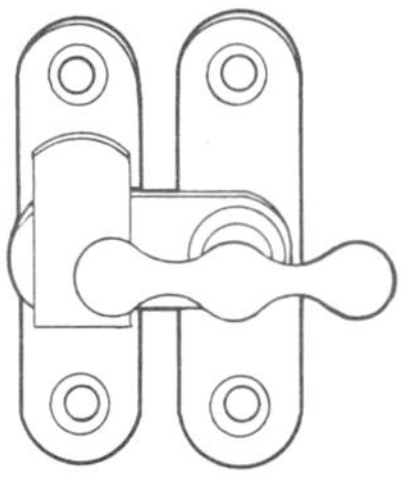

Showcase catch
For display cabinets and cupboards generally.
Fitting Surface.
When fitting to a double door, fit a bolt to the inside of one of the doors. This catch is made from brass.
Sizes 50 mm/2 in to 63 mm/2½ in.

Double roller catch
For cabinet doors.
Fitting Surface.
This catch is made from zinc-coated steel and has nylon rollers. It is also available in plated steel with rubber rollers.
Size 31 mm × 28 mm/1¼ in × 1⅛ in.

Elbow catch
For wood or metal cabinets.
Fitting Surface.
It has a simple spring action and is made from steel. The lock can be released with the elbow when neither hand is free.
Size 28 mm × 16 mm/1⅛ in × ⅝ in.

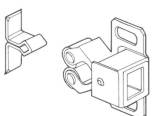

Peglock catch
For doors or flaps; can also be used to secure small detachable panels or plinths, when a panel that can be easily removed is required.
Fitting Surface.
This catch is made from nylon.
Size 38 mm × 31 mm/1½ in × 1¼ in.

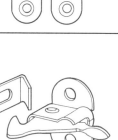

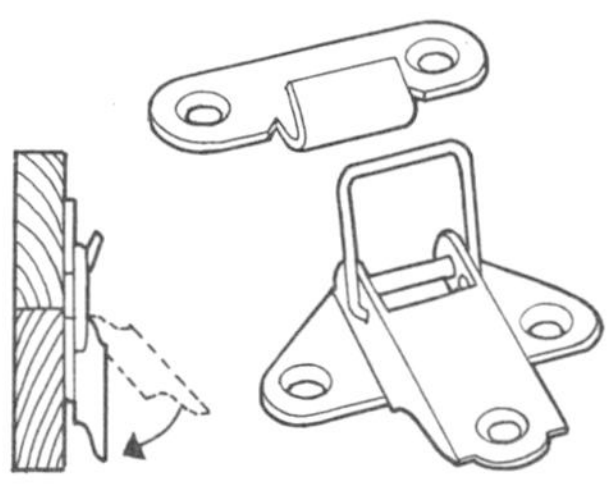

Toggle catch
For the lids of boxes and trunks where the catch presses the lid against the base.
Fitting Surface.
Both this catch and its fastening plate are available in brass- or zinc-plated steel.
Size 38 mm × 69 mm/1½ in × 2¾ in.

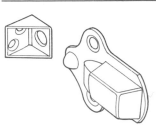

Shelves, drawers and storage fittings

With the advent of modern types of fittings and techniques, it has become possible for many woodworkers to build a piece of furniture, such as a chest of drawers or a bookshelf, that was perhaps too complicated or too time-consuming to attempt previously. With the availability of new extension fittings, it has also become feasible to construct built-in wardrobes in, for example, narrow recesses. Other modern fittings serve to utilize space more economically within the room. The extending table runner, for example, provides support for an extra work surface.

The construction of wooden drawers is perhaps one of the most difficult tasks for the woodworker. For certain applications such as bedroom, kitchen and bathroom furniture, which tend to be made from manufactured boards, this problem has been solved by the introduction of ready-made, easy to assemble plastic drawers available in kit form. No drilling or routing is required to assemble them. They are supplied with four sides, or three sides if a drawer front of another type is required to match existing furniture.

There are a number of fittings available for supporting shelves made from solid wood and manufactured boards. These fittings range from small plastic studs, 19 mm/$\frac{3}{4}$ in long that slide into sockets on the side of the carcass, to metal brackets 250 mm/10 in long that support the shelf from the back. As long as the shelves are sufficiently sturdy, that is, between 19 mm/$\frac{3}{4}$ in and 25 mm/1 in thick, and there is no more than 750 mm/30 in between supports, sagging should be minimal. The advantage of adjustable shelving is that it can be readily adapted for most storage purposes.

Shelves

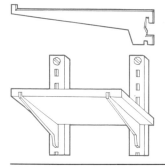

Heavy-duty shelving system
Comprises steel wall channels into which two-pronged brackets are inserted. The S-shaped bend between the prongs ensures that there is no lateral movement. Accessories include tilt brackets and bookends. Each bracket can support 40 kg/88 lb.
Size Channel 1,000 mm/40 in; bracket 150 mm/6 in and 250 mm/10 in.

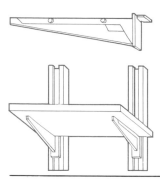

Lightweight shelving system
Comprises aluminum wall channels into which PVC locking strips are inserted to provide anchorage for the shelf brackets. Press fit each bracket in the horizontal position, then twist it until vertical. Shelf heights can be adjusted by tilting bracket up and then sliding it. Each bracket supports about 15 kg/33 lb.
Size Channel and brackets are cut to order.

Right angle shelf bracket
For shelves in which the finish is secondary to strength, for example, in workshops and greenhouses. The bracket, which is made from zinc-plated or japanned steel, offers rigid support, the long vertical arm being capable of countering much leverage.
Size 31 mm × 19 mm/1$\frac{1}{4}$ in × $\frac{3}{4}$ in.

Cantilevered bracket
For moderate shelf support. The bracket is simply fitted by pushing the pin at the back into a predrilled hole in the wall and reinforcing it with a screw driven through the rear upturned part of the bracket. Screw the shelf to the bracket, which is made from zinc-coated steel.
Size 113 mm/4$\frac{1}{2}$ in.

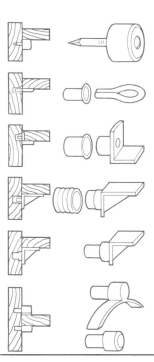

Stud shelf supports
For adjustable shelving in cabinets, bookcases and other fitted units. Fit the studs into holes about 150 mm/6 in apart along the length of the carcass sides, according to requirements. Four studs — two at each end — will support most shelves. Some of the various types of stud shelf support available are: a plastic stud with an integral nail (**A**); a coated steel or brass eye and socket for lightweight shelves (**B**); an angled, coated-steel stud and socket for supporting heavily-laden shelves (**C**); a plastic stud and bushing (**D**); a molded plastic bracket stud (**E**); and a two-piece shelf support (**F**). This last fitting comprises two plastic pegs. Fit the shelf between these two pegs so that the arms of the upper peg spread and grip the shelf. Once fitted, the shelf cannot be moved accidentally.

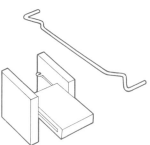

Magic wire
For fitting wooden shelves invisibly. Fit the ends of the wires into two holes positioned slightly closer than the length of the wire. Make stopped grooves in the ends of the shelf, which can then slide onto the wires. The shelf is given almost total support along its width.
Sizes 138 mm/5$\frac{1}{2}$ in to 219 mm/8$\frac{3}{4}$ in.

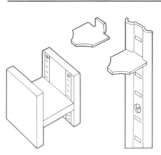

Bookcase strip
For bookcases or fitted units. Screw the strip along the length of the carcass sides and clip the studs into the horizontal slots. Both the strip and the studs are made in two finishes — zinc- or bronze-coated steel.
Size 900 mm/36 in length (can be cut shorter with a hacksaw); 19 mm/$\frac{3}{4}$ in width.

Drawers and runners

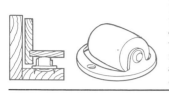

Roller
For heavily-laden drawers. Set the roller into the bottom of a drawer and fasten it with two wood screws. Form a small rabbet beneath each roller to allow for free movement.
Size 25 mm/1 in diameter.

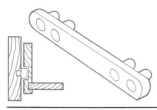

Drawer glide
For lightweight drawers. Attach the plastic glide to the inside of the carcass by inserting four wood screws into the push-in dowel supports.
Size 288 mm × 16 mm × 6 mm/11½ in × ⅝ in × ¼ in.

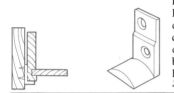

Drawer side saddle
For wooden and plastic drawers, so they open and close smoothly. Fit into the drawer aperture and adjust it by means of its slotted screw-holes.
Size 19 mm × 31 mm/¾ in × 1¼ in.

Ready to assemble drawers
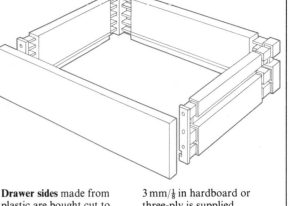

Drawer sides made from plastic are bought cut to the required size. Assemble them with corner fittings that push into the drawer ends. A bottom made of 3 mm/⅛ in hardboard or three-ply is supplied separately, as are the drawer runners.
Sizes 100 mm/4 in and 119 mm/4¾ in depth.

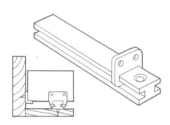

Central drawer guide
For preventing the drawer from tipping when fully extended and for securing the drawer to the center rail. Attach it to the drawer back at the top or the bottom.
Size 44 mm × 35 mm × 6 mm/1¾ in × 1⅜ in × ¼ in.

Drawer runner
For chipboard-sided drawers or for providing existing cupboards with wooden drawers. Made from steel, it is fitted with nylon wheels for smooth and quiet running.
Sizes 450 mm/18 in and 550 mm/22 in.

Storage fittings

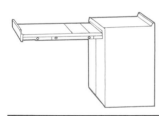

Extending table runner
For supporting a work-top extension up to 769 mm/30¾ in long. Attach this steel runner to a work top made in three sections so that it can slide down the back of the cabinet when not in use.
Size 800 mm/32 in.

Corner fitting
For providing easy access to items stored in corner cabinets. The doors help to support the shelves. It only fits inset doors as when one door is opened the other door automatically moves inside the cabinet.
Size 741 mm/29⅝ in of pole.

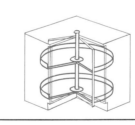

A. Rod and socket.

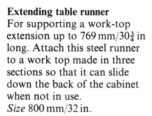

B. Guide rail.

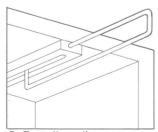

C. Extending rail.

D. Sliding rail.

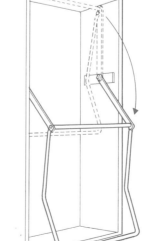

E. Pull-down rail.

Wardrobe fittings
Storage space for clothes needs to be at least 600 mm/24 in deep to allow for the coat hangers. The conventional rod and socket (A) is made from plastic-covered steel or brass. The plastic guide rail (B) runs on an aluminum track. Mounted directly on the underside of the carcass, this type of fixing prevents the sagging sometimes encountered with the rod and socket. The extending wardrobe rail (C) is used when there is insufficient depth to hang clothes sideways. The sliding wardrobe rail (D) is an alternative space-saving fitting to the extending wardrobe rail. The pull-down rail (E) utilizes the space in a floor to ceiling wardrobe, the pull-down rail giving easy access to clothes.

Glossary

Abrasive paper A general term used for paper such as garnet paper and sandpaper.
Adze A tool similar to an axe on a long shaft but with the blade turned at a right angle to the shaft axis.
Allen wrench A cranked hexagonal iron rod used for turning bolts and screws with hexagonal recessed heads.
Annual ring *see* Growth ring.
Apron rail The long rail fixed at right angles under or against a horizontal board or rail.
Arbor A spindle that is integral with an accessory for fixing into the chuck of a power tool.

Banding Strips of veneer set around central veneer panels, comprising borders and sometimes stringing.
Bare-faced Employing only one shoulder.
Batten A narrow strip of wood used for making or stiffening light frames.
Bead in To insert beading into a rabbet by gluing and occasionally pinning.
Beading A molding with a curved profile, usually semicircular.
Bearer Anything that supports and/or spreads a load.
Bevel An edge fully formed at an angle other than 90° to the face side.
Bleeding The process in which some of the substances in wood, such as soluble salts and resin, are removed, either naturally or during the manufacturing process.
Blister A defect found in laid veneer, in which the veneer is raised from the ground because there is no adhesion.
Blooming A defect resulting from a surface coating taking in moisture from the air while drying. It shows as a whitish film or mist on the surface.
Boat A piece of abrasive paper or thin cardboard folded into a container shape to hold spirit stain when coloring.
Bodying The second stage of French polishing, using a rubber.
Bolster The shoulder of a chisel blade where it meets the handle.
Bucked Crosscut.
Buff To rub with a soft absorbent cloth by hand or machine in the final stages of polishing.
Bull-nose work Planing into corners and stopped rabbets as done by the bull-nose plane.
Burl, of a tree An irregular outgrowth on a tree caused by injury to the tree and often yielding a highly ornamental figure.
Burnish *see* Buff.
Burnisher A cylindrical silver steel rod with a wooden handle for turning over the edges of a cabinet scraper.
Burr edge A wire edge such as that left on a tool after grinding or the finish on a hardboard edge after planing.
Bushing A metal or plastic lining inserted in a hole in chipboard for gripping a screw; it can be smooth or threaded. Also, a lining for a hole reducing the diameter to a required size.
Butt joint Two boards that meet side edge to side edge or side edge to end.

Calcareous deposit A chalky white stony deposit.
Capping A finishing strip of wood, often a molding.
Carpentry The structural use of wood in the building trade.
Catch plate A receiving plate on a frame for bolts or catches.
Caul Flat wood used to distribute the pressure evenly onto the veneer and ground. A double caul is two flat pieces of wood placed on either side of a counter-veneered ground to distribute the pressure evenly onto the two veneers.
Chamfer That part of an edge that is angled.
Cheeks, of a carcass The outer vertical sides of a carcass.
Cheeks, of a saw The upright pieces of a bow saw.
Chilling A surface defect in which the coating has taken in moisture from the air resulting from a solvent evaporating too quickly.
Chisel plane A bull-nose plane with the front removed leaving the blade completely exposed and projecting beyond the front of the plane.
Chuck A devise for holding wood for turning, or for gripping a bit in a brace or a twist drill in a hand drill or power drill.
Cleat A batten fastened to an accessory for holding and reinforcement purposes.
Cock beading A thin decorative beading surrounding a drawer front.
Cockle A defect, usually with dry veneer leaves, in which the material bends into ripples.
Core The center section. Blockboard, for example, has a solid wood core.
Core, of a screw The central part of the threaded portion of a screw.
Counterbore To bore a hole the diameter of the screw-head for deep sinking of the screw. This allows a shorter screw to be used.
Countersink To shape a screw-hole to allow the head of a countersunk screw to finish flush with the wood surface.
Counter veneer To veneer both sides of a ground, so balancing its tendency to warp.
Crest rail The top back rail of a chair.
Crossbeam An intermediate horizontal support.
Crosshatch To draw diagonal lines at right angles to hatch lines to distinguish one area from another.
Cross member Any member or component that crosses another or others.

Dado A groove cut across the grain of a piece of wood into which a second piece of wood is fitted. A housed joint is formed when the entire end of the second piece sits in the dado.
Datum line A reference line on a flat straight surface from which all measurements can be taken, for example, face side, face edge.
Depth The measurement from front to back or the recess measurement in from a surface.
Dowel plug A wood plug filling a cylindrical hole with its grain running perpendicular to the surrounding area.

Draw bore To insert a dowel through a mortise and tenon joint with the hole in the tenon slightly offset towards the shoulder to draw the joint up tightly.
Drawfile To draw a file horizontally backwards and forwards at right angles to a metal edge, to remove some of the marks made by filing, when the file was held at an angle with its teeth at right angles to the metal edge.
Drying in the round Drying wood in such a way as to allow air to circulate around all its surfaces.
Dulling A process in which a surface coating is rubbed with a very fine abrasive to diminish its original gloss.
Durability (seasoning) The length of time over which wood remains stable.
Dustboard A board, usually made from three-ply or hardboard, inserted between drawers to cushion their movement.

Ebonize A process in which wood is blackened to resemble ebony.
Edge The narrow surface at right angles to the sides. In this book, where it has been necessary to distinguish between the various edges of a board, the edges that run along the grain are referred to as side edges and the other two as ends.
End The cheek of a carcass or the edge that runs across the grain.

Face edge An edge planed perfectly straight and at right angles to the face side.
Face mark A mark on the face side that reaches over to the face edge to distinguish them from the other surfaces during working.
Face side The better of the two wide surfaces that is chosen to be exposed on the finished work. It must have been planed absolutely flat.
Face veneer Veneer laid on the uppermost side of the ground.
Fadding The initial stage of French polishing, using a fad or a rubber.
Fall-flap A drop leaf.
Feather To taper to a thin edge.
Fence A fixed or adjustable stop against which the work can be held or guided.
Ferrule A metal band around the handle base of a tool to prevent splitting where the shank is fitted.
Fibers The strengthening tissues in wood.
Fielded panel A panel tapered and sometimes shouldered towards its edges so it will fit into grooves in the frame.
Figure The surface appearance of the wood caused by inherent coloring substances and the arrangement of the wood tissues.
Fillet A narrow sliver of wood.
Fillister plane A general term for rabbeting planes fitted with fences.
Fishtail blade A blade with a fanned out shape.
Flush Level. When two adjacent surfaces are level with one another, they are said to be flush.
Fluting Concave grooves used for decoration.
Former The mold against which wood is held to bend it to shape.

The male former is the convex mold; and the female, the concave mold.
Full-grained finish A surface finish in which the pores between the grain are filled.

Gauge To score or mark with the spur of a gauge.
Gauged line A line marked by the spur of a marking or mortise gauge parallel with an edge of the wood.
Glue block A block of wood glued to the inside corner of a joint for additional strength.
Grain The arrangement and direction of the wood tissues. Wood may be worked in one of four ways: *with the grain*, meaning in the direction of the grain; *across the grain*, meaning at right angles to the grain; *against the grain*, meaning in the opposite direction to the grain; and *along the grain*, meaning up and down the grain.
Grain, end At the end of a piece of wood where it has been crosscut.
Grain, interlocked Grain that inclines obliquely up the tree, first in one direction and then in the other.
Grain, silver A distinctive ray figure that shows in quarter-sawed oak.
Grain, spiral Grain that follows either a left- or right-handed diagonal course up the tree.
Groove A channel worked along the grain.
Ground The base, either of solid wood or of manufactured boards, onto which veneers are laid.
Growth ring A ring produced by a layer of open earlywood followed by a layer of dense latewood.
Guide A general term for a fence or a jig.
Guide, drawer A square strip of wood glued and pinned to a drawer runner to guide the drawer in square with the carcass front.

Handed A term applied to rising butt and lift-off hinges, locks and stays. Where a door opens towards the operator, a hinge on the left side of that door is said to be left-handed; conversely, a hinge on the right side is said to be right-handed.
Hanger A piece of wood dovetailed into a carcass top rail to support a lower component, such as a drawer runner, when there is no upright running centrally from the top to the bottom of the carcass.
Hardwood A botanical term used for broad-leafed trees. The wood is not necessarily hard.
Hatch To draw diagonal lines; used to indicate waste to be cut away.
Haunch A shortened portion of a tenon to keep the joint locked at the corner.
Hinge knuckle The spine of a hinge.
Hone To sharpen a tool edge on a stone.
Horn The excess wood left on a mortise piece to prevent the mortise from splitting while the joint is being cut. The horn is sawed off after the joint is assembled.

In-cannel Where the blade of a gouge has a grinding bevel on the inside.
Inlay The insertion of wood, metal or other material into a veneered or

solid wood surface to give a decorative appearance. The inlay is flush with the surface.

Inset door A door that fits in the sides, top and bottom of a carcass or frame.

Jig A device for holding wood or a tool enabling work to be done quickly and accurately.

Joinery Fine woodwork in the building trade.

Kerf The cut made in wood by a saw blade.

Key To roughen, usually with an abrasive paper or a wood rasp, to give a surface stronger adhesive properties.

Keying The reinforcement of a mitered corner with $2\,mm/\frac{3}{32}\,in$ three-ply or veneer.

Kickback The action of a portable power saw when it jumps out of the wood towards the worker or of wood when it shoots from a fixed power saw towards the worker. This occurs when the blade cannot cut forwards either because the kerf closes around the blade, the blade hits a nail or knot, or, in the case of portable power saws, because the blade has twisted in the wood.

Kicker A device — usually a small strip of wood — fitted to a carcass rail above the drawer to prevent the drawer tipping upwards when it is withdrawn.

Knock-down fitting A modern fitting designed to allow components to be assembled and dismantled easily. It is often suitable for manufactured boards, where traditional joining or fixing is not practical.

Knot, pin A small knot.

Laminated board Solid wood or manufactured board faced with smooth hard plastic.

Lath A flexible strip of wood, thinner than a batten. It can be used for panel infilling.

Lay-on door A door that shuts against the front of a carcass rather than being framed by the carcass.

Length The measurement along the wood grain.

Loper A horizontal arm that slides out of a writing desk or an extending table to support a flap.

Louvre door A door frame with horizontal sloping wooden slats to provide both cover and air circulation.

Magic wire A wire shelf support.

Mating pieces Two pieces of wood with shaped surfaces, which when brought together fit closely.

Miter joint A joint in which the junction bisects the angle of the mating pieces, usually at 45°.

Molding A shaped wooden edge used as decoration. It can be integral (stuck) or planted on in strip form with pins and glue.

Molding, half-round A molding of semicircular section.

Molding, ogee A reverse-curved molding.

Molding, ovolo A convex molding.

Molding, quarter-round A molding of quarter-circular section.

Mortise A rectangular recess. On a workbench, a bench dog fits into a

mortise. On a joint, a tenon fits into a mortise.

Movement of wood Wood absorbs and gives off moisture depending on the ambient humidity. This causes it to expand and contract across the grain.

Muntin The dividing piece of wood between two stiles or two panels, to provide extra support to the frame or drawer bottom.

Out-cannel Where the blade of a gouge has a grinding bevel on the outside.

Oversailing A top that is planted onto the carcass cheeks and that projects on one or more sides, front and back.

Overset To set the blade of a plane or the teeth of a saw too coarsely.

Pack A bundle, especially of veneers.

Panel A thin infilling board in a frame.

Parcel A bundle, of veneers or mixed solid wood boards and manufactured boards.

Patina The valued surface finish on old furniture produced naturally by time and aided by repeated burnishing.

Pellet A slightly tapered wood plug filling a round hole with its grain running in line with the surface grain.

Pilaster A vertical piece of wood, with a rectangular section, applied to the front of a carcass.

Pilot hole A small hole drilled into wood to act as a guide for a large drill bit; it is usually made with a bradawl or a small drill bit.

Pitch The angle at which a sloping surface meets a horizontal surface.

Pitch, of a screw The ratio of screw-thread turns in relation to the distance traveled along the axis of a screw; for example, eight turns to $25\,mm/1\,in$.

Plain sawing The commonest method of slicing logs, in which they are sliced perpendicular to the growth rings. It is also called through-and-through and slash sawing.

Planted Not built into the structure but added later.

Plinth A base for a piece of furniture or ornament usually constructed separately from the rest of the carcass.

Plumb Vertical. When a component stands absolutely vertically, it is said to be plumb.

Ply A single thin sheet of veneer. An odd number of plies laid and bonded at right angles to the adjacent ply or plies is known as three-ply, five-ply or multi-ply, according to the number of plies that are used.

Pocket screw To screw through an angled slot or hole on the inside of a rail.

Post The vertical corner member in carcass construction.

Proud Uneven. Two adjacent surfaces at different levels. The surface that projects is said to be proud.

PVA Polyvinyl acetate. It is the basis of many cold glues.

PVC Polyvinyl chloride. It is a plastic sheet material.

Quarter sawing Logs first cut longitudinally into quarters, then cut again into boards.

Quirk stick A small pared piece of wood shaped to remove excess wood filler and glue from recesses and corners.

Rabbet A right-angled recess or step along the edge of a piece of wood.

Radial cut A cut from the bark towards the pith, along the rays.

Rail A horizontal piece of wood between two vertical components in a carcass or frame.

Rays Storage tissues radiating in thin lines from the pith to the phloem of a tree.

Recess An area at a lower level than the surface.

Reeds A series of small beadings worked closely together.

Relief line A line drawn around the base of a relief carving to indicate the lowest depth to which to carve.

Resin glue A chemical, vegetable or mineral glue, usually in two parts, resin and hardener, mixed together before use.

Resin, in a tree An inherent sticky substance characteristic of certain woods, particularly pines.

Reverse curve A concave curve followed by a convex curve, or vice versa.

Reviver A solution made from equal parts linseed oil, household vinegar and methylated spirit used for removing dirt, grease and wax from wood surfaces.

Ricking The distortion of a carcass out of square, usually leading to a loosening of the joints.

Rod A full-size model or sectional drawing showing constructional detail.

Roe A combination of fiddleback and ribbon figure on wood with interlocked grain.

Rubbing compound An abrasive paste rubbed onto surface coatings to dull a gloss finish.

Runner A channel or support along which the side or bottom edge of a drawer slides.

Saddle A nylon plate fitted into a drawer aperture and used in conjunction with a drawer guide. The saddle induces smooth action and alignment of the drawer.

Saddle seat The slotted part of the saddle that encompasses the drawer guide.

Scratch bead Beading formed by a scraping tool such as a scratch stock.

Screw, machine A screw with a blunt end that is unsuitable for driving into wood without pre-drilling a hole.

Screw, wood A screw with a tapered end for easy driving into wood.

Scribe To mark out against a profile with a sharp point.

Section The true shape of a piece of wood shown when cut at right angles to its length.

Set down To position below the surface.

Set in To outline features in a design by cutting vertically around them into the wood when carving.

Shank, of a bit That part of a bit or drill held in the jaws of a brace or hand drill.

Shank, of a screw That part of a screw between the head and the first thread.

Shear To break across the axis of a bolt or a screw.

Shellac A substance produced by masses of small tree insects in India. "Lakh" in Urdu means a hundred thousand. It is the basic ingredient in French polish.

Shoulder The end-grain surface perpendicular to the tenon.

Side The wide surface at right angles to the edges.

Side edge The narrow surface of a board that runs parallel with the grain.

Silver sand Fine sand.

Sizing A dilute glue used for sealing a porous surface.

Slab construction A carcass made from manufactured boards.

Slab form The form in which manufactured boards are available, that is, sheets.

Slat A wood component that has a width far greater than its thickness.

Slot screw To drive a screw through a slot instead of a hole to allow for wood shrinkage.

Slow-schedule wood Wood that takes a long time to season.

Softwood A botanical term used for cone-bearing trees with needlelike leaves. The wood is not necessarily soft.

Spacer A piece of wood or metal, either collar or block shaped, used in pairs to position the work a certain distance from the fence.

Spigot A peg or peglike protrusion, usually integral with the work, that enters a mating socket for locating or pivoting.

Spindle, in a drill press A vertical rotating shaft in a stationary housing.

Spindle, on a chair A thin, often tapered, turned piece of wood, commonly used for chair back fillings.

Spindle, on a lathe A small rotating hollow shaft on which stock is turned; one is at the tailstock and a second is at the headstock where it protrudes on both sides.

Spur, on a gauge The vertical pointed pin on a marking or mortise gauge used for marking and scoring measurement lines parallel to an edge.

Spur, on a plane The vertical cutting blade on a combination plane or multi-plane used for severing the fibers when crosscutting.

Square around To mark a continuous line around a piece of wood, using a pencil or a marking knife and a try square.

Squared line A line marked with a pencil or marking knife and a try square, the stock of which rests against a datum line or face side.

Square off To cut or plane ends or side edges at right angles to each other.

Squaring rod A device, taking many forms, for checking if two edges form a right angle.

Stable wood Wood that has been efficiently seasoned for its intended environment and is therefore liable to shrink or distort only minimally.

Steel wool Strands of very tough fine steel, used in pads for smoothing wood or surface

Glossary

coatings; also for removing surface coatings. It is available from grade 4, which is the coarsest, to grade 0000, which is the finest. Steel wool is also known as wire wool.

Stem, of a gauge The rod along which the stock of a marking, cutting or mortise gauge slides.

Stepping off To mark off a measurement, usually with dividers, along a continuous line.

Sticking and stacking A method of piling boards so air circulates around them while they are seasoning.

Stile The vertical sides in the frame of a door or a window.

Stock The body of a plane, gauge, T square or miter square.

Stock, on a lathe A roughly shaped piece of wood ready to be formed more fully on the lathe, using wood turning tools.

Stop-block, on a table saw A block of wood clamped in the required position to act as a stop and datum point when crosscutting a number of pieces of wood to the same length.

Stopped Not cut through. When a rabbet, dado or chamfer does not run to the end of the work.

Stopping The process of filling holes and cracks; also, the waxy material or resin compound used to fill the holes in wood surfaces. It is available in colored blocks or hard sticks.

Stretcher rail The horizontal crosspiece or rail connecting the lower parts of the legs of a chair, table, etc.

Striking plate A metal plate used to protect the wood surface against the bolts of locks and catches and to provide more security.

Stringing Narrow straight-grained veneer strips laid between central veneer panel(s) and borders. They can be bought or made.

Strop A strip of stiff leather dressed with a fine abrasive, used for perfecting the cutting edges of tools after they have been sharpened on an oilstone or slipstone.

Strop To perfect the cutting edge of a tool by rubbing on a leather strop or on the palm of the hand.

Stumpwood Wood cut from the base of a tree.

Sub-base A plinth. The base that raises the unit from the floor to show the design in the round and to prevent bruising the carcass when floor cleaning.

Sub-frame A subsidiary frame that is assembled separately and connected to a carcass frame.

Substrate The lower layer or layers.

Surform A brand name for a range of rasp planes. They have rasplike blades fitted into shaped holders.

Sweep The curved crosssection of a gouge.

Take off (dimension) To transfer measurements from a drawing to the work.

Tang, of a bit That part of a bit that fits into the chuck of a brace.

Tang, of a chisel The pointed end of a chisel blade that drives into the wooden handle, to hold the two parts together.

Tangential cut A cut made by plain sawing wood longitudinally.

Tannin An astringent substance found in oak and other trees that has the capacity to rust ferrous metals.

Tempered, of a tool Heat-treated to change the natural physical properties. Tempered cutting edges on tools have been treated to hold their sharp edges.

Tempered, of hardboard Hardboard soaked in oil to make it water resistant.

Template A shape or pattern cut to act as a guide in the repeat work on a design.

Thickness The sectional measurement between the two wide surfaces.

Thread, of a screw The spiral part of a screw.

Three-leaf top An extending table top in which the outside leaves can be pulled apart, allowing for the insertion of a third section.

Through From end to end; when a rabbet, dado or chamfer runs the length of the work.

Transverse cut A crosssectional cut across the trunk of the tree.

True Accurately shaped. When work is perfectly shaped it is said to be true.

True up To check that edges are straight and corners are square.

Truss The end supports for a refectory table; also, a framed structure for roof support.

Undercut To cut inside the line of the vertical.

Underframe A frame beneath a top or main frame.

Underside The opposite side to the face side.

Veneer, constructional Plain figured veneer, usually thicker than decorative veneer.

Veneer, decorative Finely-figured veneer, used for laying on the face of the ground.

Waney edge The natural edge of a tree trunk that has been retained after conversion.

Warping A general term used for cupping, bowing, in winding and diamonding in wood. Warping can be caused by underseasoning or careless storage.

Wedges, folding Pairs of wedges used as a basis for exerting pressure. As the tapered surfaces are slid together their combined thickness increases. The force generated can be applied to cramps.

Width The measurement across the grain.

Window An opening cut in a piece of veneer when doing marquetry.

Wire wool see Steel wool.

Yellow deposits Stony deposits in wood.

Credits

The publishers wish to thank the following individuals and organizations for their assistance in the preparation of this book:
Bob Adsett; E.P. Barrus Limited, Bicester, Oxfordshire; Black and Decker Limited, Maidenhead, Berkshire; Robert Brant, Hawkhurst, Kent; John Brazier, Princes Risborough, Buckinghamshire; Brine Veneer Mills Limited, London; Burgess and Galer, London; Ceka Works Limited, Pwllheli, Gwynedd; Christies South Kensington; Copydex, London; Craft Supplies, Buxton, Derbyshire; Craftsman Division, Hagemeyer (London) Limited; Geffrye Museum, London; General Woodworking Supplies, London; GKN Wood Screws, West Midlands; Harvey's Auctions Limited, London; Kity Tools UK, Leeds; James Lathan Limited, London; Ledbury Antiques, London; Jack Maynard; Peter Metcalfe, London College of Furniture; Steve Newton; Terence Morse and Son, London; Neill Tools Limited, Sheffield; Quailcraft, Working Wood magazine, Surrey; Record Ridgeway Tools Limited, Sheffield; Michael Rock; Roger's, Hertfordshire; John Sainsbury; Sandvik UK Limited; Sotheby Parke Bernet and Company; Spear and Jackson Tools Limited, Sheffield; Stanley Tools Limited, Sheffield; Statton Engineering Company, Stone, Staffordshire; TMT Design Limited, Warwick; H.G. Triggs; Victoria and Albert Museum, London; Wolf Electric Tools Limited, London.

Particular thanks are due to the following shops for the loan of tools for artists' references: E. Amette and Company Limited, London; Buck and Ryan Limited, London; Old Woodworking Tools, London; Alec Tiranti Limited, Theale, Berkshire; S. Tyzack and Son Limited, London.

Photographs 227 top left Building Research Establishment; 227 top right Bill Hallworth; 228 top International Paper Company; 228 bottom G.R. Roberts; 228/229 center National Film Board of Canada; 229 top left Malcolm Kirk, The Daily Telegraph Colour Library; 229 top right Frank Prazak, The Image Bank of Canada; 229 bottom The Commonwealth Institute; 230/235 microphotographs Building Research Establishment, Princes Risborough; 230/235 wood photographs Mike Busselle, Mitchell Beazley; 236 top left G.R. Roberts; 236/237 center Bryan and Cherry Alexander; 236 bottom left Canadian Pacific; 236 bottom right Ernest Scott; 237 G.R. Roberts; 239 top center left, top right, bottom center left, bottom center right and bottom right Mike Busselle, Mitchell Beazley.

Principal artists
Mick Gillah, Kevin Maddison, Peter Morter, Coral Mula,

Additional artwork by
Kai Choi, Harry Clow, Chris Forsey, Nigel Osborne, Charles Pickard, Mike Saunders, Del Tolton.

Title pages and section openers photographed by
Clive Corless and tools loaned by Old Woodworking Tools, London.

Additional photography by
Frances Eveleigh and Angelo Hornak.

Picture research by
Brigitte Arora assisted by Meg Price Whitlock.

Index

Where appropriate the main description for an entry is indicated in bold.

Index

Index